Teacher Edition

Eureka Math Grade 7 Module 6

Special thanks go to the Gordan A. Cain Center and to the Department of Mathematics at Louisiana State University for their support in the development of *Eureka Math*.

Published by the non-profit Great Minds

Printed in the U.S.A.
This book may be purchased from the publisher at eureka-math.org
10 9 8 7 6 5 4 3 2 1

ISBN 978-1-63255-618-9

Table of Contents[1]

Geometry

[1]Each lesson is ONE day, and ONE day is considered a 45-minute period.

Grade 7 • Module 6

Geometry

OVERVIEW

In Module 6, students delve further into several geometry topics they have been developing over the years. Grade 7 presents some of these topics (e.g., angles, area, surface area, and volume) in the most challenging form students have experienced yet. Module 6 assumes students understand the basics; the goal is to build fluency in these difficult problems. The remaining topics (i.e., working on constructing triangles and taking slices, or cross sections, of three-dimensional figures) are new to students.

In Topic A, students solve for unknown angles. The supporting work for unknown angles began in Grade 4 Module 4 (**4.MD.C.5**, **4.MD.C.6**, **4.MD.C.7**), where all of the key terms in this topic were first defined, including the following: adjacent, vertical, complementary, and supplementary angles; angles on a line; and angles at a point. In Grade 4, students used those definitions as a basis to solve for unknown angles by using a combination of reasoning (through simple number sentences and equations) and measurement (using a protractor). For example, students learned to solve for an unknown angle in a pair of supplementary angles where one angle measurement is known.

In Grade 7 Module 3, students studied how expressions and equations are an efficient way to solve problems. Two lessons were dedicated to applying the properties of equality to isolate the variable in the context of unknown angle problems. The diagrams in those lessons were drawn to scale to help students more easily make the connection between the variable and what it actually represents. Now in Module 6, the most challenging examples of unknown angle problems (both diagram-based and verbal) require students to use a synthesis of angle relationships and algebra. The problems are multi-step, requiring students to identify several layers of angle relationships and to fit them with an appropriate equation to solve. Unknown angle problems show students how to look for, and make use of, structure (MP.7). In this case, they use angle relationships to find the measurement of an angle.

Next, in Topic B, students work extensively with a ruler, compass, and protractor to construct geometric shapes, mainly triangles (**7.G.A.2**). The use of a compass is new (e.g., how to hold it and how to create equal segment lengths). Students use the tools to build triangles with given conditions such as side length and the measurement of the included angle (MP.5). Students also explore how changes in arrangement and measurement affect a triangle, culminating in a list of conditions that determine a unique triangle. Students understand two new concepts about unique triangles. They learn that under a condition that determines a unique triangle: (1) a triangle can be drawn, and (2) any two triangles drawn under the condition are identical. It is important to note that there is no mention of congruence in the CCSS until Grade 8, after a study of rigid motions. Rather, the focus of Topic B is developing students' intuitive understanding of the structure of a triangle. This includes students noticing the conditions that determine a unique triangle, more than one triangle, or no triangle (**7.G.A.2**). Understanding what makes triangles unique requires understanding what makes them identical.

Topic C introduces the idea of a slice, or cross section, of a three-dimensional figure. Students explore the two-dimensional figures that result from taking slices of right rectangular prisms and right rectangular pyramids parallel to the base and parallel to a lateral face; they also explore two-dimensional figures that result from taking skewed slices that are not parallel to either the base or a lateral face (**7.G.A.3**). The goal of the first three lessons is to get students to consider three-dimensional figures from a new perspective. One way students do this is by experimenting with an interactive website that requires students to choose how to position a three-dimensional figure so that a slice yields a particular result (e.g., how a cube should be sliced to get a pentagonal cross section).

Similar to Topic A, the subjects of area, surface area, and volume in Topics D and E are not new to students but provide opportunities for students to expand their knowledge by working with challenging applications. In Grade 6, students verified that the volume of a right rectangular prism is the same whether it is found by packing it with unit cubes or by multiplying the edge lengths of the prism (**6.G.A.2**). In Grade 7, the volume formula $V = Bh$, where B represents the area of the base, is tested on a set of three-dimensional figures that extends beyond right rectangular prisms to right prisms in general. In Grade 6, students practiced composing and decomposing two-dimensional shapes into shapes they could work with to determine area (**6.G.A.1**). Now, they learn to apply this skill to volume as well. The most challenging problems in these topics are not pure area or pure volume questions but problems that incorporate a broader mathematical knowledge such as rates, ratios, and unit conversion. It is this use of multiple skills and contexts that distinguishes real-world problems from purely mathematical ones (**7.G.B.6**).

Focus Standards

Draw, construct, and describe geometrical figures and describe the relationships between them.[2]

7.G.A.2 Draw (freehand, with ruler and protractor, and with technology) geometric shapes with given conditions. Focus on constructing triangles from three measures of angles or sides, noticing when the conditions determine a unique triangle, more than one triangle, or no triangle.

7.G.A.3 Describe the two-dimensional figures that result from slicing three-dimensional figures, as in plane sections of right rectangular prisms and right rectangular pyramids.

Solve real-life and mathematical problems involving angle measure, area, surface area, and volume.[3]

7.G.B.5 Use facts about supplementary, complementary, vertical, and adjacent angles in a multi-step problem to write and solve simple equations for an unknown angle in a figure.

7.G.B.6 Solve real-world and mathematical problems involving area, volume, and surface area of two- and three-dimensional objects composed of triangles, quadrilaterals, polygons, cubes, and right prisms.

---

[2]The balance of this cluster is taught in Modules 1 and 4.

[3]7.G.4 is taught in Module 3; 7.G.5 and 7.G.6 are introduced in Module 3.

Foundational Standards

Geometric measurement: understand concepts of angle and measure angles.

4.MD.C.7 Recognize angle measure as additive. When an angle is decomposed into non-overlapping parts, the angle measure of the whole is the sum of the angle measures of the parts. Solve addition and subtraction problems to find unknown angles on a diagram in real world and mathematical problems, e.g., by using an equation with a symbol for the unknown angle measure.

Solve real-world and mathematical problems involving area, surface area, and volume.

6.G.A.1 Find the area of right triangles, other triangles, special quadrilaterals, and polygons by composing into rectangles or decomposing into triangles and other shapes; apply these techniques in the context of solving real-world and mathematical problems.

6.G.A.2 Find the volume of a right rectangular prism with fractional edge lengths by packing it with unit cubes of the appropriate unit fraction edge lengths, and show that the volume is the same as would be found by multiplying the edge lengths of the prism. Apply the formulas $V = l\ w\ h$ and $V = b\ h$ to find volumes of right rectangular prisms with fractional edge lengths in the context of solving real-world and mathematical problems.

6.G.A.4 Represent three-dimensional figures using nets made up of rectangles and triangles, and use the nets to find the surface area of these figures. Apply these techniques in the context of solving real-world and mathematical problems.

Solve real-life and mathematical problems involving angle measure, area, surface area, and volume.

7.G.B.4 Know the formulas for area and circumference of a circle and use them to solve problems; give an informal derivation of the relationship between the circumference and area of a circle.

Focus Standards for Mathematical Practice

MP.1 **Make sense of problems and persevere in solving them**. This mathematical practice is particularly applicable for this module, as students tackle multi-step problems that require them to tie together knowledge about their current and former topics of study (i.e., a real-life composite area question that also requires proportions and unit conversion). In many cases, students have to make sense of new and different contexts and engage in significant struggle to solve problems.

MP.3 **Construct viable arguments and critique the reasoning of others.** In Topic B, students examine the conditions that determine a unique triangle, more than one triangle, or no triangle. They have the opportunity to defend and critique the reasoning of their own arguments as well as the arguments of others. In Topic C, students predict what a given slice through a three-dimensional figure yields (i.e., how to slice a three-dimensional figure for a given cross section) and must provide a basis for their predictions.

MP.5 **Use appropriate tools strategically.** In Topic B, students learn how to strategically use a protractor, ruler, and compass to build triangles according to provided conditions. An example of this is when students are asked to build a triangle provided three side lengths. Proper use of the tools helps them understand the conditions by which three side lengths determine one triangle or no triangle. Students have opportunities to reflect on the appropriateness of a tool for a particular task.

MP.7 **Look for and make use of structure.** Students must examine combinations of angle facts within a given diagram in Topic A to create an equation that correctly models the angle relationships. If the unknown angle problem is a verbal problem, such as an example that asks for the measurements of three angles on a line where the values of the measurements are consecutive numbers, students have to create an equation without a visual aid and rely on the inherent structure of the angle fact. In Topics D and E, students find area, surface area, and volume of composite figures based on the structure of two- and three-dimensional figures.

Terminology

New or Recently Introduced Terms

- **Right Rectangular Pyramid** (Given a rectangular region B in a plane E and a point V not in E, the rectangular pyramid with base B and vertex V is the union of all segments VP for any point P in B.

  It can be shown that the planar region defined by a side of the base B and the vertex V is a triangular region called a lateral face. If the vertex lies on the line perpendicular to the base at its center (i.e., the intersection of the rectangle's diagonals), the pyramid is called a right rectangular pyramid.)
- **Surface of a Pyramid** (The *surface of a pyramid* is the union of its base region and its lateral faces.)
- **Three Sides Condition** (Two triangles satisfy the *three sides condition* if there is a triangle correspondence between the two triangles such that each pair of corresponding sides are equal in length.)
- **Triangle Correspondence** (A *triangle correspondence* between two triangles is a pairing of each vertex of one triangle with one and only one vertex of the other triangle.

  A triangle correspondence also induces a correspondence between the angles of the triangles and the sides of the triangles.)

- **Triangles with Identical Measures** (Two triangles are said to have *identical measures* if there is a triangle correspondence such that all pairs of corresponding sides are equal in length and all pairs of corresponding angles are equal in measure. Two triangles with identical measures are sometimes said to be *identical.*

  Note that for two triangles to have identical measures, all six corresponding measures (i.e., 3 angle measures and 3 length measures) must be the same.)
- **Two Angles and the Included Side Condition** (Two triangles satisfy the *two angles and the included side condition* if there is a triangle correspondence between the two triangles such that two pairs of corresponding angles are each equal in measure and the pair of corresponding included sides are equal in length.)
- **Two Angles and the Side Opposite a Given Angle Condition** (Two triangles satisfy the *two angles and the side opposite a given angle condition* if there is a triangle correspondence between the two triangles such that two pairs of corresponding angles are each equal in measure and one pair of corresponding sides that are both opposite corresponding angles are equal in length.)
- **Two Sides and the Included Angle Condition** (Two triangles satisfy the *two sides and the included angle condition* if there is a triangle correspondence between the two triangles such that two pairs of corresponding sides are each equal in length and the pair of corresponding included angles are equal in measure.)

Familiar Terms and Symbols[4]

- Adjacent Angles
- Angles at a Point
- Angles on a Line
- Complementary Angles
- Right Rectangular Prism
- Supplementary Angles
- Vertical Angles

Suggested Tools and Representations

- Familiar objects and pictures to begin discussions around cross sections, such as an apple, a car, a couch, a cup, and a guitar
- A site on Annenberg Learner that illustrates cross sections: http://www.learner.org/courses/learningmath/geometry/session9/part_c/

[4]These are terms and symbols students have seen previously.

Assessment Summary

Assessment Type	Administered	Format	Standards Addressed
Mid-Module Assessment Task	After Topic B	Constructed response with rubric	7.G.B.2, 7.G.A.5
End-of-Module Assessment Task	After Topic E	Constructed response with rubric	7.G.A.2, 7.G.A.3, 7.G.B.5, 7.G.B.6

Topic A

Unknown Angles

7.G.B.5

Focus Standard:	7.G.B.5	Use facts about supplementary, complementary, vertical, and adjacent angles in a multi-step problem to write and solve simple equations for an unknown angle in a figure.
Instructional Days:	4	
Lesson 1:	Complementary and Supplementary Angles (P)[1]	
Lessons 2–4:	Solving for Unknown Angles Using Equations (P, P, P)	

The topic of unknown angles was first introduced to students in Grade 4, where they determined unknown angle values by measuring with a protractor and by setting up and solving simple equations. Though the problems in Grade 7 are more sophisticated, the essential goal remains the same. The goal is to model the angle relationship with an equation and find the value that makes the equation true, which can be confirmed by measuring any diagram with a protractor, similar to Grade 4. There are more lines in any given diagram than in Grade 4, which means there are more angle relationships to assess in order to find the unknown angle, such as a case where vertical angles and angles on a line are in combination. In contrast to typical procedural instruction, students are encouraged to examine complex diagrams thoroughly to best understand them. In Lesson 1, students revisit key vocabulary and then tackle problems—both diagram-based and verbal—that focus on complementary and supplementary angles. Lessons 2–4 broaden in scope to include the angle facts of angles on a line and angles at a point, and the problems become progressively more challenging. The goal is to present students with a variety of problems so they can practice analyzing the structure of the diagrams and translating their observations into an equation (MP.7).

[1]Lesson Structure Key: **P**-Problem Set Lesson, **M**-Modeling Cycle Lesson, **E**-Exploration Lesson, **S**-Socratic Lesson

Lesson 1: Complementary and Supplementary Angles

Student Outcomes

- Students solve for unknown angles in word problems and in diagrams involving complementary and supplementary angles.

Lesson Notes

Students review key terminology regarding angles before attempting several examples. In Lessons 1–4, students practice identifying the angle relationship(s) in each problem and then model the relationship with an equation that yields the unknown value. By the end of the four unknown angle lessons, students should be fluent in this topic and should be able to solve for unknown angles in problems that incorporate multiple angle relationships with mastery.

Solving angle problems has been a part of instruction for many years. Traditionally, the instructions are simply to solve for an unknown angle. Consider using these problems to highlight MP.4 and MP.2 to build conceptual understanding. Students model the geometric situation with equations and can readily explain the connection between the symbols they use and the situation being modeled. Ask students consistently to explain the geometric relationships involved and how they are represented by the equations they write. A final component of deep conceptual understanding is for students to assess the reasonableness of their solutions, both algebraically and geometrically.

Classwork

Opening Exercise (6 minutes)

In the tables below, students review key definitions and angle facts (statements). Students first saw this information in Grade 4, so these definitions and abbreviations should be familiar. Though abbreviations are presented, teachers may choose to require students to state complete definitions. Consider reading each statement with the *blank* and challenging students to name the missing word.

Opening Exercise

As we begin our study of unknown angles, let us review key definitions.

Term	Definition
Adjacent	Two angles, $\angle AOC$ and $\angle COB$, with a common side $\overrightarrow{OC}$, are __________ *angles* if C is in the interior of $\angle AOB$.
Vertical; vertically opposite	When two lines intersect, any two non-adjacent angles formed by those lines are called __________ *angles*, or __________ __________ *angles*.
Perpendicular	Two lines are __________ if they intersect in one point, and any of the angles formed by the intersection of the lines is $90°$. Two segments or rays are __________ if the lines containing them are __________ lines.

Teachers should note that the *Angles on a Line* fact is stated with a diagram of two angles on a line, but the definition can be expanded to include two or more angles on a line. When two angles are on a line, they are also referred to as a *linear pair.*

Complete the missing information in the table below. In the *Statement* column, use the illustration to write an equation that demonstrates the angle relationship; use all forms of angle notation in your equations.

Angle Relationship	Abbreviation	Statement	Illustration
Adjacent Angles	*∠s add*	The measurements of adjacent angles add. $a + b = c$	B, C, A, D, a°, b°, c°
Vertical Angles	*vert. ∠s*	Vertical angles have equal measures. $a = b$	D, G, C, F, E, a°, b°
Angles on a Line	*∠s on a line*	If the vertex of a ray lies on a line but the ray is not contained in that line, then the sum of measurements of the two angles formed is $180°$. $a + b = 180$	C, A, B, D, a°, b°
Angles at a Point	*∠s at a point*	Suppose three or more rays with the same vertex separate the plane into angles with disjointed interiors. Then, the sum of all the measurements of the angles is $360°$. $a + b + c = 360$	B, A, D, C, a°, b°, c°

Note that the distinction between *an angle* and *the measurement of an angle* is not made in this table or elsewhere in the module. Students study congruence in Grade 8, where stating two angles as equal will mean that a series of rigid motions exists such that one angle will map exactly onto the other. For now, the notion of two angles being equal is a function of each angle having the same degree measurement as the other. Furthermore, while students were exposed to the m notation preceding the measurement of an angle in Module 3, continue to expose them to different styles of notation so they learn to discern meaning based on context of the situation.

Discussion (7 minutes)

Lead students through a discussion regarding the *Angles on a Line* fact that concludes with definitions of supplementary and complementary angles.

- *Angles on a Line* refers to two angles on a line whose measurements sum to 180°. Could we use this fact if the two angles are not adjacent, but their measurements still sum to 180°? Could we still call them angles on a line?
  - *No, it would not be appropriate because the angles no longer sit on a line.*
- It would be nice to have a term for non-adjacent angles that sum to 180° since 180° is such a special value in geometry. We can make a similar argument for two angles whose measurements sum to 90°. It would be nice to have a term that describes two angles that may or may not be adjacent to each other whose measurements sum to 90°.

Before students examine the following table, have them do the following exercise in pairs. Ask students to generate a pair of values that sum to 180° and a pair of values that sum to 90°. Ask one student to draw angles with these measurements that are adjacent and the other student to draw angles with these measurements that are not adjacent.

Direct students to the examples of supplementary and complementary angle pairs, and ask them to try to develop their own definitions of the terms. Discuss a few of the student-generated definitions as a class, and record formal definitions in the table.

> *Scaffolding:*
> It may be helpful to create a prominent, permanent visual display to remind students of these definitions.

The diagrams in the lessons depict line segments only but should be interpreted as rays and lines.

Angle Relationship	Definition	Diagram
Complementary Angles	***If the sum of the measurements of two angles is 90°, then the angles are called complementary; each angle is called a complement of the other.***	30° 60° 30° 60°
Supplementary Angles	***If the sum of the measurements of two angles is 180°, then the angles are called supplementary; each angle is called a supplement of the other.***	120° 60° 120° 60°

Exercise 1 (4 minutes)

Students set up and solve an equation for the unknown angle based on the relevant angle relationship in the diagram.

Exercise 1

1. **In a complete sentence, describe the relevant angle relationships in the diagram. Write an equation for the angle relationship shown in the figure, and solve for x. Confirm your answers by measuring the angle with a protractor.**

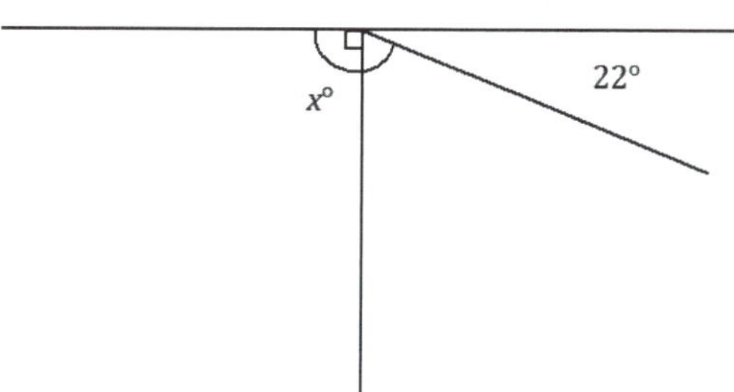

The angles $x°$ and $22°$ are supplementary and sum to $180°$.

$$x + 22 = 180$$
$$x + 22 - 22 = 180 - 22$$
$$x = 158$$

The measure of the angle is $158°$.

Example 1 (5 minutes)

Students set up and solve an equation for the unknown angle based on the relevant angle relationship described in the word problem.

Optional Discussion: The following four questions are a summative exercise that teachers may or may not need to do with students before Example 1 and Exercises 2–4. Review terms that describe arithmetic operations.

- *More than, increased by, exceeds by, and greater than*
- *Less than, decreased by, and fewer than*
- *Times, twice, doubled, and product of*
- *Half of* (or any fractional term, such as *one-third of*), *out of, and ratio of*

- What does it mean for two angle measurements to be in a ratio of $1:4$? Explain by using a tape diagram.
  - *Two angle measurements in a ratio of* $1:4$ *can be represented as follows.*

x	☐			
$4x$	☐	☐	☐	☐

Once we have expressions to represent the angle measurements, they can be used to create an equation depending on the problem. Encourage students to draw a tape diagram for Example 1.

Example 1

The measures of two supplementary angles are in the ratio of $2:3$. Find the measurements of the two angles.

$$2x + 3x = 180$$
$$5x = 180$$
$$\left(\frac{1}{5}\right)5x = \left(\frac{1}{5}\right)180$$
$$x = 36$$

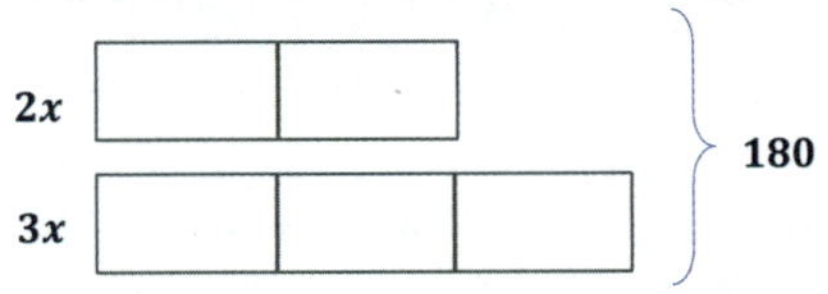

Angle 1 $= 2(36)° = 72°$

Angle 2 $= 3(36)° = 108°$

Exercises 2–4 (12 minutes)

Students set up and solve an equation for the unknown angle based on the relevant angle relationship described in the word problem.

Exercises 2–4

2. **In a pair of complementary angles, the measurement of the larger angle is three times that of the smaller angle. Find the measurements of the two angles.**

$$x + 3x = 90$$
$$4x = 90$$
$$\left(\frac{1}{4}\right)4x = \left(\frac{1}{4}\right)90$$
$$x = 22.5$$

Angle 1 $= 22.5°$

Angle 2 $= 3(22.5)° = 67.5°$

3. **The measure of a supplement of an angle is $6°$ more than twice the measure of the angle. Find the measurement of the two angles.**

$$x + (2x + 6) = 180$$
$$3x + 6 = 180$$
$$3x + 6 - 6 = 180 - 6$$
$$3x = 174$$
$$\left(\frac{1}{3}\right)3x = \left(\frac{1}{3}\right)174$$
$$x = 58$$

Angle 1 $= 58°$

Angle 2 $= 2(58)° + 6° = 122°$

4. The measure of a complement of an angle is 32° more than three times the angle. Find the measurement of the two angles.

$$x + (3x + 32) = 90$$
$$4x + 32 - 32 = 90 - 32$$
$$4x = 58$$
$$\left(\frac{1}{4}\right)4x = \left(\frac{1}{4}\right)58$$
$$x = 14.5$$

Angle 1 $= 14.5^\circ$

Angle 2 $= 3(14.5)^\circ + 32^\circ = 75.5^\circ$

Scaffolding solutions:

a. *The x° angle and the 22° angle are supplementary and sum to 180°.*

b. $x + 22 = 180$

c. *The equation shows that an unknown value, x (which is the unknown angle in the diagram) plus 22 is equal to 180. The 22 represents the angle that we know, which is 22 degrees.*

d.
$$x + 22 = 180$$
$$x + 22 - 22 = 180 - 22$$
$$x = 158$$

$x = 158$ *means that in the diagram, the missing angle measures 158°.*

e. *The answer of 158° is correct. If we substitute 158 for x, we get $158 + 22 = 180$, which is a true number sentence. This is reasonable because when we look at the diagram, we would expect the angle to be obtuse.*

Scaffolding:

Students struggling to organize their solutions may benefit from the following:

- In a complete sentence, describe the relevant angle relationships in the diagram. Write an equation that represents this relationship (MP.4). Explain how the parts of the equation describe the situation (MP.2). Solve the equation, and interpret the solution. Check your answer. Is it reasonable?

Example 2 (5 minutes)

Example 2

Two lines meet at a point that is also the vertex of an angle. Set up and solve an appropriate equation for x and y.

$$16 + y = 90$$
$$16 - 16 + y = 90 - 16$$
$$y = 74$$

Complementary angles

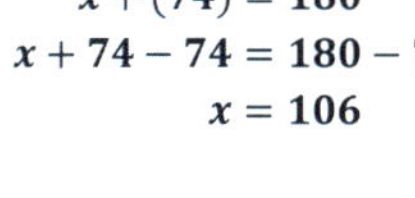

$$x + (74) = 180$$
$$x + 74 - 74 = 180 - 74$$
$$x = 106$$

Supplementary angles

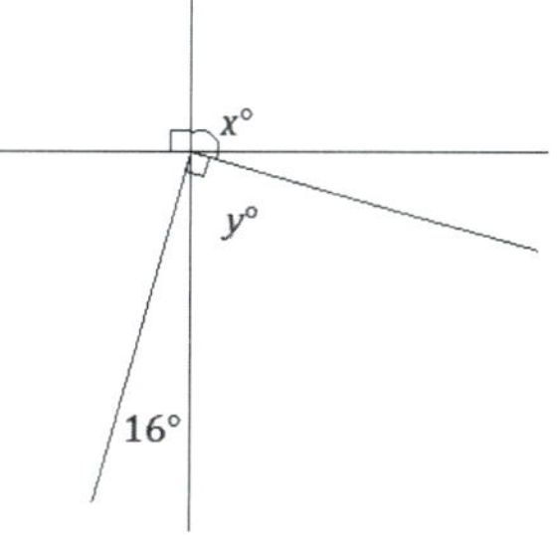

Closing (1 minute)

- To determine the measurement of an unknown angle, we must identify the angle relationship(s) and then model the relationship with an equation that yields the unknown value.
- If the sum of the measurements of two angles is $90°$, the angles are complementary angles, and one is the complement of the other.
- If the sum of the measurements of two angles is $180°$, the angles are supplementary angles, and one is the supplement of the other.

Lesson Summary

- **Supplementary angles are two angles whose measurements sum to $180°$.**
- **Complementary angles are two angles whose measurements sum to $90°$.**
- **Once an angle relationship is identified, the relationship can be modeled with an equation that will find an unknown value. The unknown value may be used to find the measure of the unknown angle.**

Exit Ticket (5 minutes)

Name ______________________________ Date ______________

Lesson 1: Complementary and Supplementary Angles

Exit Ticket

1. Set up and solve an equation for the value of x. Use the value of x and a relevant angle relationship in the diagram to determine the measurement of $\angle EAF$.

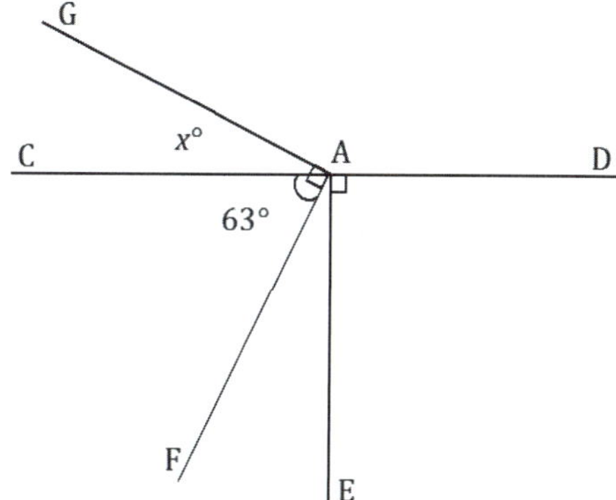

2. The measurement of the supplement of an angle is 39° more than half the angle. Find the measurement of the angle and its supplement.

Exit Ticket Sample Solutions

1. **Set up and solve an equation for the value of x. Use the value of x and a relevant angle relationship in the diagram to determine the measurement of $\angle EAF$.**

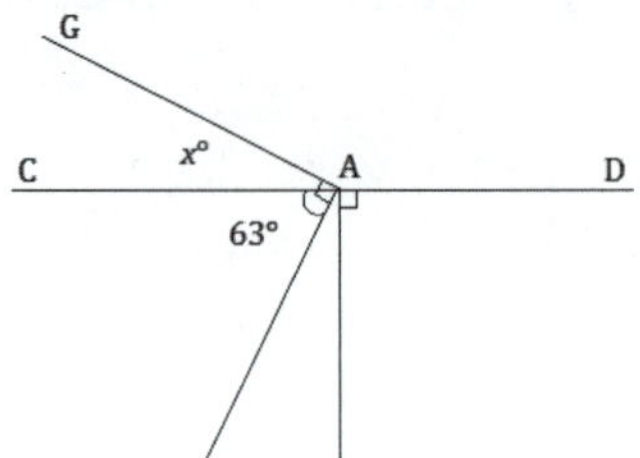

$$x + 63 = 90$$
$$x + 63 - 63 = 90 - 63$$
$$x = 27$$

$\angle CAG$ and $\angle EAF$ are the complements of $63°$. The measurement of $\angle CAG$ is $27°$; therefore, the measurement of $\angle EAF$ is also $27°$.

2. **The measurement of the supplement of an angle is $39°$ more than half the angle. Find the measurement of the angle and its supplement.**

$$x + \left(\frac{1}{2}x + 39\right) = 180$$
$$1.5x + 39 = 180$$
$$1.5x + 39 - 39 = 180 - 39$$
$$1.5x = 141$$
$$1.5x \div 1.5 = 141 \div 1.5$$
$$x = 94$$

The measurement of the angle is $94°$.

The measurement of the supplement is $\frac{1}{2}(94)° + 39° = 86°$.

OR

$$x + \left(\frac{1}{2}x + 39\right) = 180$$
$$\frac{3}{2}x + 39 = 180$$
$$\frac{3}{2}x + 39 - 39 = 180 - 39$$
$$\frac{3}{2}x = 141$$
$$\left(\frac{2}{3}\right)\left(\frac{3}{2}x\right) = \left(\frac{2}{3}\right)141$$
$$x = 94$$

The measurement of the angle is $94°$.

The measurement of the supplement is $\frac{1}{2}(94)° + 39° = 86°$.

Problem Set Sample Solutions

1. Two lines meet at a point that is also the endpoint of a ray. Set up and solve the appropriate equations to determine x and y.

$x + 55 = 90$ *Complementary angles*
$x + 55 - 55 = 90 - 55$
$x = 35$

$55 + y = 180$ *Supplementary angles*
$55 - 55 + y = 180 - 55$
$y = 125$

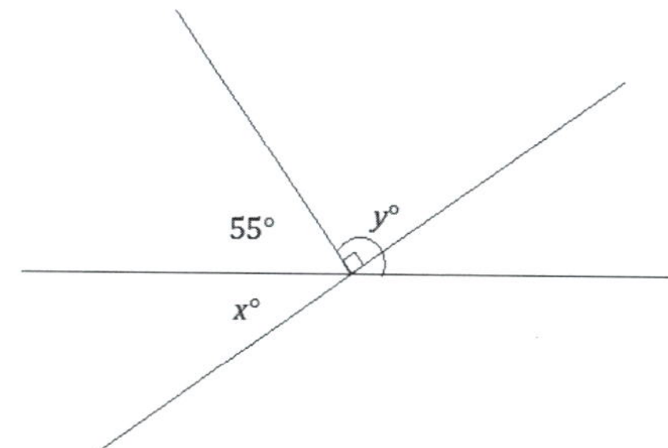

> *Scaffolding:*
> As shown in Exercise 4, some students may benefit from a scaffolded task. Use the five-part scaffold to help organize the question for those students who might benefit from it.

2. Two lines meet at a point that is also the vertex of an angle. Set up and solve the appropriate equations to determine x and y.

$y + x = 90$ *Complementary angles*

$x + 32 = 90$ *Complementary angles*
$x + 32 - 32 = 90 - 32$
$x = 58$

$y + (58) = 90$ *Complementary angles*
$y + 58 - 58 = 90 - 58$
$y = 32$

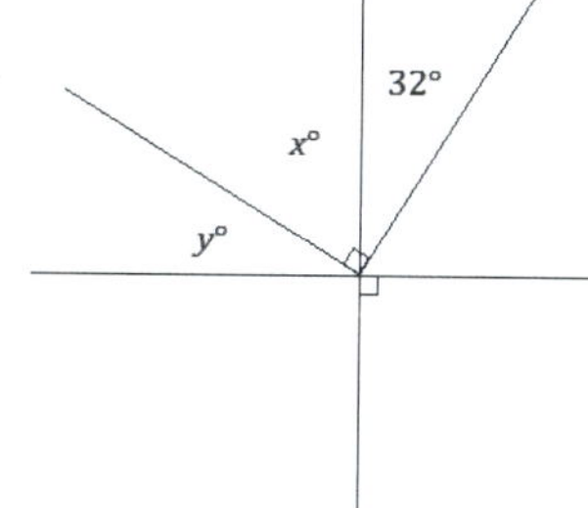

3. Two lines meet at a point that is also the vertex of an angle. Set up and solve an appropriate equation for x and y.

$x + y = 180$ *Supplementary angles*

$28 + y = 90$ *Complementary angles*
$28 - 28 + y = 90 - 28$
$y = 62$

$x + (62) = 180$ *Supplementary angles*
$x + 62 - 62 = 180 - 62$
$x = 118$

x° y° 28°

4. Set up and solve the appropriate equations for s and t.

$$79 + t = 90$$ *Complementary angles*
$$79 - 79 + t = 90 - 79$$
$$t = 11$$

$$19 + (11) + 79 + s = 180$$ *Angles on a line*
$$109 + s = 180$$
$$109 - 109 + s = 180 - 109$$
$$s = 71$$

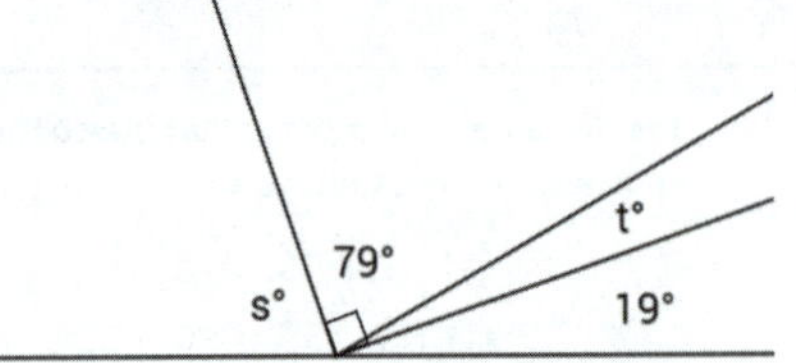

5. Two lines meet at a point that is also the endpoint of two rays. Set up and solve the appropriate equations for m and n.

$$43 + m = 90$$ *Complementary angles*
$$43 - 43 + m = 90 - 43$$
$$m = 47$$

$$38 + 43 + (47) + n = 180$$ *Angles on a line*
$$128 + n = 180$$
$$128 - 128 + n = 180 - 128$$
$$n = 52$$

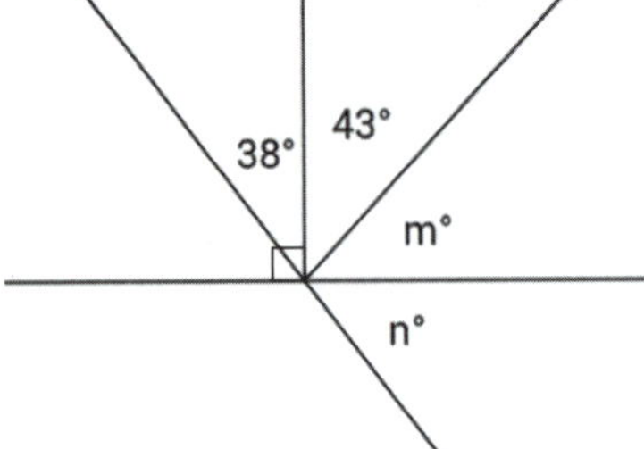

6. The supplement of the measurement of an angle is $16°$ less than three times the angle. Find the measurement of the angle and its supplement.

$$x + (3x - 16) = 180$$
$$4x - 16 + 16 = 180 + 16$$
$$4x = 196$$
$$\left(\frac{1}{4}\right)4x = \left(\frac{1}{4}\right)196$$
$$x = 49$$

Angle $= 49°$

Supplement $= 3(49)° - 16° = 131°$

7. The measurement of the complement of an angle exceeds the measure of the angle by 25%. Find the measurement of the angle and its complement.

$$x + \left(x + \frac{1}{4}x\right) = 90$$
$$x + \frac{5}{4}x = 90$$
$$\frac{9}{4}x = 90$$
$$\left(\frac{4}{9}\right)\frac{9}{4}x = \left(\frac{4}{9}\right)90$$
$$x = 40$$

Angle $= 40°$

Complement $= \frac{5}{4}(40)° = 50°$

8. The ratio of the measurement of an angle to its complement is $1:2$. Find the measruement of the angle and its complement.

$$x + 2x = 90$$
$$3x = 90$$
$$\left(\frac{1}{3}\right)3x = \left(\frac{1}{3}\right)90$$
$$x = 30$$

Angle $= 30°$

Complement $= 2(30)° = 60°$

9. The ratio of the measurement of an angle to its supplement is $3:5$. Find the measurement of the angle and its supplement.

$$3x + 5x = 180$$
$$8x = 180$$
$$\left(\frac{1}{8}\right)8x = \left(\frac{1}{8}\right)180$$
$$x = 22.5$$

Angle $= 3(22.5)° = 67.5°$

Supplement $= 5(22.5)° = 112.5°$

10. Let x represent the measurement of an acute angle in degrees. The ratio of the complement of x to the supplement of x is $2:5$. Guess and check to determine the value of x. Explain why your answer is correct.

Solutions will vary; $x = 30°$.

The complement of $30°$ *is* $60°$. *The supplement of* $30°$ *is* $150°$. *The ratio of* 60 *to* 150 *is equivalent to* $2:5$.

Lesson 2: Solving for Unknown Angles Using Equations

Student Outcomes

- Students solve for unknown angles in word problems and in diagrams involving complementary, supplementary, vertical, and adjacent angles.

Classwork

Opening Exercise (5 minutes)

MP.6

Opening Exercise

Two lines meet at a point. In a complete sentence, describe the relevant angle relationships in the diagram. Find the values of r, s, and t.

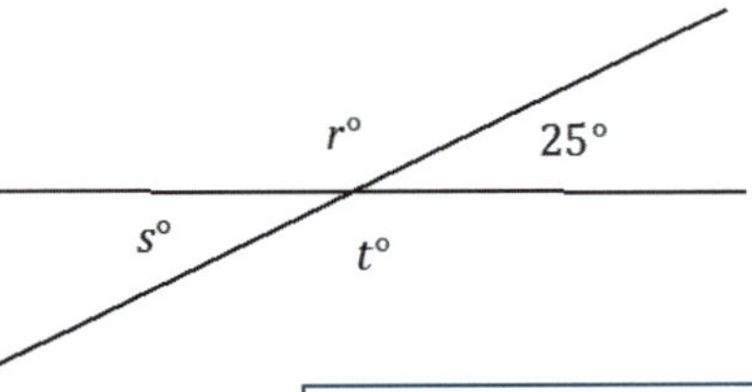

The two intersecting lines form two pairs of vertical angles; $s = 25$, and $r° = t°$. Angles $s°$ and $r°$ are angles on a line and sum to $180°$.

$$s = 25$$
$$r + 25 = 180$$
$$r + 25 - 25 = 180 - 25$$
$$r = 155,$$
$$t = 155$$

Scaffolding:

Students may benefit from repeated practice drawing angle diagrams from verbal descriptions. For example, tell them "Draw a diagram of two supplementary angles, where one has a measure of 37°." Students struggling to organize their solution to a problem may benefit from the five-part process of the Exit Ticket in Lesson 1, including writing an equation, explaining the connection between the equation and the situation, and assessing whether an answer is reasonable. This builds conceptual understanding.

In the following examples and exercises, students set up and solve an equation for the unknown angle based on the relevant angle relationships in the diagram. Model the good habit of always stating the geometric reason when you use one. This is a requirement in high school geometry.

Example 1 (4 minutes)

Example 1

Two lines meet at a point that is also the endpoint of a ray. In a complete sentence, describe the relevant angle relationships in the diagram. Set up and solve an equation to find the value of p and r.

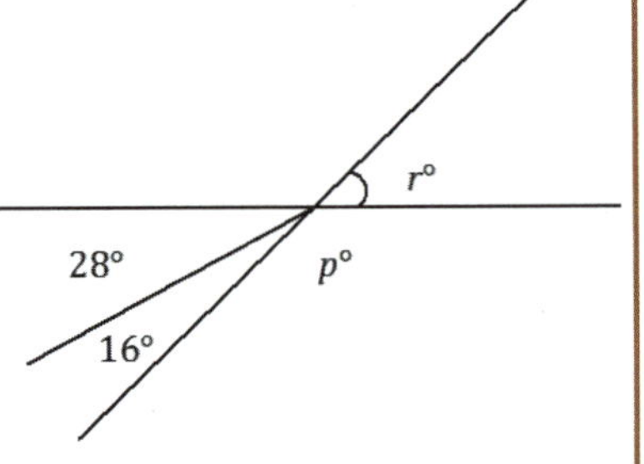

The angle $r°$ is vertically opposite from and equal to the sum of the angles with measurements $28°$ and $16°$, or a sum of $44°$. Angles $r°$ and $p°$ are angles on a line and sum to $180°$.

$$r = 28 + 16 \quad \textit{Vert. } \angle s$$
$$r = 44$$
$$p + (44) = 180 \quad \angle s \textit{ on a line}$$
$$p + 44 - 44 = 180 - 44$$
$$p = 136$$

Take the opportunity to distinguish the correct usage of *supplementary* versus *angles on a line* in this example. Remind students that *supplementary* should be used in reference to two angles, whereas *angles on a line* can be used for two *or more* angles.

Exercise 1 (4 minutes)

Exercise 1

Three lines meet at a point. In a complete sentence, describe the relevant angle relationship in the diagram. Set up and solve an equation to find the value of a.

The two $a°$ angles and the angle $144°$ are angles on a line and sum to $180°$.

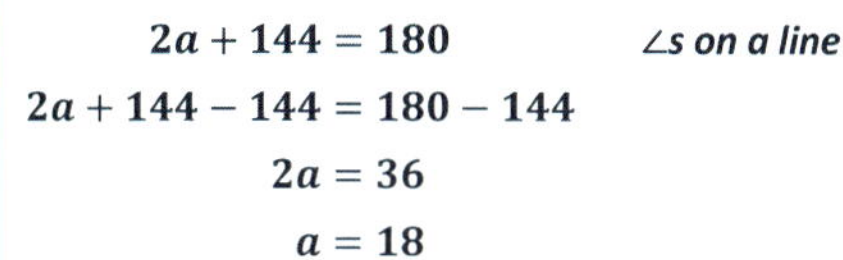

∠s on a line

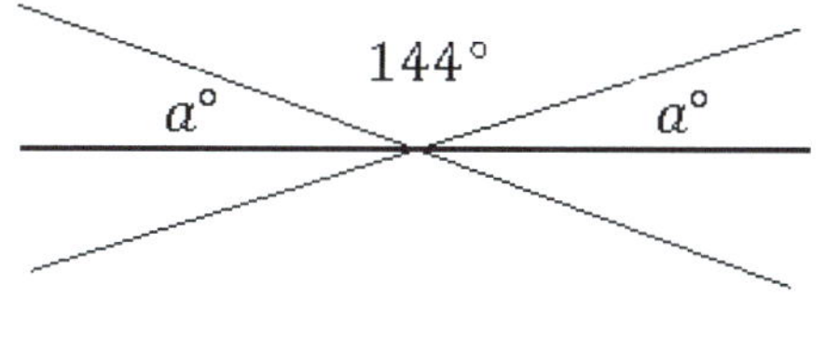

Example 2 (4 minutes)

Encourage students to label diagrams as needed to facilitate their solutions. In this example, the label $y°$ is added to the diagram to show the relationship of $z°$ with $19°$. This addition allows for methodical progress toward the solution.

Example 2

Three lines meet at a point. In a complete sentence, describe the relevant angle relationships in the diagram. Set up and solve an equation to find the value of z.

Let $y°$ be the angle vertically opposite and equal in measurement to $19°$.
The angles $z°$ and $y°$ are complementary and sum to $90°$.

$$z + y = 90$$
$$z + 19 = 90$$
$$z + 19 - 19 = 90 - 19$$
$$z = 71$$

Complementary ∠s

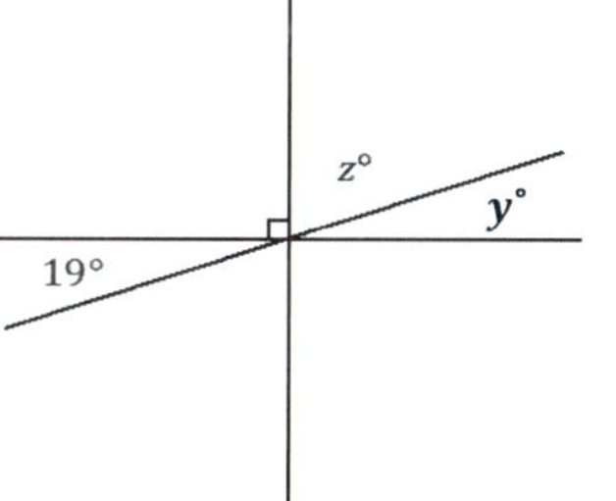

Exercise 2 (4 minutes)

Exercise 2

Three lines meet at a point; $\angle AOF = 144°$. In a complete sentence, describe the relevant angle relationships in the diagram. Set up and solve an equation to determine the value of c.

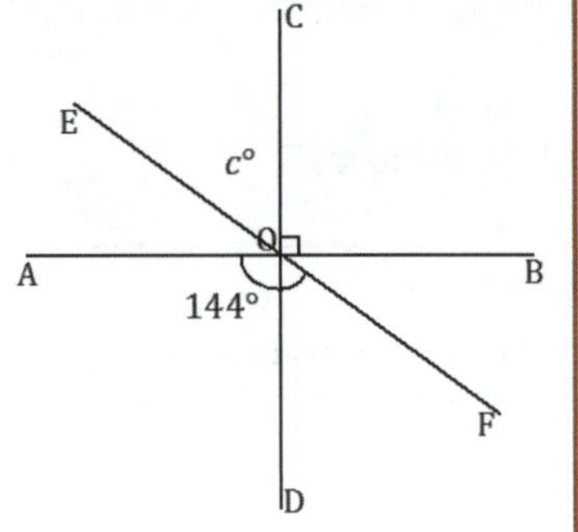

$\angle EOB$, formed by adjacent angles $\angle EOC$ and $\angle COB$, is vertical to and equal in measurement to $\angle AOF$.
The measurement of $\angle EOB$ is $c° + 90°$ ($\angle$s add).

$c + 90 = 144$ *Vert. $\angle$s*

$c + 90 - 90 = 144 - 90$

$c = 54$

Example 3 (4 minutes)

Example 3

Two lines meet at a point that is also the endpoint of a ray. The ray is perpendicular to one of the lines as shown. In a complete sentence, describe the relevant angle relationships in the diagram. Set up and solve an equation to find the value of t.

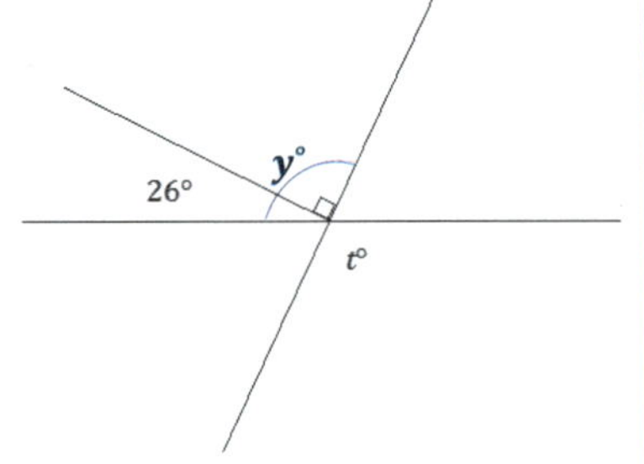

The measurement of the angle formed by adjacent angles of $26°$ and $90°$ is the sum of the adjacent angles. This angle is vertically opposite and equal in measurement to the angle $t°$.

Let $y°$ be the measure of the indicated angle.

$y = 116$ *$\angle$s add*

$t = (y)$ *Vert. $\angle$s*

$t = 116$

Exercise 3 (4 minutes)

Exercise 3

Two lines meet at a point that is also the endpoint of a ray. The ray is perpendicular to one of the lines as shown. In a complete sentence, describe the relevant angle relationships in the diagram. You may add labels to the diagram to help with your description of the angle relationship. Set up and solve an equation to find the value of v.

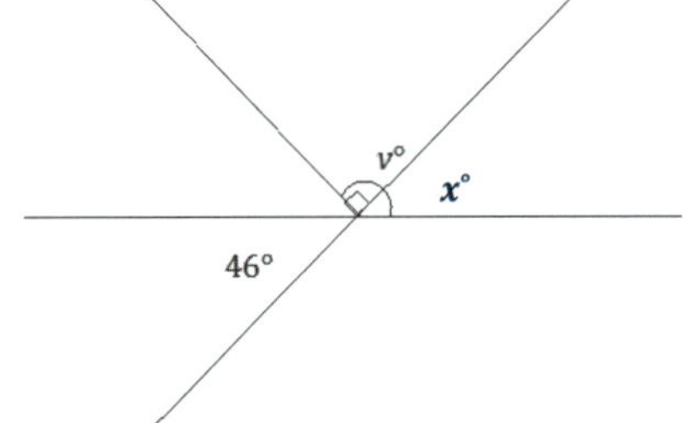

One possible response: Let $x°$ be the angle vertically opposite and equal in measurement to $46°$. The angles $x°$ and $v°$ are adjacent angles, and the angle they form together is equal to the sum of their measurements.

$x = 46$ *Vert. $\angle$s*

$v = 90 + 46$ *$\angle$s add*

$v = 136$

Example 4 (4 minutes)

Example 4

Three lines meet at a point. In a complete sentence, describe the relevant angle relationships in the diagram. Set up and solve an equation to find the value of x. Is your answer reasonable? Explain how you know.

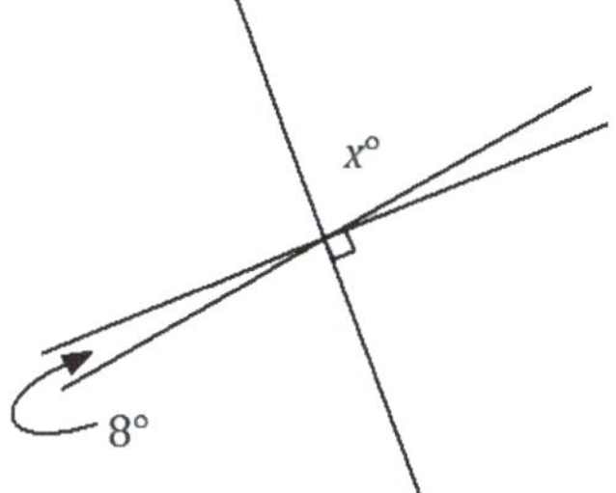

The angle $x°$ is vertically opposite from the angle formed by the right angle that contains and shares a common side with an $8°$ angle.

$x = 90 - 8$ *∠s add and vert. ∠s*

$x = 82$

The answer is reasonable because the angle marked by $x°$ is close to appearing as a right angle.

Exercise 4 (4 minutes)

Exercise 4

Two lines meet at a point that is also the endpoint of two rays. In a complete sentence, describe the relevant angle relationships in the diagram. Set up and solve an equation to find the value of x. Find the measurements of $\angle AOB$ and $\angle BOC$.

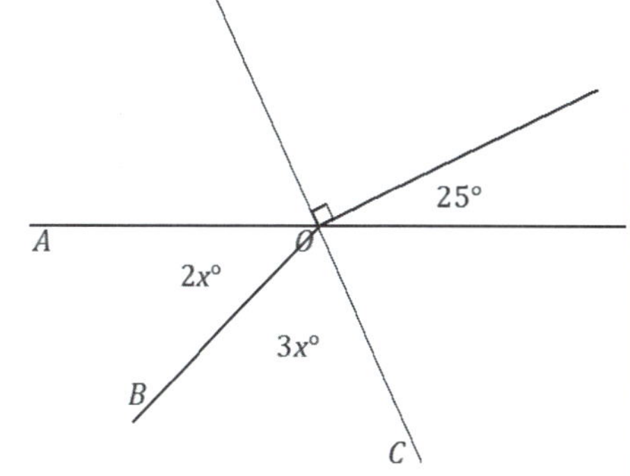

$\angle AOC$ is vertically opposite from the angle formed by adjacent angles $90°$ and $25°$.

$2x + 3x = 90 + 25$ *∠s add and vert. ∠s*

$5x = 115$

$x = 23$

$\angle AOC = 2(23)° = 46°$

$\angle BOC = 3(23)° = 69°$

Exercise 5 (4 minutes)

Exercise 5

a. **In a complete sentence, describe the relevant angle relationships in the diagram. Set up and solve an equation to find the value of x. Find the measurements of $\angle AOB$ and $\angle BOC$.**

$\angle AOB$ and $\angle BOC$ are complementary and sum to $90°$.

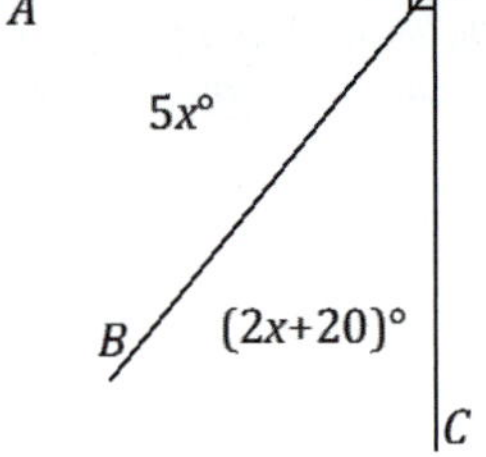

$$5x + (2x + 20) = 90 \quad \text{complementary } \angle\text{s}$$
$$7x + 20 = 90$$
$$7x + 20 - 20 = 90 - 20$$
$$7x = 70$$
$$x = 10$$

$\angle AOB = 5(10)° = 50°$

$\angle BOC = 2(10)° + 20° = 40°$

MP.7

b. **Katrina was solving the problem above and wrote the equation $7x + 20 = 90$. Then, she rewrote this as $7x + 20 = 70 + 20$. Why did she rewrite the equation in this way? How does this help her to find the value of x?**

She grouped the quantity on the right-hand side of the equation similarly to that of the left-hand side. This way, it is clear that the quantity $7x$ on the left-hand side must be equal to the quantity 70 on the right-hand side.

Closing (1 minute)

- In every unknown angle problem, it is important to identify the angle relationship(s) correctly in order to set up an equation that yields the unknown value.
- Check your answer by substituting and/or measuring to be sure it is correct.

Lesson Summary

- **To solve an unknown angle problem, identify the angle relationship(s) first to set up an equation that will yield the unknown value.**
- **Angles on a line and supplementary angles are not the same relationship. *Supplementary* angles are two angles whose angle measures sum to $180°$ whereas *angles on a line* are two or more adjacent angles whose angle measures sum to $180°$.**

Exit Ticket (3 minutes)

Name ______________________________ Date ______________

Lesson 2: Solving for Unknown Angles Using Equations

Exit Ticket

Two lines meet at a point that is also the vertex of an angle. Set up and solve an equation to find the value of x. Explain why your answer is reasonable.

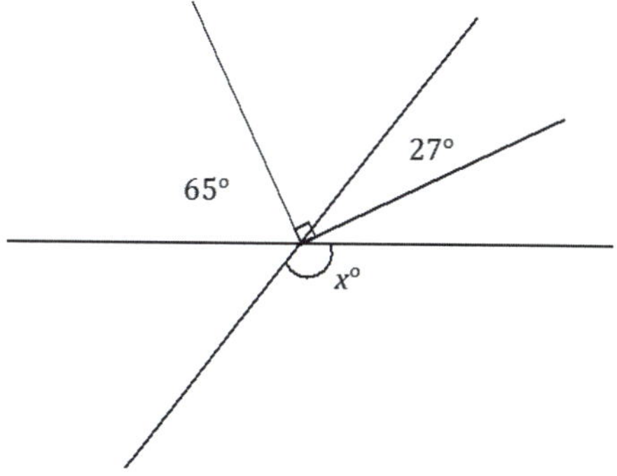

Exit Ticket Sample Solutions

Two lines meet at a point that is also the vertex of an angle. Set up and solve an equation to find the value of x. Explain why your answer is reasonable.

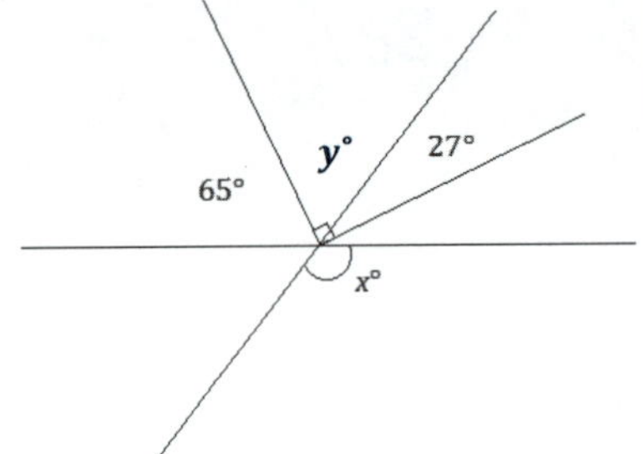

$$65 + (90 - 27) = x$$
$$x = 128$$

OR

$$y + 27 = 90$$
$$y + 27 - 27 = 90 - 27$$
$$y = 63$$

$$65 + y = x$$
$$65 + (63) = x$$
$$x = 128$$

The answers seem reasonable because a rounded value of y as 60 and a rounded value of its adjacent angle 65 as 70 yields a sum of 130, which is close to the calculated answer.

Problem Set Sample Solutions

Note: Arcs indicating unknown angles begin to be dropped from the diagrams. It is necessary for students to determine the specific angle whose measure is being sought. Students should draw their own arcs.

1. **Two lines meet at a point that is also the endpoint of a ray. Set up and solve an equation to find the value of c.**

$$c + 90 + 17 = 180 \quad \text{∠s on a line}$$
$$c + 107 = 180$$
$$c + 107 - 107 = 180 - 107$$
$$c = 73$$

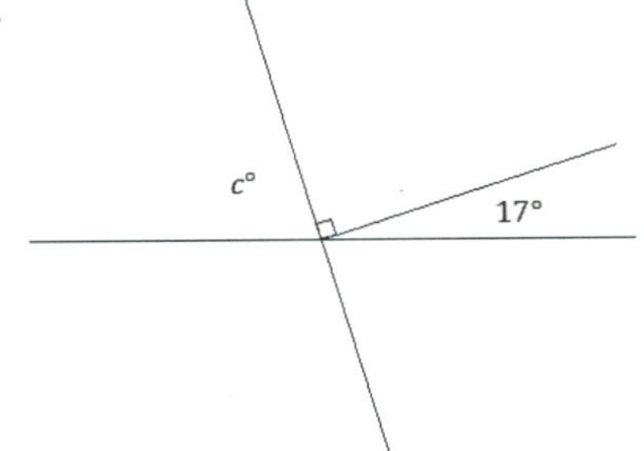

Scaffolded solutions:

a. *Use the equation above.*

b. *The angle marked $c°$, the right angle, and the angle with measurement $17°$ are angles on a line, and their measurements sum to $180°$.*

c. *Use the solution above. The answer seems reasonable because it looks like it has a measurement a little less than a $90°$ angle.*

> *Scaffolding:*
> Students struggling to organize their solution may benefit from prompts such as the following: Write an equation to model this situation. Explain how your equation describes the situation. Solve and interpret the solution. Is it reasonable?

2. Two lines meet at a point that is also the endpoint of a ray. Set up and solve an equation to find the value of a. Explain why your answer is reasonable.

$a + 33 = 49$ *∠s add and vert. ∠s*

$a + 33 - 33 = 49 - 33$

$a = 16$

The answers seem reasonable because a rounded value of a as 20 and a rounded value of its adjacent angle 33 as 30 yields a sum of 50, which is close to the rounded value of the measurement of the vertical angle.

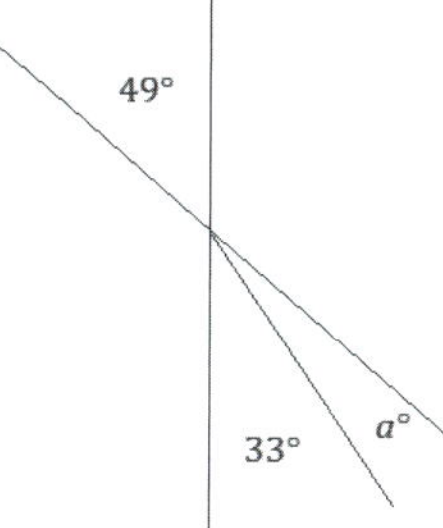

3. Two lines meet at a point that is also the endpoint of a ray. Set up and solve an equation to find the value of w.

$w + 90 = 125$ *∠s add and vert. ∠s*

$w + 90 - 90 = 125 - 90$

$w = 35$

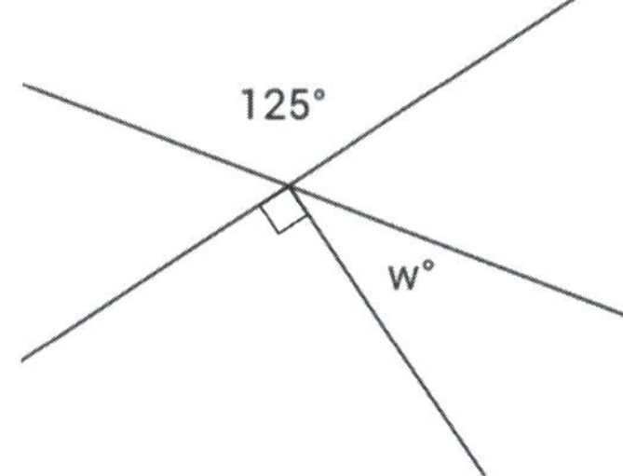

4. Two lines meet at a point that is also the vertex of an angle. Set up and solve an equation to find the value of m.

$(90 - 68) + 24 = m$ *∠s add and vert. ∠s*

$m = 46$

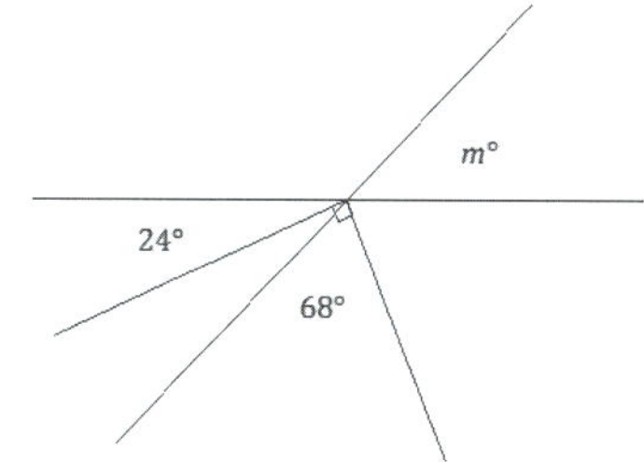

5. Three lines meet at a point. Set up and solve an equation to find the value of r.

$r + 122 + 34 = 180$ *∠s on a line and vert. ∠s*

$r + 156 = 180$

$r + 156 - 156 = 180 - 156$

$r = 24$

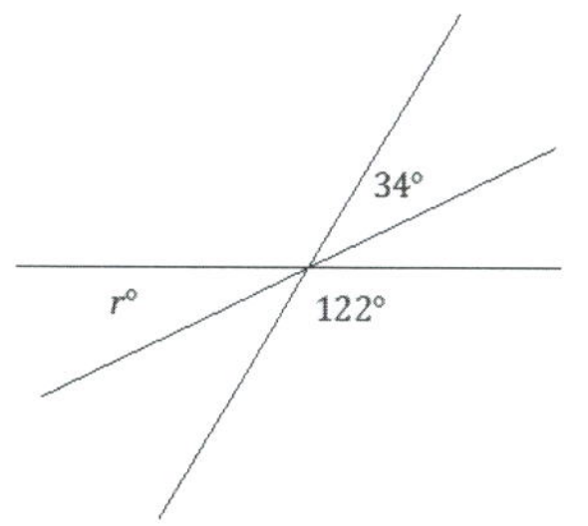

6. **Three lines meet at a point that is also the endpoint of a ray. Set up and solve an equation to find the value of each variable in the diagram.**

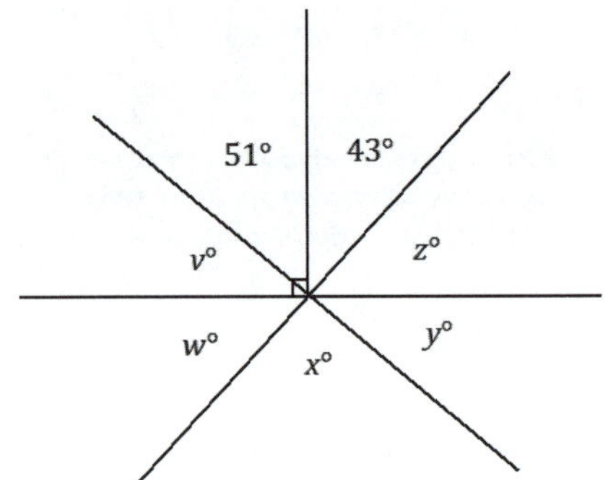

$v = 90 - 51$ *Complementary ∠s*
$v = 39$

$w + 39 + 51 + 43 = 180$ *∠s on a line*
$w + 133 = 180$
$w + 133 - 133 = 180 - 133$
$w = 47$

$x = 51 + 43$ *Vert. ∠s*
$x = 94$

$y = 39$ *Vert. ∠s*

$z = 47$ *Vert. ∠s*

7. **Set up and solve an equation to find the value of x. Find the measurement of $\angle AOB$ and of $\angle BOC$.**

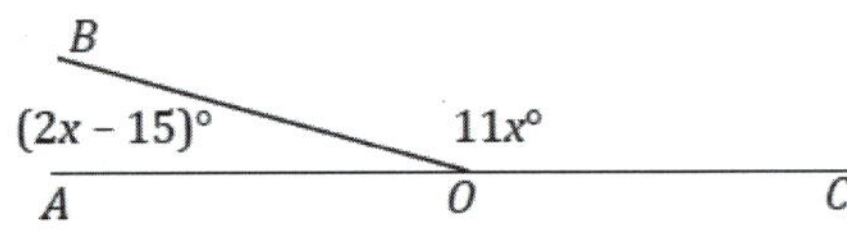

$(2x - 15) + 11x = 180$ *Supplementary ∠s*
$13x - 15 = 180$
$13x - 15 + 15 = 180 + 15$
$13x = 195$
$x = 15$

The measurement of $\angle AOB$: $2(15)^\circ - 15^\circ = 15^\circ$

The measurement of $\angle BOC$: $11(15)^\circ = 165^\circ$

Scaffolded solutions:

a. *Use the equation above.*

b. *The marked angles are angles on a line, and their measurements sum to* 180°.

c. *Once 15 is substituted for* x, *then the measurement of* $\angle AOB$ *is* 15° *and the measurement of* $\angle BOC$ *is* 165°. *These answers seem reasonable since* $\angle AOB$ *is acute and* $\angle BOC$ *is obtuse.*

> *Scaffolding:*
>
> Students struggling to organize their solution may benefit from prompts such as the following:
>
> - Write an equation to model this situation. Explain how your equation describes the situation. Solve and interpret the solution. Is it reasonable?

8. **Set up and solve an equation to find the value of x. Find the measurement of $\angle AOB$ and of $\angle BOC$.**

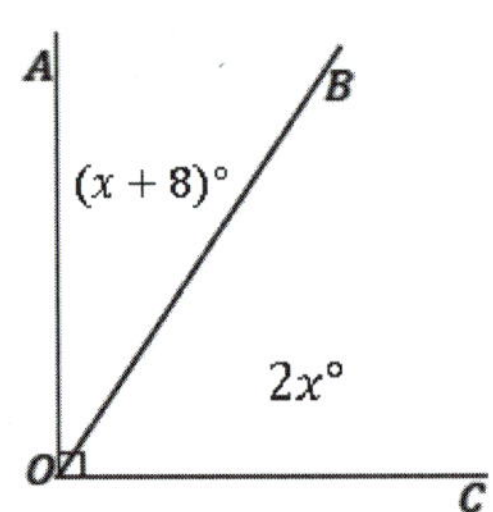

$x + 8 + 2x = 90$ *Complementary ∠s*
$3x + 8 = 90$
$3x + 8 - 8 = 90 - 8$
$3x = 82$
$x = 27\frac{1}{3}$

The measurement of $\angle AOB$: $\left(27\frac{1}{3}\right)^\circ + 8^\circ = 35\frac{1}{3}^\circ$

The measurement of $\angle BOC$: $2\left(27\frac{1}{3}\right)^\circ = 54\frac{2}{3}^\circ$

9. **Set up and solve an equation to find the value of x. Find the measurement of $\angle AOB$ and of $\angle BOC$.**

$$4x + 5 + 5x + 22 = 180 \quad \text{∠s on a line}$$
$$9x + 27 = 180$$
$$9x + 27 - 27 = 180 - 27$$
$$9x = 153$$
$$x = 17$$

The measurement of $\angle AOB$: $4(17)^\circ + 5^\circ = 73^\circ$

The measurement of $\angle BOC$: $5(17)^\circ + 22^\circ = 107^\circ$

10. **Write a verbal problem that models the following diagram. Then, solve for the two angles.**

One possible response: Two angles are supplementary. The measurement of one angle is five times the measurement of the other. Find the measurements of both angles.

$$10x + 2x = 180 \quad \text{Supplementary ∠s}$$
$$12x = 180$$
$$x = 15$$

The measurement of Angle 1: $10(15)^\circ = 150^\circ$

The measurement of Angle 2: $2(15)^\circ = 30^\circ$

Lesson 3: Solving for Unknown Angles Using Equations

Student Outcomes

- Students solve for unknown angles in word problems and in diagrams involving all learned angle facts.

Classwork

Opening Exercise (5 minutes)

> *Scaffolding:*
>
> Encourage students to redraw parts of the diagram to emphasize relationships. For example, line AB and ray OE could be redrawn to see the relationship $y°$ and the angle whose measure is $134°$.

Opening Exercise

Two lines meet at a point that is also a vertex of an angle; the measurement of $\angle AOF$ is $134°$. Set up and solve an equation to find the values of x and y. Are your answers reasonable? How do you know?

$$14 + x + 90 = 134 \qquad \textit{∠s add}$$
$$x + 104 = 134$$
$$x + 104 - 104 = 134 - 104$$
$$x = 30$$

$$y + 134 = 180 \qquad \textit{∠s on a line}$$
$$y + 134 - 134 = 180 - 134$$
$$y = 46$$

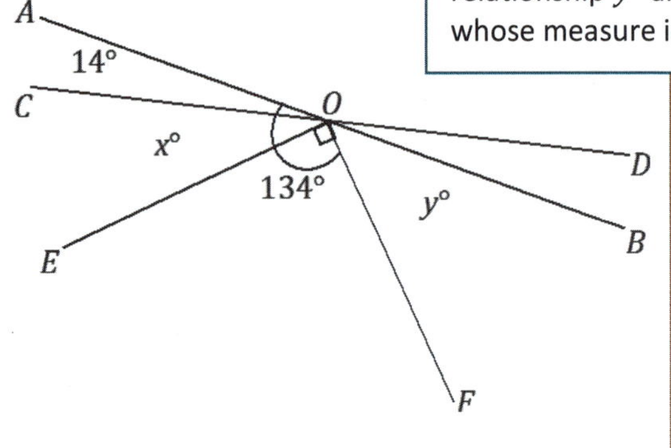

The answers are reasonable because the angle marked $y°$ appears to be approximately half the measurement of a right angle, and the angle marked $x°$ appears to be approximately double in measurement of $\angle AOC$.

In the following examples and exercises, students set up and solve an equation for the unknown angle based on the relevant angle relationships in the diagram. Encourage students to note the appropriate angle fact abbreviation for any step that depends on an angle relationship.

> *Scaffolding:*
>
> Remind students that a full rotation or turn through a circle is $360°$.
>
> 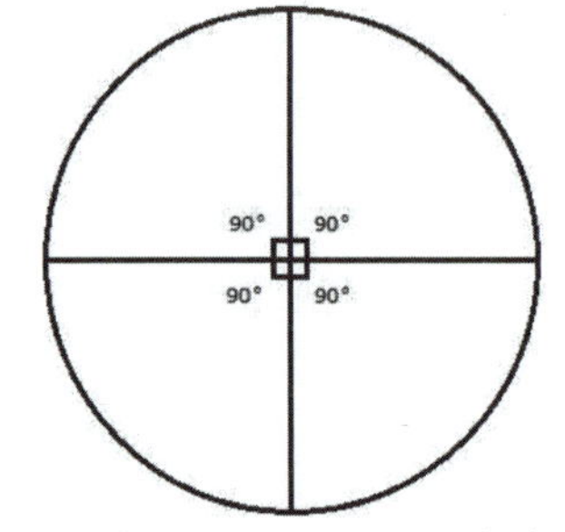
>
>
> A circular protractor may help to demonstrate this.

Example 1 (4 minutes)

Example 1

Set up and solve an equation to find the value of x.

$$x + 90 + 123 = 360 \qquad \textit{∠s at a point}$$
$$x + 213 = 360$$
$$x + 213 - 213 = 360 - 213$$
$$x = 147$$

Exercise 1 (4 minutes)

Exercise 1

Five rays meet at a common endpoint. In a complete sentence, describe the relevant angle relationships in the diagram. Set up and solve an equation to find the value of a.

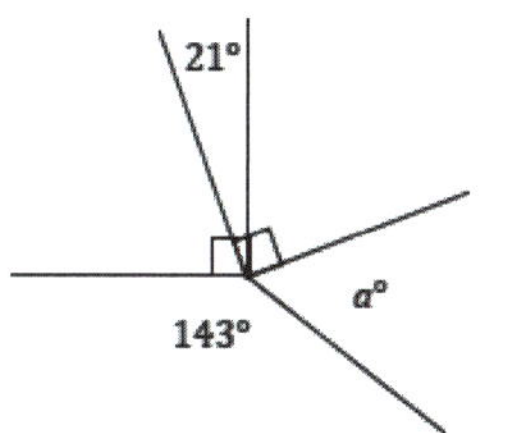

The sum of angles at a point is $360°$.

$90 + (90 - 21) + a + 143 = 360$ *∠s at a point*

$302 + a = 360$

$302 - 302 + a = 360 - 302$

$a = 58$

Example 2 (4 minutes)

Example 2

Four rays meet at a common endpoint. In a complete sentence, describe the relevant angle relationships in the diagram. Set up and solve an equation to find the value of x. Find the measurements of $\angle BAC$ and $\angle DAE$.

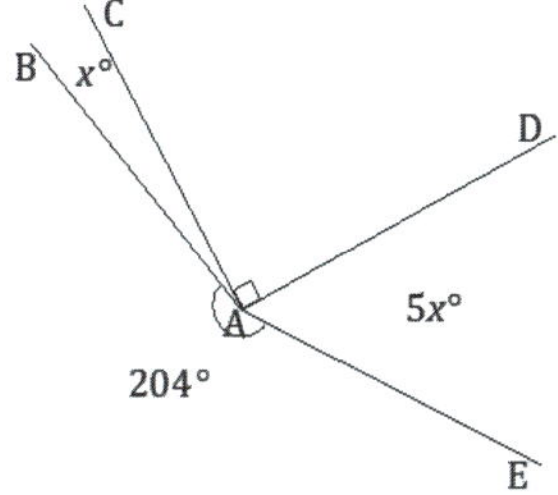

The sum of the degree measurements of $\angle BAC$, $\angle CAD$, $\angle DAE$ and the arc that measures $204°$ is $360°$.

$x + 90 + 5x + 204 = 360$ *∠s at a point*

$6x + 294 = 360$

$6x + 294 - 294 = 360 - 294$

$6x = 66$

$\left(\frac{1}{6}\right)6x = \left(\frac{1}{6}\right)66$

$x = 11$

The measurement of $\angle BAC$: $11°$

The measurement of $\angle DAE$: $5(11)° = 55°$

Exercise 2 (4 minutes)

Exercise 2

Four rays meet at a common endpoint. In a complete sentence, describe the relevant angle relationships in the diagram. Set up and solve an equation to find the value of x. Find the measurement of $\angle CAD$.

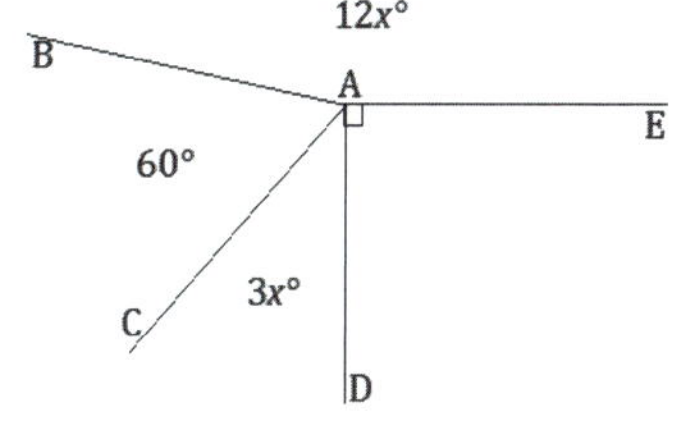

$\angle BAC$, $\angle CAD$, $\angle DAE$, and $\angle EAB$ are angles at a point and sum to $360°$.

$3x + 60 + 12x + 90 = 360$ *∠s at a point*

$15x + 150 = 360$

$15x + 150 - 150 = 360 - 150$

$15x = 210$

$\left(\frac{1}{15}\right)15x = \left(\frac{1}{15}\right)210$

$x = 14$

The measurement of $\angle CAD$: $3(14)° = 42°$

Example 3 (4 minutes)

Example 3

Two lines meet at a point that is also the endpoint of two rays. In a complete sentence, describe the relevant angle relationships in the diagram. Set up and solve an equation to find the value of x. Find the measurements of $\angle BAC$ and $\angle BAH$.

$\angle DAE$ is formed by adjacent angles $\angle EAF$ and $\angle FAD$; the measurement of $\angle DAE$ is equal to the sum of the measurements of the adjacent angles. This is also true for the measurement of $\angle CAH$, formed by adjacent angles $\angle CAB$ and $\angle BAH$. $\angle CAH$ is vertically opposite from and equal in measurement to $\angle DAE$.

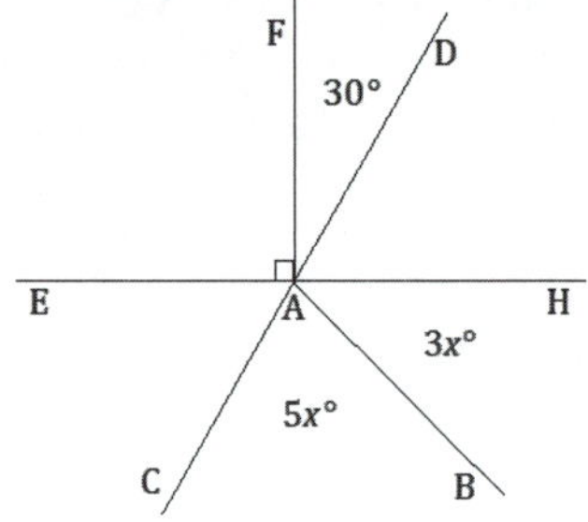

$90 + 30 = 120$ ***$\angle DAE$, $\angle$s add***

$5x + 3x = 8x$ ***$\angle CAH$, $\angle$s add***

$8x = 120$ ***Vert. $\angle$s***

$\left(\frac{1}{8}\right)8x = \left(\frac{1}{8}\right)120$

$x = 15$

The measurement of $\angle BAC$: $5(15)° = 75°$

The measurement of $\angle BAH$: $3(15)° = 45°$

Exercise 3 (4 minutes)

Exercise 3

Lines AB and EF meet at a point which is also the endpoint of two rays. In a complete sentence, describe the relevant angle relationships in the diagram. Set up and solve an equation to find the value of x. Find the measurements of $\angle DHF$ and $\angle AHD$.

The measurement of $\angle AHF$, formed by adjacent angles $\angle AHD$ and $\angle DHF$, is equal to the sum of the measurements of the adjacent angles. This is also true for the measurement of $\angle EHB$, which is formed by adjacent angles $\angle EHC$ and $\angle CHB$. $\angle AHF$ is vertically opposite from and equal in measurement to $\angle EHB$.

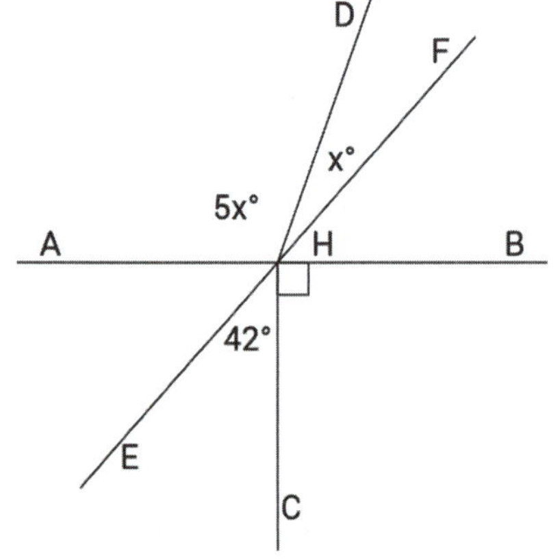

$5x + x = 6x$ ***$\angle AHF$, $\angle$s add***

$42 + 90 = 132$ ***$\angle EHB$, $\angle$s add***

$6x = 132$ ***Vert. $\angle$s***

$\left(\frac{1}{6}\right)6x = \left(\frac{1}{6}\right)132$

$x = 22$

The measurement of $\angle DHF$: $22°$

The measurement of $\angle AHD$: $5(22)° = 110°$

The following examples are designed to highlight MP.7 by helping students to see the connection between an angle diagram and the equation used to model it. Solving equations with variables on both sides is a topic in Grade 8. Teachers may choose to show the solution method if they so choose.

Example 4 (6 minutes)

Example 4

Two lines meet at a point. Set up and solve an equation to find the value of x. Find the measurement of one of the vertical angles.

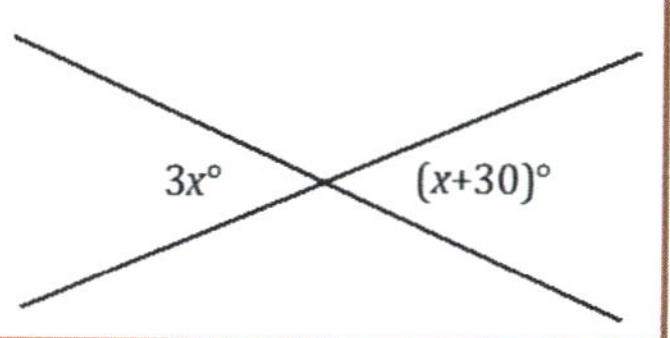

Students use information in the figure and a protractor to solve for x.

i) Students measure a $30°$ angle as shown; the remaining portion of the angle must be $x°$ (∠s add).

ii) Students can use their protractor to find the measurement of $x°$ and use this measurement to partition the other angle in the vertical pair.

As a check, students should substitute the measured x value into each expression and evaluate; each angle of the vertical pair should equal the other. Students can also use their protractor to measure each angle of the vertical angle pair.

MP.7

With a modified figure, students can write an algebraic equation that they have the skills to solve.

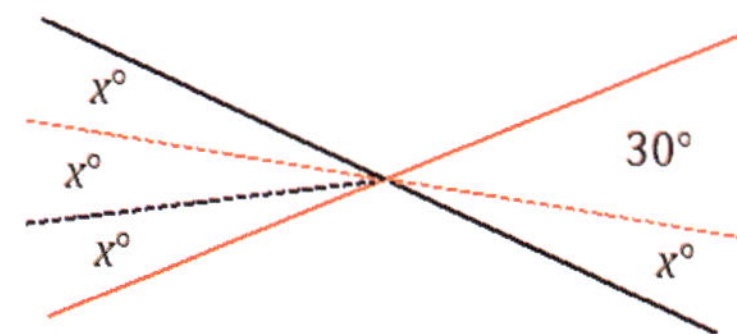

$$2x = 30 \qquad \textit{Vert. ∠s}$$
$$\left(\frac{1}{2}\right)2x = \left(\frac{1}{2}\right)30$$
$$x = 15$$

Measurement of each angle in the vertical pair: $3(15)° = 45°$

Extension: The algebra steps above are particularly helpful as a stepping-stone in demonstrating how to solve the equation that takes care of the problem in one step as follows:

$$3x = x + 30 \qquad \textit{Vert. ∠s}$$
$$3x - x = x - x + 30$$
$$2x = 30$$
$$\left(\frac{1}{2}\right)2x = \left(\frac{1}{2}\right)30$$
$$x = 15$$

Measurement of each angle in the vertical pair: $3(15)° = 45°$

Students understand the first line of this solution because of their knowledge of vertical angles. In fact, the only line they are not familiar with is the second line of the solution, which is a skill that they learn in Grade 8. Showing students this solution is simply a preview.

Exercise 4 (4 minutes)

Exercise 4

Set up and solve an equation to find the value of x. Find the measurement of one of the vertical angles.

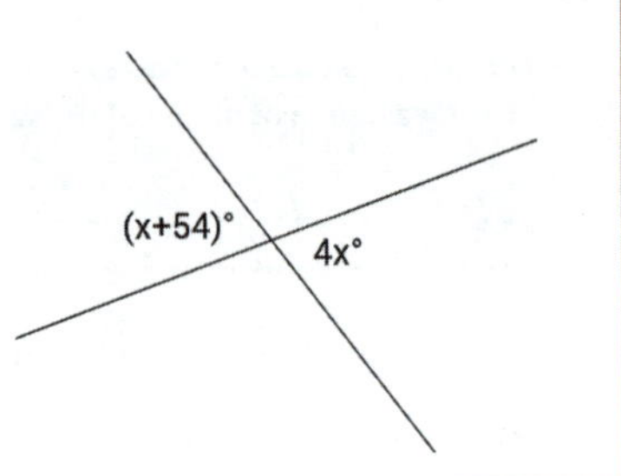

Students use information in the figure and a protractor to solve for x.

i) Students measure a $54°$ angle as shown; the remaining portion of the angle must be x ($\angle$s add).

ii) Students can use their protractors to find the measurement of x and use this measurement to partition the other angle in the vertical pair.

Students should perform a check as in Example 4 before solving an equation that matches the modified figure.

$$54 = 3x \qquad \textit{Vert. } \angle s$$
$$\left(\frac{1}{3}\right)54 = \left(\frac{1}{3}\right)3x$$
$$x = 18$$

Measurement of each vertical angle: $4(18)° = 72°$

Extension:

$$x + 54 = 4x \qquad \textit{Vert. } \angle s$$
$$x - x + 54 = 4x - x$$
$$54 = 3x$$
$$\left(\frac{1}{3}\right)54 = \left(\frac{1}{3}\right)3x$$
$$x = 18$$

Closing (1 minute)

- In every unknown angle problem, it is important to identify the angle relationship(s) correctly in order to set up an equation that yields the unknown value.
- Check your answer by substituting and/or measuring to be sure it is correct.

Lesson Summary

Steps to Solving for Unknown Angles

- Identify the angle relationship(s).
- Set up an equation that will yield the unknown value.
- Solve the equation for the unknown value.
- Substitute the answer to determine the angle(s).
- Check and verify your answer by measuring the angle with a protractor.

Exit Ticket (5 minutes)

Name ______________________________ Date ______________

Lesson 3: Solving for Unknown Angles Using Equations

Exit Ticket

1. Two rays have a common endpoint on a line. Set up and solve an equation to find the value of z. Find the measurements of $\angle AYC$ and $\angle DYB$.

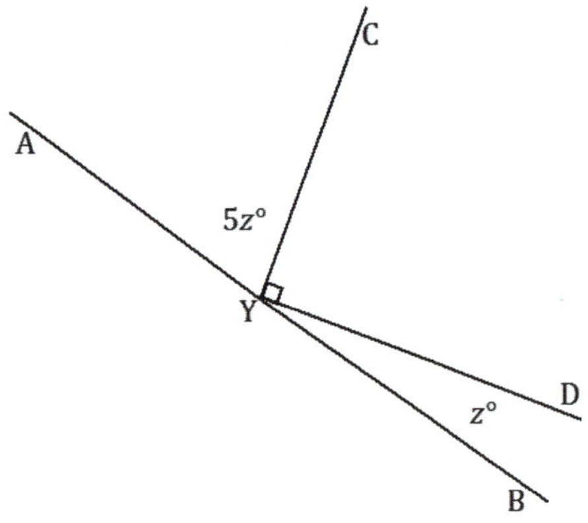

2. Two lines meet at a point that is also the vertex of an angle. Set up and solve an equation to find the value of x. Find the measurements of $\angle CAH$ and $\angle EAG$.

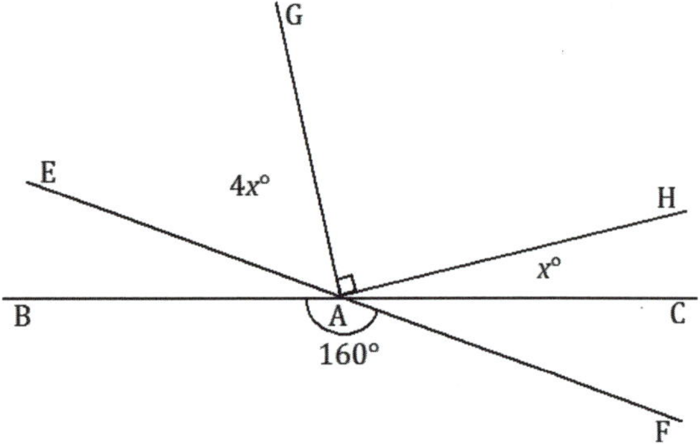

Exit Ticket Sample Solutions

1. Two rays have a common endpoint on a line. Set up and solve an equation to find the value of z. Find the measurements of $\angle AYC$ and $\angle DYB$.

$$5z + 90 + z = 180 \quad \textit{∠s on a line}$$
$$6z + 90 = 180$$
$$6z + 90 - 90 = 180 - 90$$
$$6z = 90$$
$$\left(\frac{1}{6}\right)6z = \left(\frac{1}{6}\right)90$$
$$z = 15$$

The measurement of $\angle AYC$*:* $5(15)° = 75°$

The measurement of $\angle DYB$*:* $15°$

Scaffolded solutions:

a. *Use the equation above.*

b. *The angle marked* $z°$*, the right angle, and the angle with measurement* $5z°$ *are angles on a line. Their measurements sum to* $180°$*.*

c. *The answers seem reasonable because once* 15 *is substituted in for* z*, the measurement of* $\angle AYC$ *is* $75°$*, which is slightly smaller than a right angle, and the measurement of* $\angle DYB$ *is* $15°$*, which is an acute angle.*

Scaffolding:

Students struggling to organize their solution may benefit from prompts such as the following: Write an equation to model this situation. Explain how your equation describes the situation. Solve and interpret the solution. Is it reasonable?

2. Two lines meet at a point that is also the vertex of an angle. Set up and solve an equation to find the value of x. Find the measurements of $\angle CAH$ and $\angle EAG$.

$$4x + 90 + x = 160 \quad \textit{vert. ∠s}$$
$$5x + 90 = 160$$
$$5x + 90 - 90 = 160 - 90$$
$$5x = 70$$
$$\left(\frac{1}{5}\right)5x = \left(\frac{1}{5}\right)70$$
$$x = 14$$

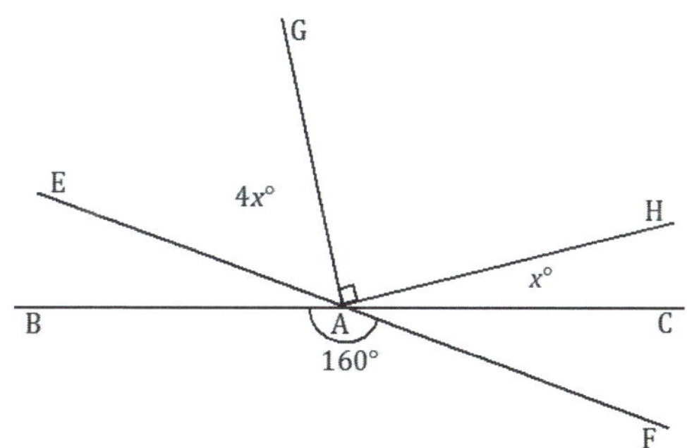

The measurement of $\angle CAH$*:* $14°$

The measurement of $\angle EAG$*:* $4(14)° = 56°$

Problem Set Sample Solutions

Set up and solve an equation for the unknown angle based on the relevant angle relationships in the diagram. Add labels to diagrams as needed to facilitate their solutions. List the appropriate angle fact abbreviation for any step that depends on an angle relationship.

1. Two lines meet at a point. Set up and solve an equation to find the value of x.

$$x + 15 = 72 \quad \textit{Vert. ∠s}$$
$$x + 15 - 15 = 72 - 15$$
$$x = 57$$

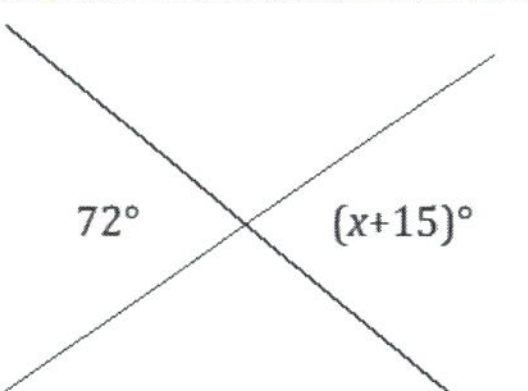

2. Three lines meet at a point. Set up and solve an equation to find the value of a. Is your answer reasonable? Explain how you know.

Let $b = a$. *Vert. ∠s*

∠s on a line

$$78 + b + 52 = 180$$
$$b + 130 = 180$$
$$b + 130 - 130 = 180 - 130$$
$$b = 50$$

Since $b = a$, $a = 50$.

The answer seems reasonable since it is similar in magnitude to the $52°$ angle.

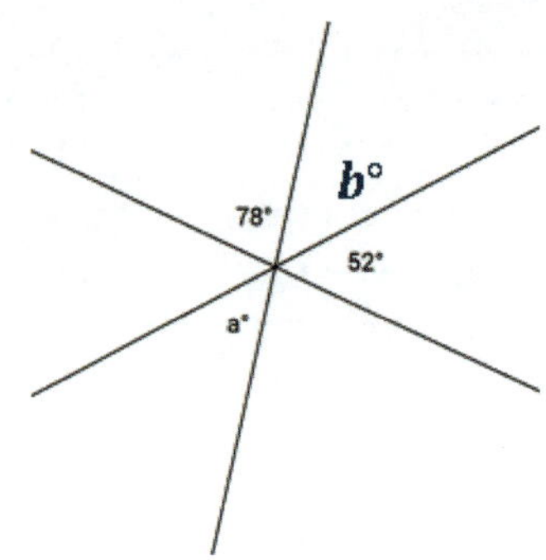

3. Two lines meet at a point that is also the endpoint of two rays. Set up and solve an equation to find the values of a and b.

$$a + 32 + 90 = 180 \quad \text{∠s on a line}$$
$$a + 122 = 180$$
$$a + 122 - 122 = 180 - 122$$
$$a = 58$$

$$a + b + 90 = 180 \quad \text{∠s on a line}$$
$$58 + b + 90 = 180$$
$$b + 148 = 180$$
$$b + 148 - 148 = 180 - 148$$
$$b = 32$$

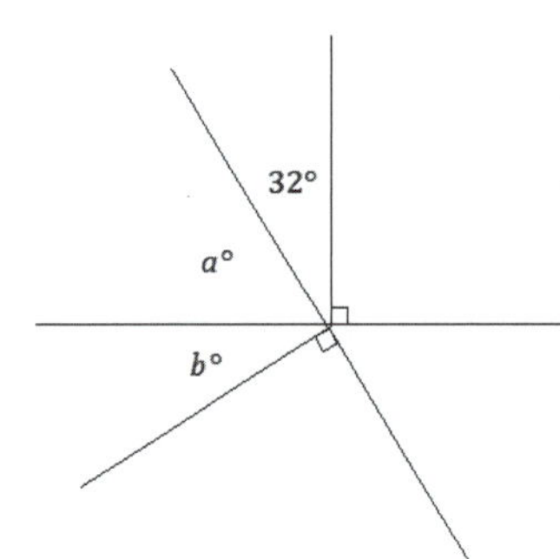

Scaffolding:

Students struggling to organize their solution may benefit from prompts such as the following: Write an equation to model this situation. Explain how your equation describes the situation. Solve and interpret the solution. Is it reasonable?

Scaffolded solutions:

a. *Use the equation above.*

b. *The angle marked $a°$, the angle with measurement $32°$, and the right angle are angles on a line. Their measurements sum to $180°$.*

c. *The answers seem reasonable because once the values of a and b are substituted, it appears that the two angles ($a°$ and $b°$) form a right angle. We know those two angles should form a right angle because the angle adjacent to it is a right angle.*

4. Three lines meet at a point that is also the endpoint of a ray. Set up and solve an equation to find the values of x and y.

$$x + 39 + 90 = 180 \quad \text{∠s on a line}$$
$$x + 129 = 180$$
$$x + 129 - 129 = 180 - 129$$
$$x = 51$$

$$y + x + 90 = 180 \quad \text{∠s on a line}$$
$$y + 51 + 90 = 180$$
$$y + 141 = 180$$
$$y + 141 - 141 = 180 - 141$$
$$y = 39$$

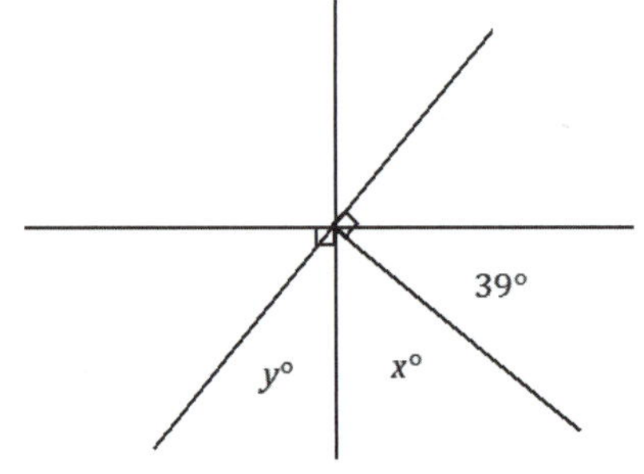

5. Two lines meet at a point. Find the measurement of one of the vertical angles. Is your answer reasonable? Explain how you know.

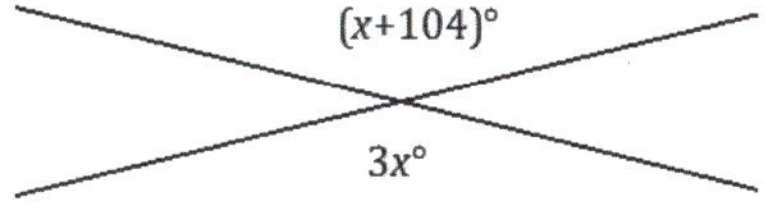

$$2x = 104 \quad \textit{vert.} \angle s$$
$$\left(\frac{1}{2}\right)2x = \left(\frac{1}{2}\right)104$$
$$x = 52$$

Measurement of each vertical angle: $3(52)° = 156°$

The answer seems reasonable because a rounded value of 50 *would make the numeric value of each expression* 150 *and* 154*, which are reasonably close for a check.*

A solution can include a modified diagram, as shown, and the supporting algebra work.

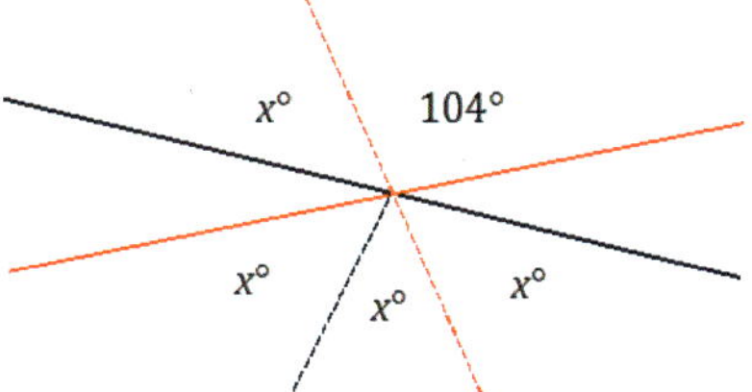

Solutions may also include the full equation and solution:

$$3x = x + 104 \quad \textit{Vert.} \angle s$$
$$3x - x = x - x + 104$$
$$2x = 104$$
$$\left(\frac{1}{2}\right)2x = \left(\frac{1}{2}\right)104$$
$$x = 52$$

6. Three lines meet at a point that is also the endpoint of a ray. Set up and solve an equation to find the value of y.

Let $x°$ *and* $z°$ *be the measurements of the indicated angles.*

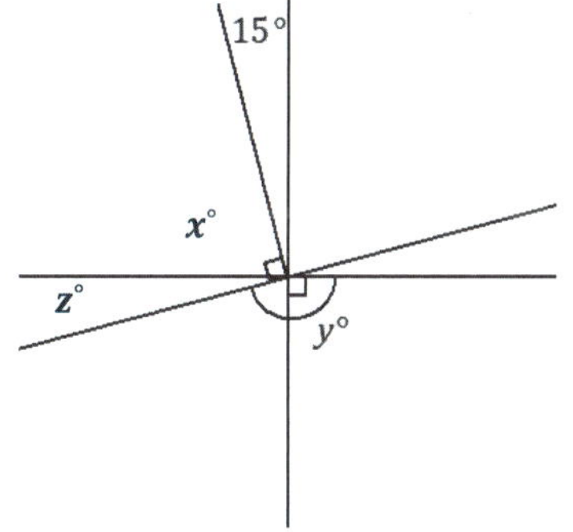

$$x + 15 = 90 \quad \textit{Vert.} \angle s$$
$$x + 15 - 15 = 90 - 15$$
$$x = 75$$

$$x + z = 90 \quad \textit{Complementary} \angle s$$
$$75 + z = 90$$
$$75 - 75 + z = 90 - 75$$
$$z = 15$$

$$z + y = 180 \quad \angle s \textit{ on a line}$$
$$15 + y = 180$$
$$15 - 15 + y = 180 - 15$$
$$y = 165$$

7. Three adjacent angles are at a point. The second angle is $20°$ more than the first, and the third angle is $20°$ more than the second angle.

a. Find the measurements of all three angles.

$$x + (x + 20) + (x + 20 + 20) = 360 \quad \text{∠s at a point}$$
$$3x + 60 = 360$$
$$3x + 60 - 60 = 360 - 60$$
$$3x = 300$$
$$\left(\frac{1}{3}\right)3x = \left(\frac{1}{3}\right)300$$
$$x = 100$$

Angle 1: $100°$

Angle 2: $100° + 20° = 120°$

Angle 3: $100° + 20° + 20° = 140°$

MP.2 & MP.7

b. Compare the expressions you used for the three angles and their combined expression. Explain how they are equal and how they reveal different information about this situation.

By the commutative and associative laws, $x + (x + 20) + (x + 20 + 20)$ *is equal to* $(x + x + x) + (20 + 20 + 20)$, *which is equal to* $3x + 60$. *The first expression,* $x + (x + 20) + (x + 20 + 20)$, *shows the sum of three unknown numbers, where the second is* 20 *more than the first, and the third is* 20 *more than the second. The expression* $3x + 60$ *shows the sum of three times an unknown number with* 60.

8. Four adjacent angles are on a line. The measurements of the four angles are four consecutive even numbers. Determine the measurements of all four angles.

$$x + (x + 2) + (x + 4) + (x + 6) = 180 \quad \text{∠s on a line}$$
$$4x + 12 = 180$$
$$4x + 12 - 12 = 180 - 12$$
$$4x = 168$$
$$\left(\frac{1}{4}\right)4x = \left(\frac{1}{4}\right)168$$
$$x = 42$$

The four angle measures are $42°$, $44°$, $46°$, *and* $48°$.

Scaffolding:

Teachers may need to review the term *consecutive* for students to successfully complete Problem Set 8.

9. Three adjacent angles are at a point. The ratio of the measurement of the second angle to the measurement of the first angle is $4:3$. The ratio of the measurement of the third angle to the measurement of the second angle is $5:4$. Determine the measurements of all three angles.

Let the smallest measure of the three angles be $3x°$. *Then, the measure of the second angle is* $4x°$, *and the measure of the third angle is* $5x°$.

$$3x + 4x + 5x = 360 \quad \text{∠s at a point}$$
$$12x = 360$$
$$\left(\frac{1}{12}\right)12x = \left(\frac{1}{12}\right)360$$
$$x = 30$$

Angle 1: $3(30)° = 90°$

Angle 2: $4(30)° = 120°$

Angle 3: $5(30)° = 150°$

10. Four lines meet at a point. Solve for x and y in the following diagram.

$2x + 18 + 90 = 180$ *∠s on a line*

$2x + 108 = 180$

$2x + 108 - 108 = 180 - 108$

$2x = 72$

$\left(\frac{1}{2}\right)2x = \left(\frac{1}{2}\right)72$

$x = 36$

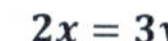

$2x = 3y$ *Vert. ∠s*

$2(36) = 3y$

$72 = 3y$

$\left(\frac{1}{3}\right)72 = \left(\frac{1}{3}\right)3y$

$y = 24$

Lesson 4: Solving for Unknown Angles Using Equations

Student Outcomes

- Students solve for unknown angles in word problems and in diagrams involving all learned angle facts.

Classwork

Opening Exercise (5 minutes)

> **Opening Exercise**
>
> **The complement of an angle is four times the measurement of the angle. Find the measurement of the angle and its complement.**
>
> $x + 4x = 90$ *Complementary ∠s*
>
> $5x = 90$
>
> $\left(\frac{1}{5}\right)5x = \left(\frac{1}{5}\right)90$
>
> $x = 18$
>
> *The measurement of the angle is* $18°$.
>
> *The measurement of the complement of the angle is* $72°$.

> *Scaffolding:*
>
> As in earlier lessons, tasks such as the Opening Exercise can be scaffolded into parts as follows:
>
> - Explain the angle relationships in the diagram. Write the equation. Explain how the equation represents the diagram, including particular parts. Solve the equation. Interpret the solution, and determine if it is reasonable.

In the following examples and exercises, students set up and solve an equation for the unknown angle based on the relevant angle relationships in the diagram. Encourage students to list the appropriate angle fact abbreviation for any step that depends on an angle relationship.

Example 1 (4 minutes)

Two options are provided here for Example 1. The second is more challenging than the first.

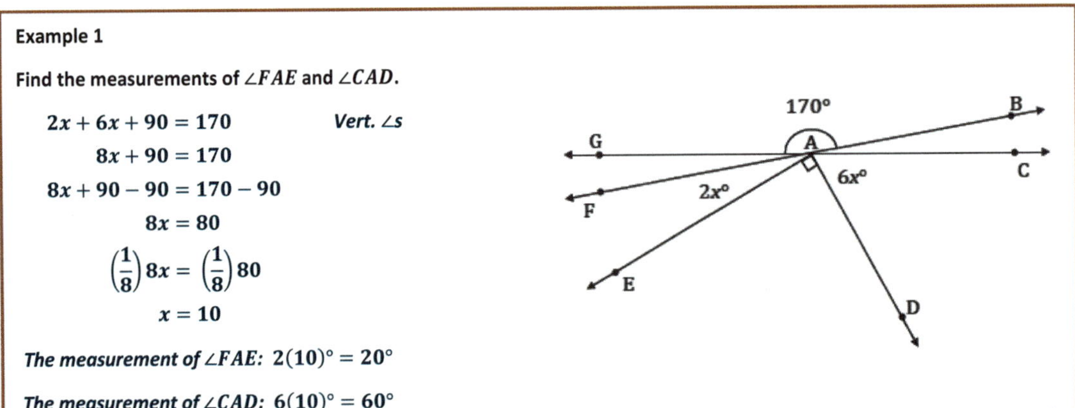

> **Example 1**
>
> **Find the measurements of $\angle FAE$ and $\angle CAD$.**
>
> $2x + 6x + 90 = 170$ *Vert. ∠s*
>
> $8x + 90 = 170$
>
> $8x + 90 - 90 = 170 - 90$
>
> $8x = 80$
>
> $\left(\frac{1}{8}\right)8x = \left(\frac{1}{8}\right)80$
>
> $x = 10$
>
> *The measurement of* $\angle FAE$: $2(10)° = 20°$
>
> *The measurement of* $\angle CAD$: $6(10)° = 60°$

Two lines meet at a point. List the relevant angle relationship in the diagram. Set up and solve an equation to find the value of x. Find the measurement of one of the vertical angles.

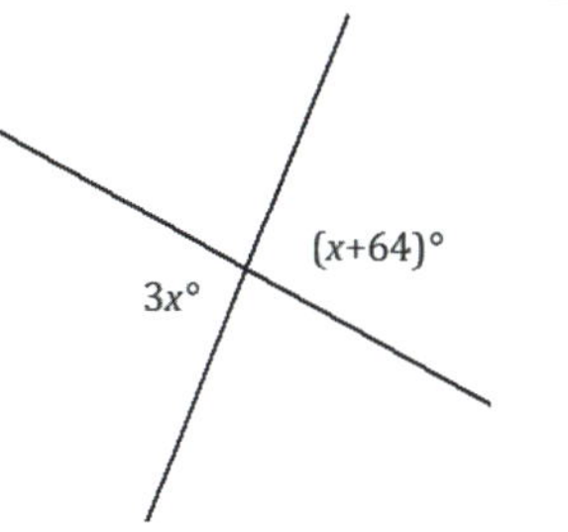

Students use information in the figure and a protractor to solve for x.

i) Students measure a 64° angle as shown; the remaining portion of the angle must be x° ($\angle$s add).

ii) Students can use their protractors to find the measurement of x° and use this measurement to partition the other angle in the vertical pair.

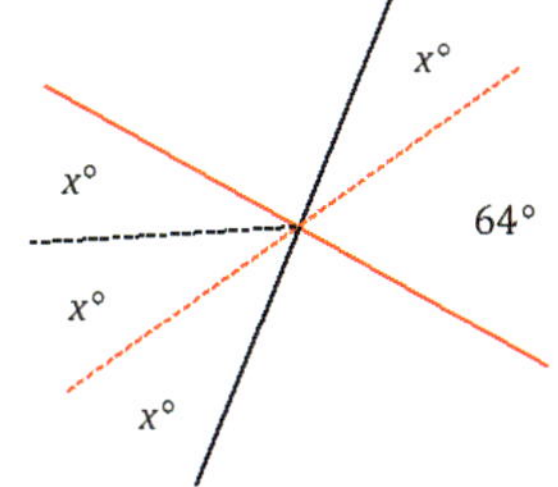

As a check, students should substitute the measured x value into each expression and evaluate; each angle of the vertical pair should be equal to the other. Students can also use their protractor to measure each angle of the vertical angle pair.

MP.7

With a modified figure, students can write an algebraic equation that they have the skills to solve.

$$2x = 64 \qquad \textit{Vert. } \angle s$$
$$\left(\frac{1}{2}\right)2x = \left(\frac{1}{2}\right)64$$
$$x = 32$$

Measurement of each angle in the vertical pair: $3(32)^\circ = 96^\circ$

Extension:

$$3x = x + 64 \qquad \textit{vert. } \angle s$$
$$3x - x = x - x + 64$$
$$2x = 64$$
$$\left(\frac{1}{2}\right)2x = \left(\frac{1}{2}\right)64$$
$$x = 32$$

Measurement of each angle in the vertical pair: $3(32)^\circ = 96^\circ$

Exercise 1 (4 minutes)

Exercise 1

Set up and solve an equation to find the value of x. List the relevant angle relationship in the diagram. Find the measurement of one of the vertical angles.

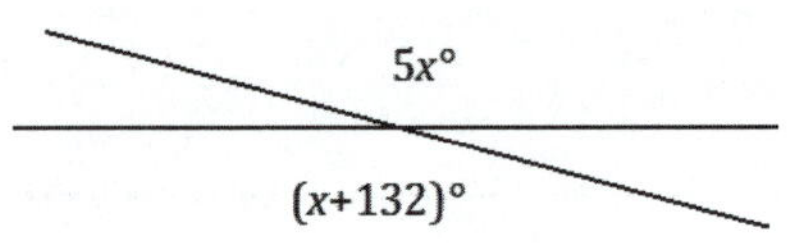

Students use information in the figure and a protractor to solve for x.

i) Measure a $132°$ angle as shown; the remaining portion of the original angle must be $x°$ (∠s add).

ii) Partition the other angle in the vertical pair into equal angles of $x°$.

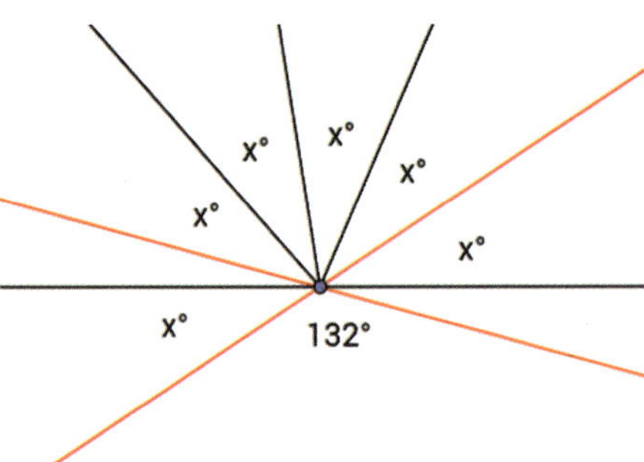

Students should perform a check (as in Example 1) before solving an equation that matches the modified figure.

Extension:

$$4x = 132 \qquad \textit{vert. } \angle s$$

$$\left(\frac{1}{4}\right)4x = \left(\frac{1}{4}\right)132$$

$$x = 33$$

Measurement of each vertical angle: $5(33)° = 165°$

Note: Students can check their answers for any question by measuring each unknown angle with a protractor, as all diagrams are drawn to scale.

Example 2 (4 minutes)

Example 2

Three lines meet at a point. List the relevant angle relationships in the diagram. Set up and solve an equation to find the value of b.

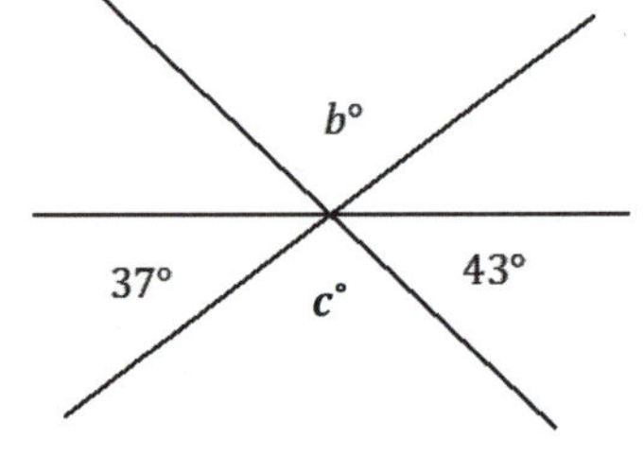

Let $b = c$.

$$c + 37 + 43 = 180 \qquad \angle s \textit{ on a line}$$

$$c + 80 = 180$$

$$c + 80 - 80 = 180 - 80$$

$$c = 100$$

Since $b = c$, $b = 100$.

Exercise 2 (4 minutes)

Students set up and solve an equation for the unknown angle based on the relevant angle relationships in the diagram. List the appropriate angle fact abbreviation in the initial equation.

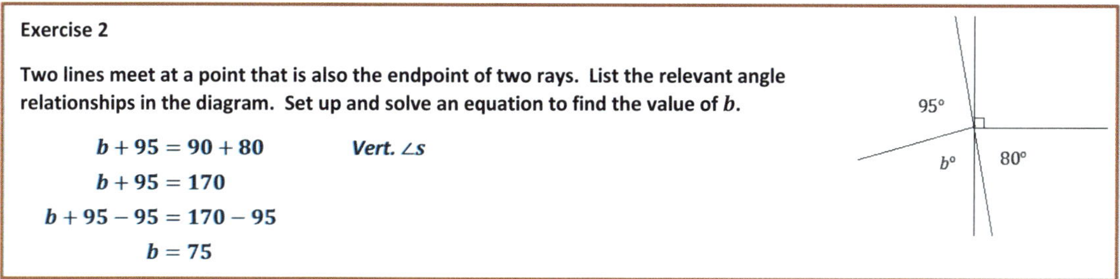

Exercise 2

Two lines meet at a point that is also the endpoint of two rays. List the relevant angle relationships in the diagram. Set up and solve an equation to find the value of b.

$$b + 95 = 90 + 80 \qquad \textit{Vert. } \angle s$$
$$b + 95 = 170$$
$$b + 95 - 95 = 170 - 95$$
$$b = 75$$

Example 3 (6 minutes)

Students set up and solve an equation for the unknown angle based on the relevant angle relationship in the question. In this case, suggest that students use the words *angle* and *supplement* as placeholders in their equations. Students can use a tape diagram to solve for the unknown angles.

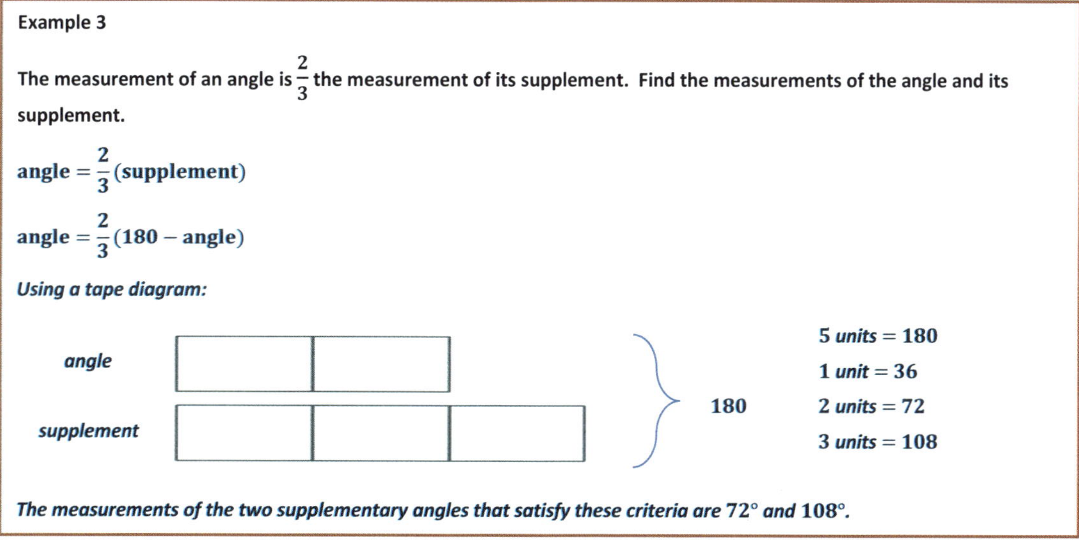

Example 3

The measurement of an angle is $\frac{2}{3}$ the measurement of its supplement. Find the measurements of the angle and its supplement.

$$\text{angle} = \frac{2}{3}(\text{supplement})$$

$$\text{angle} = \frac{2}{3}(180 - \text{angle})$$

Using a tape diagram:

5 units $= 180$
1 unit $= 36$
2 units $= 72$
3 units $= 108$

The measurements of the two supplementary angles that satisfy these criteria are $72°$ and $108°$.

The tape diagram model is an ideal strategy for this question. If students are not familiar with the tape diagram model, use a Guess and Check table with them. Here is an example of such a table with two entries for guesses that did not result in a correct answer.

Guess	$\frac{2}{3}(\text{Guess})$	Sum (should be $180°$)
60	$\frac{2}{3}(60) = 40$	$60 + 40 = 100$; not the answer
90	$\frac{2}{3}(90) = 60$	$90 + 60 = 150$; not the answer

Exercise 3 (5 minutes)

Students set up and solve an equation for the unknown angle based on the relevant angle relationship in the question. In this case, suggest that students use the words *angle* and *complement* as placeholders in their equations. Students can use a tape diagram to solve for the unknown angles.

Exercise 3

The measurement of an angle is $\frac{1}{4}$ the measurement of its complement. Find the measurements of the two complementary angles.

$$\text{angle} = \frac{1}{4}(\text{complement})$$

$$\text{angle} = \frac{1}{4}(90 - \text{angle})$$

Using a tape diagram:

5 *units* $= 90$

1 *unit* $= 18$

4 *units* $= 72$

The measurements of the two complementary angles that satisfy these criteria are $18°$ and $72°$.

Example 4 (4 minutes)

Example 4

Three lines meet at a point that is also the endpoint of a ray. List the relevant angle relationships in the diagram. Set up and solve an equation to find the value of z.

Let $x°$ be the measurement of the indicated angle.

$x + 90 + 29 = 180$ *∠s on a line*

$x + 119 = 180$

$x + 119 - 119 = 180 - 119$

$x = 61$

$z = x + 90$ *∠s add*

$z = 61 + 90$

$z = 151$

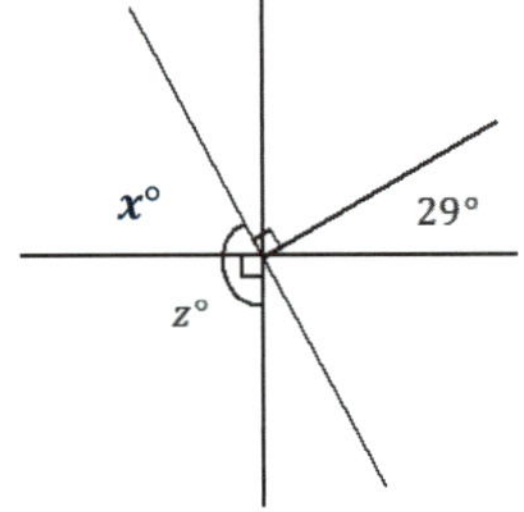

Exercise 4 (4 minutes)

Exercise 4

Two lines meet at a point that is also the vertex of an angle. Set up and solve an equation to find the value of x. Find the measurements of $\angle GAF$ and $\angle BAC$.

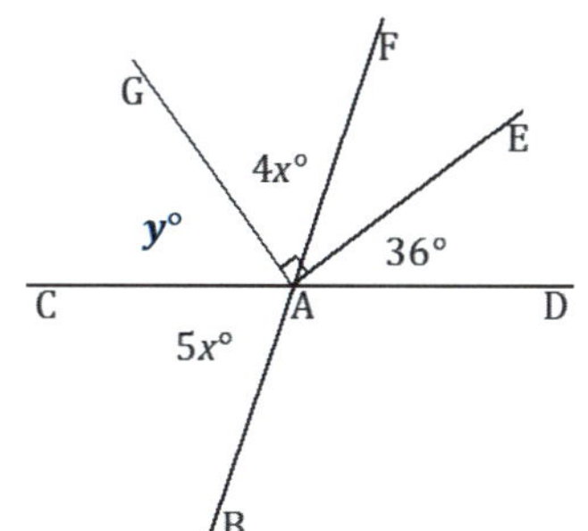

Let $y°$ be the measurement of the indicated angle.

$$y = 180 - (90 + 36) \quad \angle\text{s on a line}$$
$$y = 54$$

$$4x + y + 5x = 180 \quad \angle\text{s on a line}$$
$$4x + 54 + 5x = 180$$
$$9x + 54 = 180$$
$$9x + 54 - 54 = 180 - 54$$
$$9x = 126$$
$$\left(\frac{1}{9}\right)9x = \left(\frac{1}{9}\right)126$$
$$x = 14$$

The measurement of $\angle GAF$: $4(14)° = 56°$

The measurement of $\angle BAC$: $5(14)° = 70°$

Closing (1 minute)

- In every unknown angle problem, it is important to identify the angle relationship(s) correctly in order to set up an equation that yields the unknown value.
- Check your answer by substituting and/or measuring to be sure it is correct.

Lesson Summary

Steps to Solving for Unknown Angles

- **Identify the angle relationship(s).**
- **Set up an equation that will yield the unknown value.**
- **Solve the equation for the unknown value.**
- **Substitute the answer to determine the measurement of the angle(s).**
- **Check and verify your answer by measuring the angle with a protractor.**

Exit Ticket (4 minutes)

Name ______________________________ Date ______________

Lesson 4: Solving for Unknown Angles Using Equations

Exit Ticket

Lines BC and EF meet at A. Rays AG and AD form a right angle. Set up and solve an equation to find the values of x and w.

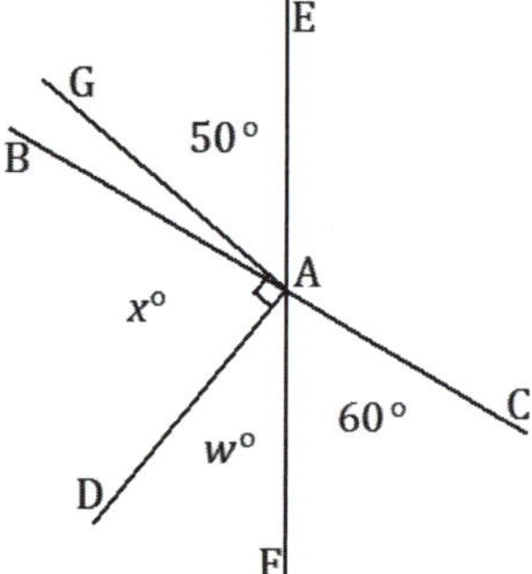

Exit Ticket Sample Solutions

Lines BC and EF meet at A. Rays AG and AD form a right angle. Set up and solve an equation to find the values of x and w.

$\angle BAE = 60$ *Vert. ∠s*

$\angle BAG = 10$ *∠s add*

$x + \angle BAG = 90$ *Complementary ∠s*

$x + 10 = 90$

$x + 10 - 10 = 90 - 10$

$x = 80$

$x + w + 60 = 180$ *∠s on a line*

$80 + w + 60 = 180$

$140 + w = 180$

$140 + w - 140 = 180 - 140$

$w = 40$

Problem Set Sample Solutions

Set up and solve an equation for the unknown angle based on the relevant angle relationships in the diagram. Add labels to diagrams as needed to facilitate their solutions. List the appropriate angle fact abbreviation for any step that depends on an angle relationship.

> *Scaffolding:*
>
> Some students may need to use the corner of a piece of paper to confirm which rays form the right angle:
>
> $59 + d = 90$ or
>
> $59 + d + c = 90$.

1. **Four rays have a common endpoint on a line. Set up and solve an equation to find the value of c.**

$59 + d = 90$ *Complementary ∠s*

$59 - 59 + d = 90 - 59$

$d = 31$

$d + c + 140 = 180$ *∠s on a line*

$31 + c + 140 = 180$

$c + 171 = 180$

$c + 171 - 171 = 180 - 171$

$c = 9$

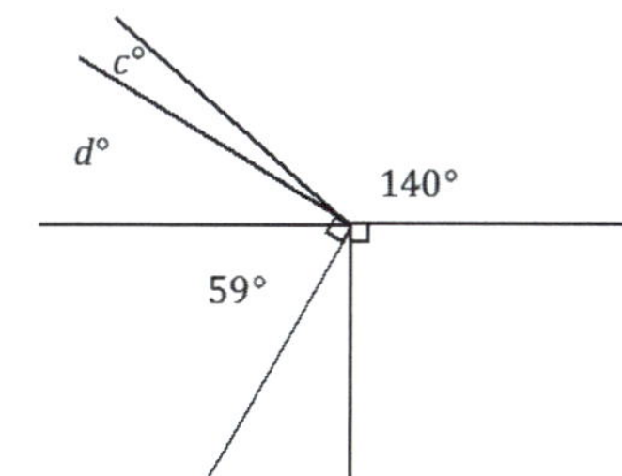

2. Lines BC and EF meet at A. Set up and solve an equation to find the value of x. Find the measurements of $\angle EAH$ and $\angle HAC$.

$$\angle BAE + 57 = 90$$ *Complementary ∠s*
$$\angle BAE + 57 - 57 = 90 - 57$$
$$\angle BAE = 33$$

$$\angle BAE + 3x + 4x = 180$$ *∠s on a line*
$$33 + 3x + 4x = 180$$
$$33 + 7x = 180$$
$$33 - 33 + 7x = 180 - 33$$
$$7x = 147$$
$$\left(\frac{1}{7}\right)7x = \left(\frac{1}{7}\right)147$$
$$x = 21$$

The measurement of $\angle EAH$: $3(21)° = 63°$

The measurement of $\angle HAC$: $4(21)° = 84°$

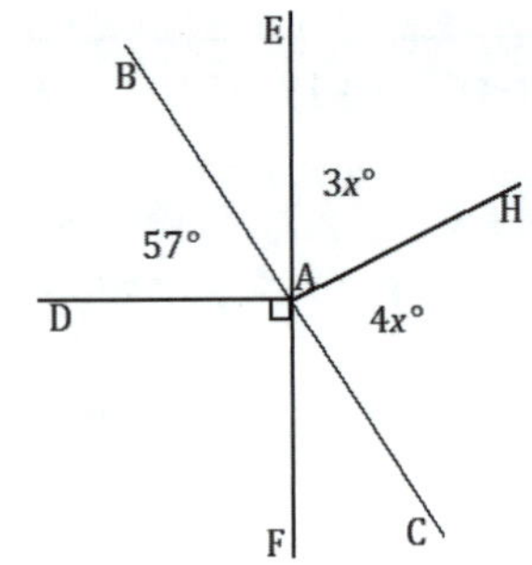

> *Scaffolding:*
>
> Students struggling to organize their solution may benefit from prompts such as the following:
>
> - Write an equation to model this situation. Explain how your equation describes the situation. Solve and interpret the solution. Is it reasonable?

3. Five rays share a common endpoint. Set up and solve an equation to find the value of x. Find the measurements of $\angle DAG$ and $\angle GAH$.

$$2x + 36.5 + 36.5 + 2x + 3x = 360$$ *∠s at a point*
$$7x + 73 = 360$$
$$7x + 73 - 73 = 360 - 73$$
$$7x = 287$$
$$\left(\frac{1}{7}\right)7x = \left(\frac{1}{7}\right)287$$
$$x = 41$$

The measurement of $\angle EAF$: $2(41)° = 82°$

The measurement of $\angle GAH$: $3(41)° = 123°$

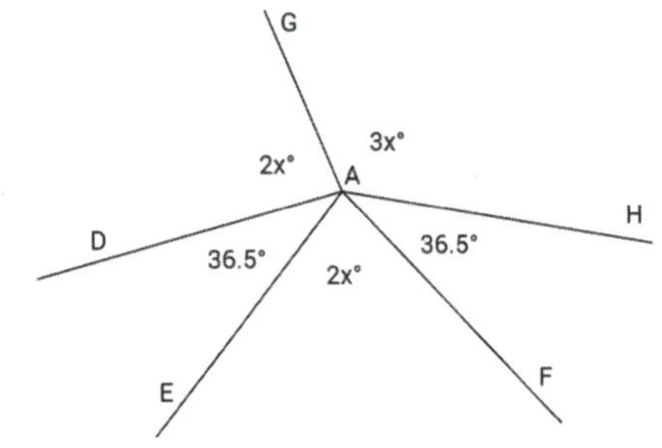

4. Four lines meet at a point which is also the endpoint of three rays. Set up and solve an equation to find the values of x and y.

$$2y + 12 + 15 + 90 = 180$$ *∠s on a line*
$$2y + 117 = 180$$
$$2y + 117 - 117 = 180 - 117$$
$$2y = 63$$
$$\left(\frac{1}{2}\right)2y = \left(\frac{1}{2}\right)63$$
$$y = 31.5$$

$$3x = 2y$$ *Vert. ∠s*
$$3x = 2(31.5)$$
$$3x = 63$$
$$\left(\frac{1}{3}\right)3x = \left(\frac{1}{3}\right)63$$
$$x = 21$$

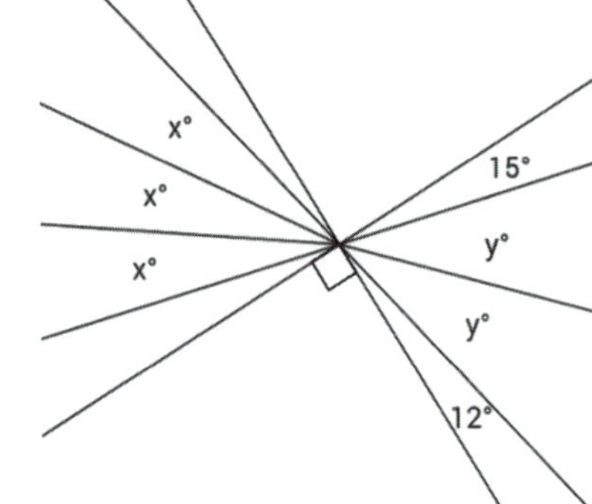

5. Two lines meet at a point that is also the vertex of a right angle. Set up and solve an equation to find the value of x. Find the measurements of $\angle CAE$ and $\angle BAG$.

$\angle DAB = 4x$ *vert. ∠s*

$\angle DAG = 90 + 15 = 105$ *∠s add*

$4x + 3x = 105$ *∠s add*

$7x = 105$

$\left(\frac{1}{7}\right)7x = \left(\frac{1}{7}\right)105$

$x = 15$

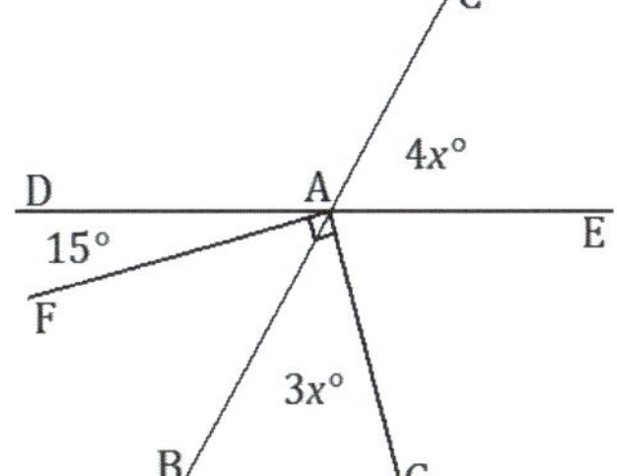

The measurement of $\angle CAE$: $4(15)° = 60°$

The measurement of $\angle BAG$: $3(15)° = 45°$

6. Five angles are at a point. The measurement of each angle is one of five consecutive, positive whole numbers.

a. Determine the measurements of all five angles.

$$x + (x+1) + (x+2) + (x+3) + (x+4) = 360$$
$$5x + 10 = 360$$
$$5x + 10 - 10 = 360 - 10$$
$$5x = 350$$
$$\left(\frac{1}{5}\right)5x = \left(\frac{1}{5}\right)350$$
$$x = 70$$

Angle measures are $70°$, $71°$, $72°$, $73°$, *and* $74°$.

MP.2 & MP.7

b. Compare the expressions you used for the five angles and their combined expression. Explain how they are equivalent and how they reveal different information about this situation.

By the commutative and associative laws, $x + (x+1) + (x+2) + (x+3) + (x+4)$ *is equal to* $(x + x + x + x + x) + (1 + 2 + 3 + 4)$, *which is equal to* $5x + 10$. *The first expression,* $x + (x+1) + (x+2) + (x+3) + (x+4)$, *shows the sum of five unknown numbers where the second is* 1 *degree more than the first, the third is* 1 *degree more than the second, and so on. The expression* $5x + 10$ *shows the sum of five times an unknown number with* 10.

7. Let $x°$ be the measurement of an angle. The ratio of the measurement of the complement of the angle to the measurement of the supplement of the angle is $1:3$. The measurement of the complement of the angle and the measurement of the supplement of the angle have a sum of $180°$. Use a tape diagram to find the measurement of this angle.

$(90 - x):(180 - x) = 1:3$

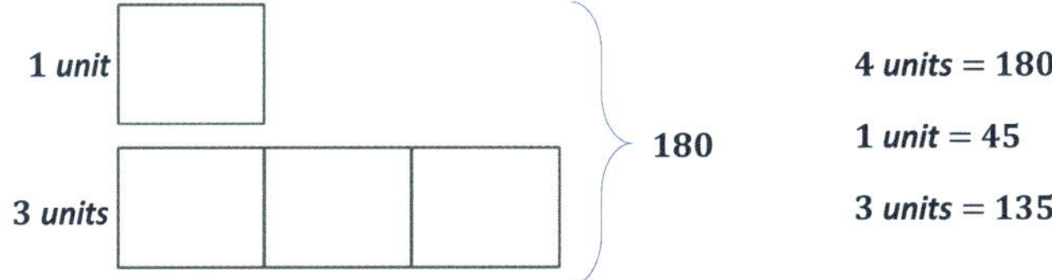

4 units $= 180$

1 unit $= 45$

3 units $= 135$

The measurement of the angle that satisfies these criteria is $45°$.

8. Two lines meet at a point. Set up and solve an equation to find the value of x. Find the measurement of one of the vertical angles.

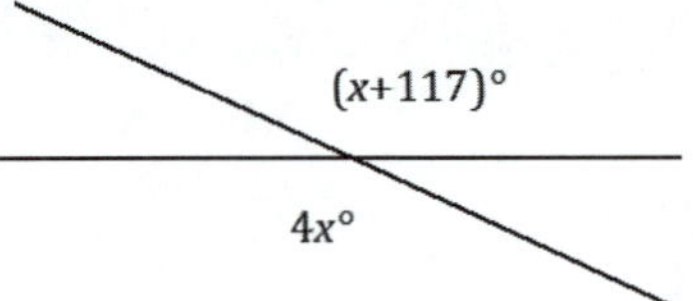

A solution can include a modified diagram (as shown) and the supporting algebra work:

$$3x = 117 \quad \textit{vert. } \angle s$$
$$\left(\frac{1}{3}\right)3x = \left(\frac{1}{3}\right)117$$
$$x = 39$$

Each vertical angle: $4(39)° = 156°$

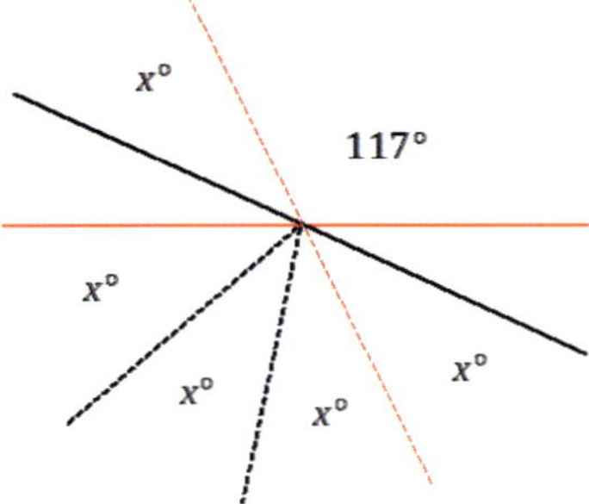

Solutions may also include the full equation and solution:

$$4x = x + 117$$
$$4x - x = x - x + 117$$
$$3x = 117$$
$$\left(\frac{1}{3}\right)3x = \left(\frac{1}{3}\right)117$$
$$x = 39$$

9. The difference between three times the measurement of the complement of an angle and the measurement of the supplement of that angle is $20°$. What is the measurement of the angle?

$$3(90 - x) - (180 - x) = 20$$
$$270 - 3x - 180 + x = 20$$
$$90 - 2x = 20$$
$$90 - 90 - 2x = 20 - 90$$
$$-2x = -70$$
$$\left(-\frac{1}{2}\right)(-2x) = \left(-\frac{1}{2}\right)(-70)$$
$$x = 35$$

The measurement of the angle is $35°$.

Topic B

Constructing Triangles

7.G.A.2

Focus Standard:	7.G.A.2	Draw (freehand, with ruler and protractor, and with technology) geometric shapes with given conditions. Focus on constructing triangles from three measures of angles or sides, noticing when the conditions determine a unique triangle, more than one triangle, or no triangle.
Instructional Days:	11	
Lesson 5:	Identical Triangles (S)[1]	
Lesson 6:	Drawing Geometric Shapes (E)	
Lesson 7:	Drawing Parallelograms (P)	
Lesson 8:	Drawing Triangles (E)	
Lesson 9:	Conditions for a Unique Triangle—Three Sides and Two Sides and the Included Angle (E)	
Lesson 10:	Conditions for a Unique Triangle—Two Angles and a Given Side (E)	
Lesson 11:	Conditions on Measurements That Determine a Triangle (E)	
Lesson 12:	Unique Triangles—Two Sides and a Non-Included Angle (E)	
Lessons 13–14:	Checking for Identical Triangles (P, P)	
Lesson 15:	Using Unique Triangles to Solve Real-World and Mathematical Problems (P)	

Lesson 5 provides the foundation for almost every other lesson in Topic B. Students learn how to label two triangles as identical or different by understanding triangle correspondence and learning the relevant notation and terminology pertaining to it. In Lesson 6, students practice using a ruler, protractor, and compass to construct geometric shapes set by given conditions (e.g., constructing circles of radius 5 cm and 12 cm or constructing a triangle so that one angle is $100°$). Students use a new tool, a set square, to draw parallelograms in Lesson 7. With an understanding of how to use the construction tools, students focus next on drawing triangles. Exercises in Lesson 8 demonstrate how a given set of conditions determines how many different triangles can be drawn. For example, the number of triangles that can be drawn with a requirement of a $90°$ angle is different from the number that can be drawn with a requirement of side lengths 4 cm, 5 cm, and 6 cm.

[1]Lesson Structure Key: **P**-Problem Set Lesson, **M**-Modeling Cycle Lesson, **E**-Exploration Lesson, **S**-Socratic Lesson

Next, students consider whether the triangles they construct are identical. In fact, standard **7.G.A.2** asks students to construct triangles from three measures of angles or sides, "noticing when the conditions determine a unique triangle, more than one triangle, or no triangle." The guiding question is then: What conditions determine a unique triangle (i.e., the construction always yields identical triangles), more than one triangle (i.e., the construction leads to non-identical triangles), or no triangle (i.e., a triangle cannot be formed by the construction)? In Lessons 9–10, students explore the conditions that determine a unique triangle. Note that the discussion regarding the conditions that determine a unique triangle is distinct from the discussion regarding whether two figures are congruent, which requires a study of rigid motions (Grade 8 Module 2). However, the study of what constitutes uniqueness is inextricably linked to the notion of identical figures. In Lesson 11, students discover the side-length conditions and angle-measurement conditions that determine whether or not a triangle can be formed. Lesson 12 focuses on the conditions that do not guarantee a unique triangle. With all these conditions covered, Lessons 13 and 14 ask students to practice constructing viable arguments to explain whether provided information determines a unique triangle, more than one triangle, or no triangle. Finally, in Lesson 15, students solve real-world and mathematical problems by applying their understanding of the correspondences that exist between identical triangles.

Lesson 5: Identical Triangles

Student Outcomes

- Students use a triangle correspondence to recognize when two triangles match identically.
- Students use notation to denote a triangle correspondence and use the triangle correspondence to talk about corresponding angles and sides.
- Students are able to label equal angles and sides of triangles with multiple arcs or tick marks.

Lesson Notes

This lesson provides a basis for identifying two triangles as *identical*. To clearly define triangles as identical, students must understand what a triangle correspondence is and be able to manipulate the relevant notation and terminology. Once this is understood, students have the means, specifically the language, to discuss what makes a triangle unique in Lesson 7 and forward. Exercise 7 in the Problem Set is designed as an exploratory challenge; do not expect students to develop an exact answer at this level.

Classwork

Opening (2 minutes)

Opening

When studying triangles, it is essential to be able to communicate about the parts of a triangle without any confusion. The following terms are used to identify particular angles or sides:

- **between**
- **adjacent to**
- **opposite to**
- **included [side/angle]**

Exercises 1–7 (15 minutes)

Exercises 1–7

Use the figure $\triangle ABC$ to fill in the following blanks.

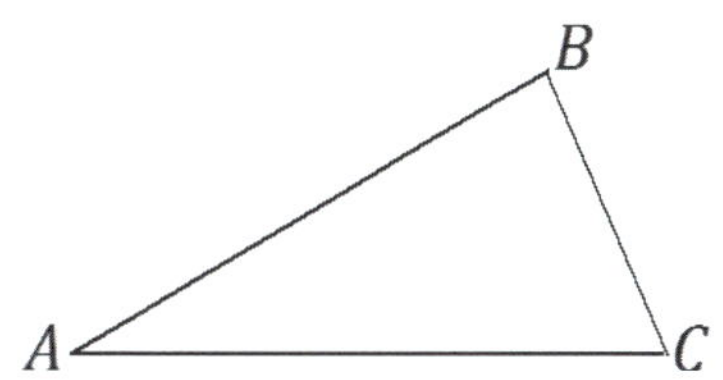

1. **$\angle A$ is *between* sides $\overline{AB}$ and $\overline{AC}$.**
2. **$\angle B$ is *adjacent to* side $\overline{AB}$ and to side $\overline{BC}$.**
3. **Side $\overline{AB}$ is *opposite to* $\angle C$.**
4. **Side *$\overline{BC}$* is the included side of $\angle B$ and $\angle C$.**

MP.6

5. $\angle B$ is opposite to side $\overline{AC}$.

6. Side $\overline{AB}$ is between $\angle A$ and $\angle B$.

7. What is the included angle of sides $\overline{AB}$ and $\overline{BC}$? $\angle B$.

Now that we know what to call the parts within a triangle, we consider how to discuss two triangles. We need to compare the parts of the triangles in a way that is easy to understand. To establish some alignment between the triangles, we pair up the vertices of the two triangles. We call this a *correspondence*. Specifically, a correspondence between two triangles is a pairing of each vertex of one triangle with one (and only one) vertex of the other triangle. A correspondence provides a systematic way to compare parts of two triangles.

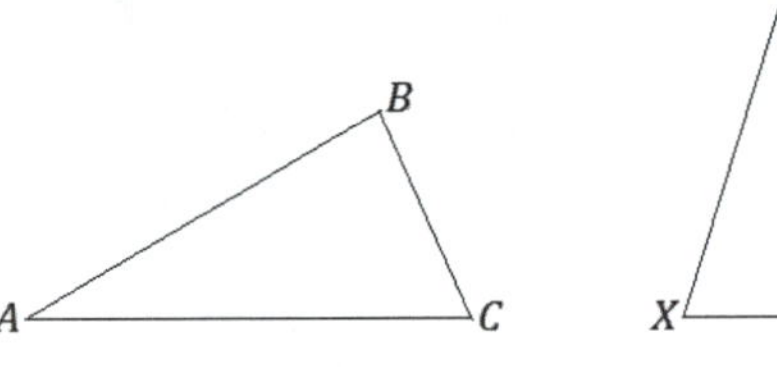

Figure 1

In Figure 1, we can choose to assign a correspondence so that A matches to X, B matches to Y, and C matches to Z. We notate this correspondence with double arrows: $A \leftrightarrow X$, $B \leftrightarrow Y$, and $C \leftrightarrow Z$. This is just one of six possible correspondences between the two triangles. Four of the six correspondences are listed below; find the remaining two correspondences.

$A \leftrightarrow X$, $B \leftrightarrow Y$, $C \leftrightarrow Z$

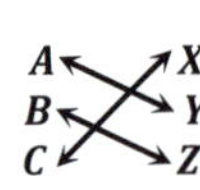

$A \leftrightarrow Z$, $B \leftrightarrow Y$, $C \leftrightarrow X$

$A \leftrightarrow Y$, $B \leftrightarrow X$, $C \leftrightarrow Z$

$A \leftrightarrow Y$, $B \leftrightarrow Z$, $C \leftrightarrow X$

$A \leftrightarrow Z$, $B \leftrightarrow X$, $C \leftrightarrow Y$

A simpler way to indicate the triangle correspondences is to let the order of the vertices define the correspondence (i.e., the first corresponds to the first, the second to the second, and the third to the third). The correspondences above can be written in this manner. Write the remaining two correspondences in this way.

$\triangle ABC \leftrightarrow \triangle XYZ$ $\triangle ABC \leftrightarrow \triangle XZY$ $\triangle ABC \leftrightarrow \triangle ZYX$

$\triangle ABC \leftrightarrow \triangle YXZ$ $\triangle ABC \leftrightarrow \triangle YZX$ $\triangle ABC \leftrightarrow \triangle ZXY$

Students have already seen a correspondence without knowing the formal use of the word. The correspondence between numerical coordinates and geometric points allows methods from algebra to be applied to geometry. The correspondence between a figure in a scale drawing and the corresponding figure in a scale drawing allows students to compute actual lengths and areas from the scale drawing. In Grade 8, students learn that figures are congruent when there is a transformation that makes a correspondence between the two figures. The idea of a correspondence lays the foundation for the understanding of functions discussed in Algebra I.

Discussion (5 minutes)

Review the remaining two correspondences that students filled out.

- Why do we take time to set up a correspondence?
  - *A correspondence provides a systematic way to compare parts of two triangles. Without a correspondence, it would be difficult to discuss the parts of a triangle because we would have no way of referring to particular sides, angles, or vertices.*

EUREKA MATH

- Assume the correspondence $\triangle ABC \leftrightarrow \triangle YZX$. What can we conclude about the vertices?
  - *Vertex A corresponds to Y, B corresponds to Z, and C corresponds to X.*
- How is it possible for any two triangles to have a total of six correspondences?
  - *We can match the first vertex in one triangle with any of the three vertices in the second triangle. Then, the second vertex of one triangle can be matched with any of the remaining two vertices in the second triangle.*

With a correspondence in place, comparisons can be made about corresponding sides and corresponding angles. The following are corresponding vertices, angles, and sides for the triangle correspondence $\triangle ABC \leftrightarrow \triangle YXZ$. Complete the missing correspondences.

Vertices:	$A \leftrightarrow Y$	$\boldsymbol{B \leftrightarrow X}$	$\boldsymbol{C \leftrightarrow Z}$
Angles:	$\angle A \leftrightarrow \angle Y$	$\boldsymbol{\angle B \leftrightarrow \angle X}$	$\boldsymbol{\angle C \leftrightarrow \angle Z}$
Sides:	$\overline{AB} \leftrightarrow \overline{YX}$	$\boldsymbol{\overline{BC} \leftrightarrow \overline{XZ}}$	$\boldsymbol{\overline{CA} \leftrightarrow \overline{ZY}}$

Example 1 (5 minutes)

Example 1

Given the following triangle correspondences, use double arrows to show the correspondence between vertices, angles, and sides.

Triangle Correspondence	$\triangle ABC \leftrightarrow \triangle STR$
Correspondence of Vertices	$A \longleftrightarrow S$ $B \longleftrightarrow T$ $C \longleftrightarrow R$
Correspondence of Angles	$\angle A \longleftrightarrow \angle S$ $\angle B \longleftrightarrow \angle T$ $\angle C \longleftrightarrow \angle R$
Correspondence of Sides	$\overline{AB} \longleftrightarrow \overline{ST}$ $\overline{BC} \longleftrightarrow \overline{TR}$ $\overline{CA} \longleftrightarrow \overline{RS}$

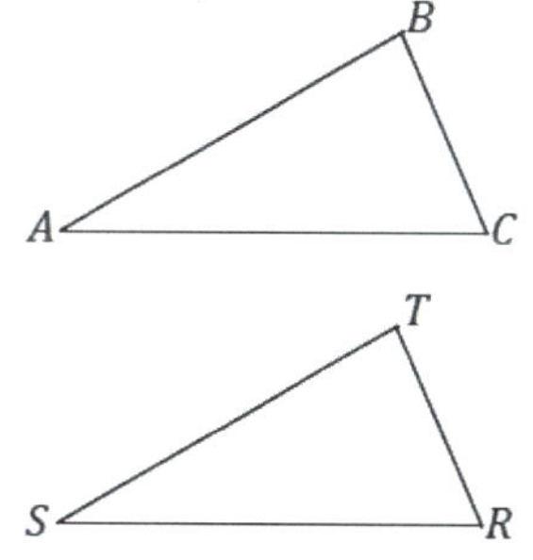

Examine Figure 2. By simply looking, it is impossible to tell the two triangles apart unless they are labeled. They look exactly the same (just as identical twins look the same). One triangle could be picked up and placed on top of the other.

Two triangles are identical if there is a triangle correspondence so that corresponding sides and angles of each triangle are equal in measurement. In Figure 2, there is a correspondence that will match up equal sides and equal angles, $\triangle ABC \leftrightarrow \triangle XYZ$; we can conclude that $\triangle ABC$ is identical to $\triangle XYZ$. This is not to say that we cannot find a correspondence in Figure 2 so that unequal sides and unequal angles are matched up, but there certainly is one correspondence that will match up angles with equal measurements and sides of equal lengths, making the triangles identical.

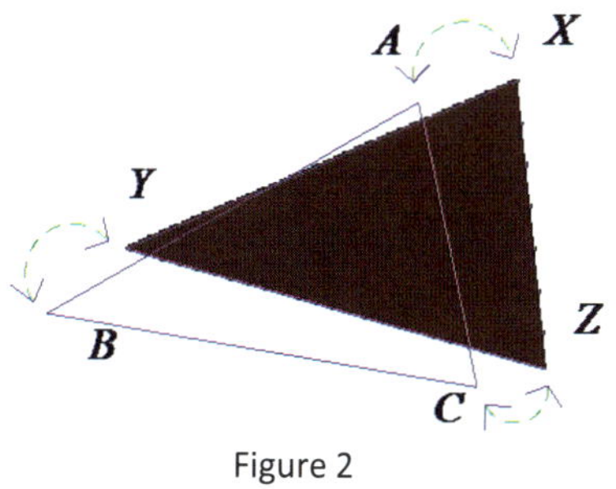

Figure 2

Discussion (5 minutes)

In Figure 2, $\triangle ABC$ **is identical to** $\triangle XYZ$.

- Which side is equal in measurement to $\overline{XZ}$? Justify your response.
  - *The length of* $\overline{AC}$ *is equal to the length of* $\overline{XZ}$ *because it is known that the triangle correspondence* $\triangle ABC \leftrightarrow \triangle XYZ$ ***matches equal sides and equal angles.***
- Which angle is equal in measurement to $\angle B$? Justify your response.
  - The measurement of $\angle Y$ *is equal to the measurement of* $\angle B$ *because it is known that the triangle correspondence* $\triangle ABC \leftrightarrow \triangle XYZ$ ***matches equal sides and equal angles.***

In discussing identical triangles, it is useful to have a way to indicate those sides and angles that are equal. We mark sides with tick marks and angles with arcs if we want to draw attention to them. If two angles or two sides have the same number of marks, it means they are equal.

In this figure, $AC = DE = EF$, **and** $\angle B = \angle E$.

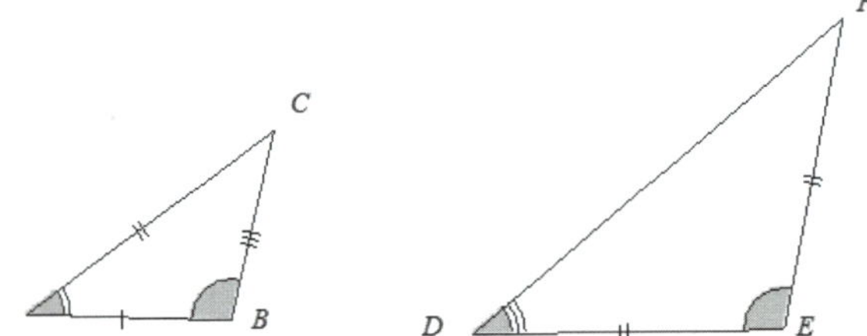

Example 2 (3 minutes)

Example 2

Two identical triangles are shown below. Give a triangle correspondence that matches equal sides and equal angles.

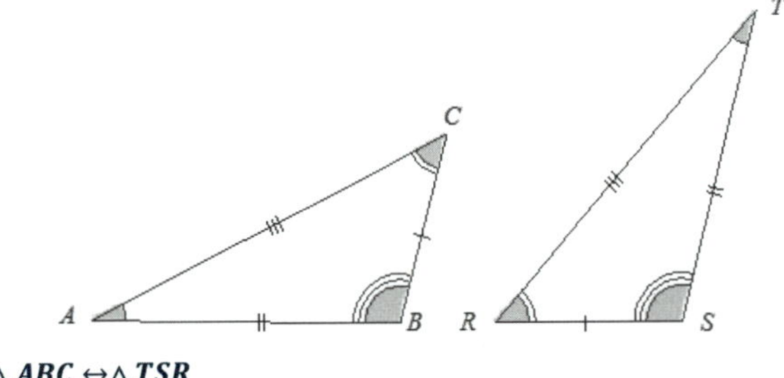

$\triangle ABC \leftrightarrow \triangle TSR$

Scaffolding:

Triangle cardstock cutouts may facilitate the process of determining a correspondence. Find this image in the supplement to Lesson 5 in a size large enough to cut out.

Exercise 8 (3 minutes)

Exercise 8

8. **Sketch two triangles that have a correspondence. Describe the correspondence in symbols or words. Have a partner check your work.**

   Answers will vary. Encourage students to check for correct use of notation and correctly made correspondences.

Vocabulary

CORRESPONDENCE: A *correspondence* between two triangles is a pairing of each vertex of one triangle with one (and only one) vertex of the other triangle.

If $A \leftrightarrow X$, $B \leftrightarrow Y$, and $C \leftrightarrow Z$ is a correspondence between two triangles (written $\triangle ABC \leftrightarrow \triangle XYZ$), then $\angle A$ matches $\angle X$, side $\overline{AB}$ matches side $\overline{XY}$, and so on.

Closing (2 minutes)

> **Lesson Summary**
>
> - **Two triangles and their respective parts can be compared once a correspondence has been assigned to the two triangles. Once a correspondence is selected, corresponding sides and corresponding angles can also be determined.**
> - **Double arrows notate corresponding vertices. Triangle correspondences can also be notated with double arrows.**
> - **Triangles are identical if there is a correspondence so that corresponding sides and angles are equal.**
> - **An equal number of tick marks on two different sides indicates the sides are equal in measurement. An equal number of arcs on two different angles indicates the angles are equal in measurement.**

Exit Ticket (5 minutes)

Name ________________________________ Date ____________________

Lesson 5: Identical Triangles

Exit Ticket

1. The following triangles are identical and have the correspondence $\triangle ABC \leftrightarrow \triangle YZX$. Find the measurements for each of the following sides and angles. Figures are not drawn to scale.

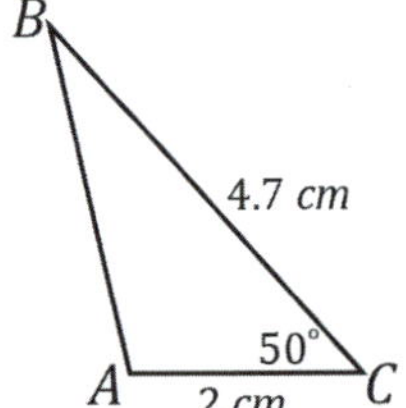

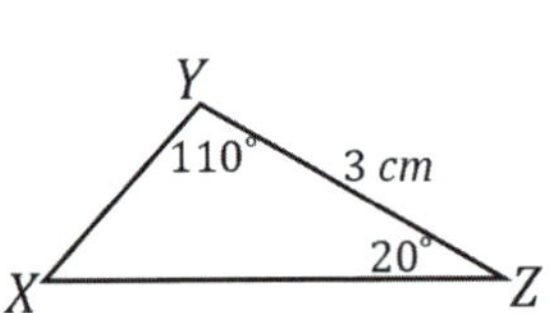

$AB =$ ________

________ $= ZX$

________ $= XY$

$\angle A =$ ________

$\angle B =$ ________

________ $= \angle X$

2. Explain why correspondences are useful.

Exit Ticket Sample Solutions

1. The following triangles are identical and have the correspondence $\triangle ABC \leftrightarrow \triangle YZX$. Find the measurements for each of the following sides and angles. Figures are not drawn to scale.

$AB = 3$ cm

4.7 cm $= ZX$

2 cm $= XY$

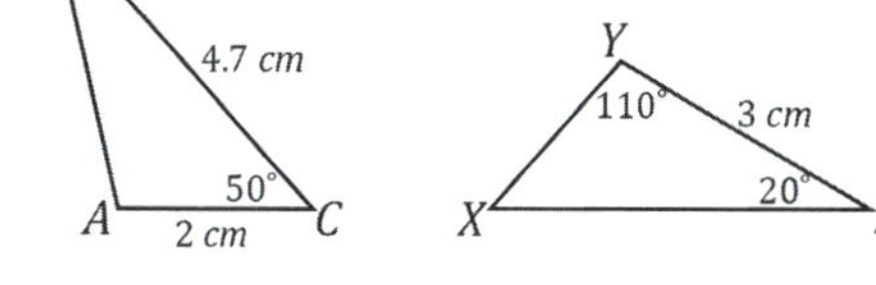

$\angle A = 110°$

$\angle B = 20°$

$50° = \angle X$

2. Explain why correspondences are useful.

A correspondence offers a systematic way to compare parts of two triangles. We can make statements about similarities or differences between two triangles using a correspondence, whereas without one, we would not have a reference system to make such comparisons.

Problem Set Sample Solutions

Given the following triangle correspondences, use double arrows to show the correspondence between vertices, angles, and sides.

1.

Triangle Correspondence	$\triangle ABC \leftrightarrow \triangle RTS$
Correspondence of Vertices	$A \longleftrightarrow R$ $B \longleftrightarrow T$ $C \longleftrightarrow S$
Correspondence of Angles	$\angle A \longleftrightarrow \angle R$ $\angle B \longleftrightarrow \angle T$ $\angle C \longleftrightarrow \angle S$
Correspondence of Sides	$\overline{AB} \longleftrightarrow \overline{RT}$ $\overline{BC} \longleftrightarrow \overline{TS}$ $\overline{CA} \longleftrightarrow \overline{SR}$

2.

Triangle Correspondence	$\triangle ABC \leftrightarrow \triangle FGE$
Correspondence of Vertices	$A \longleftrightarrow F$ $B \longleftrightarrow G$ $C \longleftrightarrow E$
Correspondence of Angles	$\angle A \longleftrightarrow \angle F$ $\angle B \longleftrightarrow \angle G$ $\angle C \longleftrightarrow \angle E$
Correspondence of Sides	$\overline{AB} \longleftrightarrow \overline{FG}$ $\overline{BC} \longleftrightarrow \overline{GE}$ $\overline{CA} \longleftrightarrow \overline{EF}$

3.

Triangle Correspondence	$\triangle QRP \leftrightarrow \triangle WYX$
Correspondence of Vertices	$Q \longleftrightarrow W$ $R \longleftrightarrow Y$ $P \longleftrightarrow X$
Correspondence of Angles	$\angle Q \longleftrightarrow \angle W$ $\angle R \longleftrightarrow \angle Y$ $\angle P \longleftrightarrow \angle X$
Correspondence of Sides	$\overline{QR} \longleftrightarrow \overline{WY}$ $\overline{RP} \longleftrightarrow \overline{YX}$ $\overline{PQ} \longleftrightarrow \overline{XW}$

Name the angle pairs and side pairs to find a triangle correspondence that matches sides of equal length and angles of equal measurement.

4. $DE = ZX$ $XY = EF$ $DF = ZY$

$\angle E = \angle X$ $\angle Z = \angle D$ $\angle F = \angle Y$

$\triangle DEF \leftrightarrow \triangle ZXY$

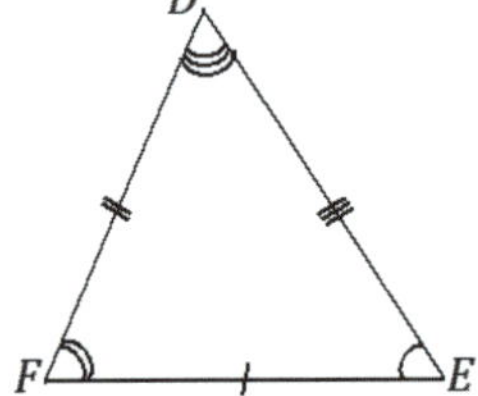

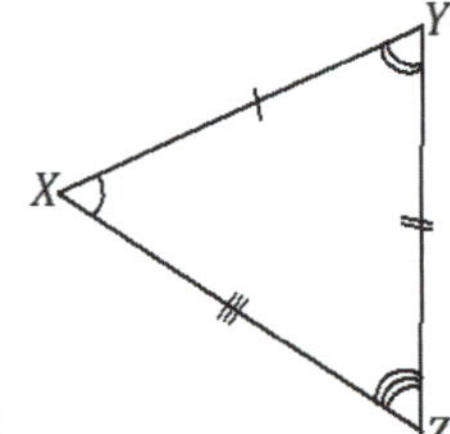

5.

$JK = WX$ $YX = LK$ $LJ = YW$

$\angle Y = \angle L$ $\angle J = \angle W$ $\angle K = \angle X$

$\triangle JKL \leftrightarrow \triangle WXY$

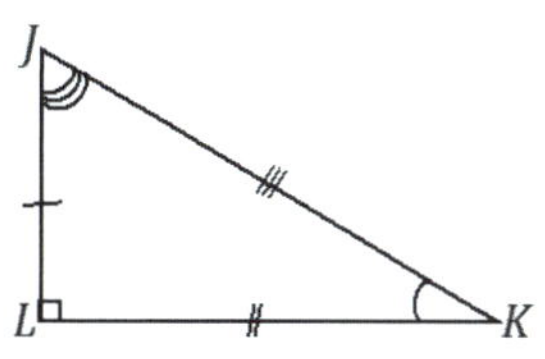

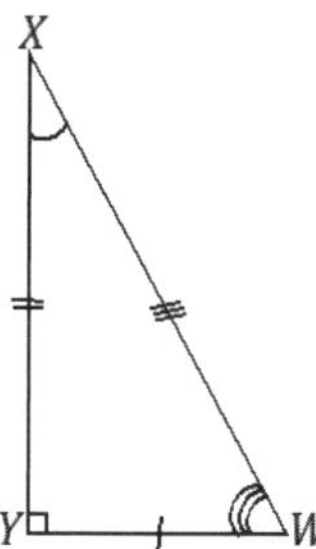

6.

$PQ = UT$ $TV = QR$ $RP = VU$

$\angle Q = \angle T$ $\angle U = \angle P$ $\angle R = \angle V$

$\triangle PQR \leftrightarrow \triangle UTV$

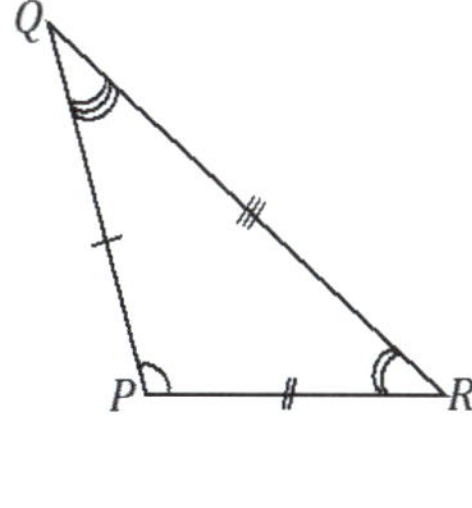

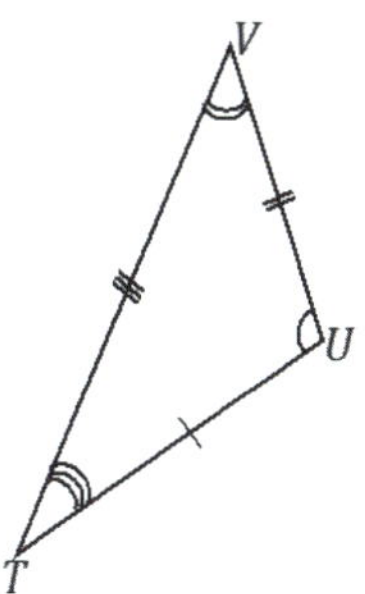

7. Consider the following points in the coordinate plane.

a. How many different (non-identical) triangles can be drawn using any three of these six points as vertices?

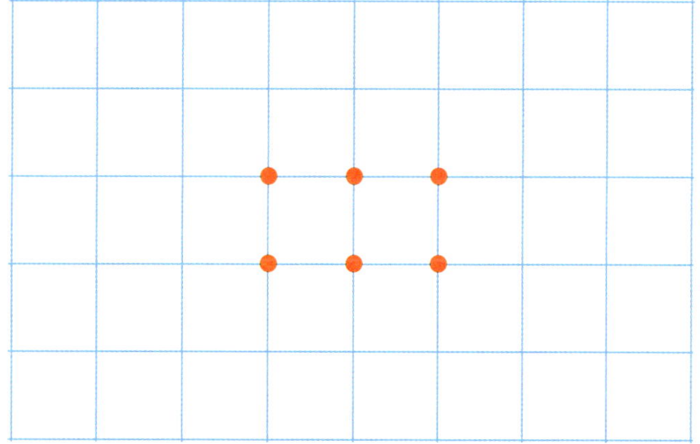

There is a total of 18 triangles but only 4 different triangles. Each triangle is identical with one of these four:

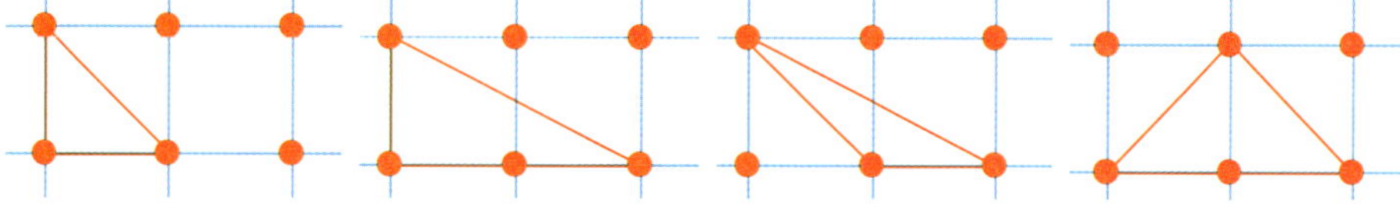

b. How can we be sure that there are no more possible triangles?

Any other triangle will have a correspondence so that equal sides and angles of equal measurement can be lined up (i.e., one can be laid over another, and the two triangles will match).

8. Quadrilateral $ABCD$ is identical with quadrilateral $WXYZ$ with a correspondence $A \leftrightarrow W$, $B \leftrightarrow X$, $C \leftrightarrow Y$, and $D \leftrightarrow Z$.

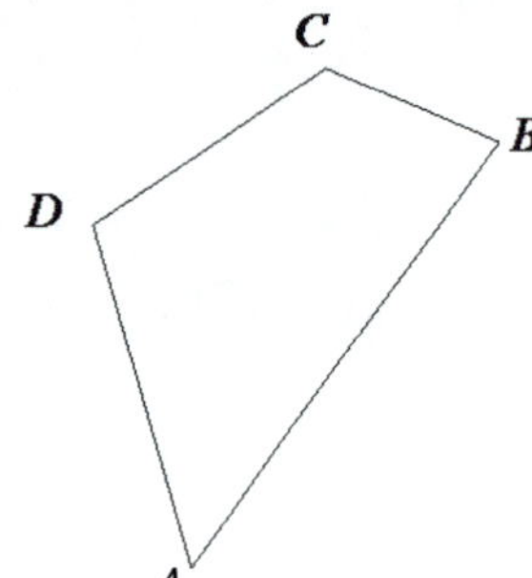

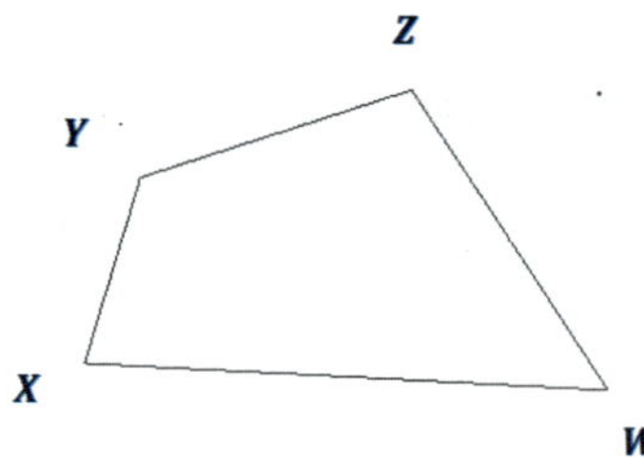

a. In the figure above, label points W, X, Y, and Z on the second quadrilateral.

b. Set up a correspondence between the side lengths of the two quadrilaterals that matches sides of equal length.

$\overline{AB} \leftrightarrow \overline{WX}$, $\overline{BC} \leftrightarrow \overline{XY}$, $\overline{CD} \leftrightarrow \overline{YZ}$, and $\overline{AD} \leftrightarrow \overline{WZ}$

c. Set up a correspondence between the angles of the two quadrilaterals that matches angles of equal measure.

$\angle A \leftrightarrow \angle W$, $\angle B \leftrightarrow \angle X$, $\angle C \leftrightarrow \angle Y$, and $\angle D \leftrightarrow \angle Z$

Example 2: Scaffolding Supplement

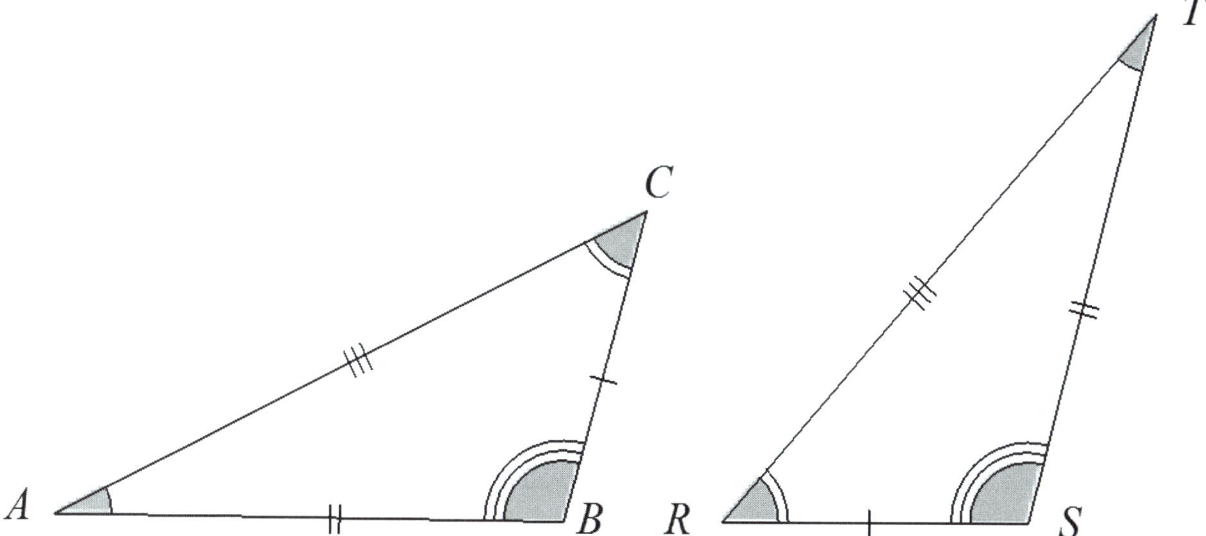

Lesson 6: Drawing Geometric Shapes

Student Outcomes

- Students use a compass, protractor, and ruler to draw geometric shapes based on given conditions.

Lesson Notes

The following sequence of lessons is based on standard **7.G.A.2**: Draw (freehand, with a ruler and protractor, and with technology) geometric shapes with given conditions. Focus on constructing triangles from three measurements of angles or sides, noticing when the conditions determine a unique triangle, more than one triangle, or no triangle. Instruction including the use of a compass is included to develop deeper understanding of geometric relationships and also as preparation for high school geometry.

By the close of Lesson 6, students should be able to use a ruler, protractor, and compass to draw simple figures. Although students have previously worked with a ruler and protractor, they have negligible experience using a compass before Grade 7. Therefore, they need to be given some time for exercises that demonstrate how to accurately use a compass. Practice through simple constructions will prepare students for Lessons 7–12, which require drawing triangles according to the conditions that determine a unique triangle, more than one triangle, or no triangle. For example, by constructing a triangle with three given side lengths, students gain a better understanding of how the construction is done and why this condition always yields a unique triangle.

As always, teachers should gauge how students handle the early problems and proceed with their students' performance in mind. If students struggle with problems, such as Exploratory Challenge Problems 2 and 3, try giving them variations of the same questions before moving forward. The goal of the lesson is to draw geometric shapes using a compass, protractor, and ruler. If students struggle to develop a facility with these tools, spend more time on the exercises with fewer instructions.

Students should have some experience with freehand sketches. Consider doing some exercises twice, once with a tool and once without. To highlight Mathematical Practice 5, students should have the opportunity to compare creating a diagram with a tool versus doing it freehand.

Classwork

Discussion (5 minutes)

The purpose of this lesson is to develop a familiarity with the tools of construction through problems. Students construct several basic geometric shapes and composite figures. Discuss what a compass is and how to use it.

A compass is a tool for drawing circles. The point where the needle of the compass sits represents the center of the circle, and its radius can be adjusted by widening or narrowing the two arms of the compass.

> *Scaffolding:*
>
> Ten problems are provided, but teachers may choose how many to work through based on students' abilities.

Tips to drawing circles with a thumbscrew compass:

- Adjust the compass to the intended radius length.
- Using one hand, place weight on the point of the compass, and let the pencil-end be relatively loose.
- Angle the compass relative to the paper; holding the compass perpendicular to the paper makes it difficult to maneuver.

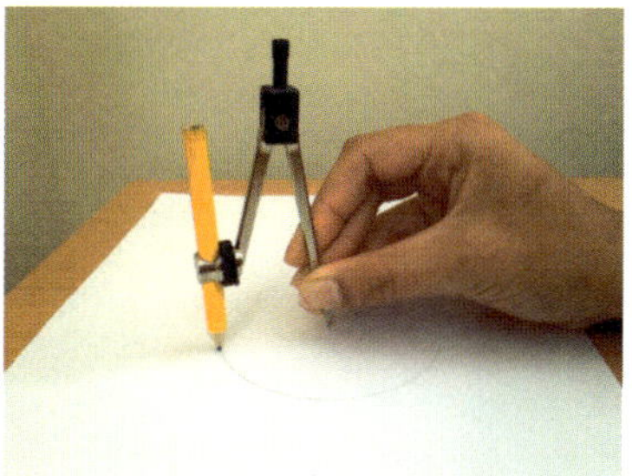

Holding the compass perpendicular to the paper makes it difficult to maneuver.

Angling the compass relative to the paper makes it easier to rotate.

What makes using the compass difficult?

- Students using traditional metal compasses might have difficulty with the following: keeping weight on the point while drawing with the pencil, dealing with a pencil coming loose and falling out, and making a hole in the paper with the point. Students using safety compasses might have difficulty keeping weight on the center of the compass, moving the slider around unintentionally, and keeping track of the radius adjustment.

Have students try drawing a few circles of their own before working on the exercises.

There are alternatives to using the kind of compass shown above; here are two examples:

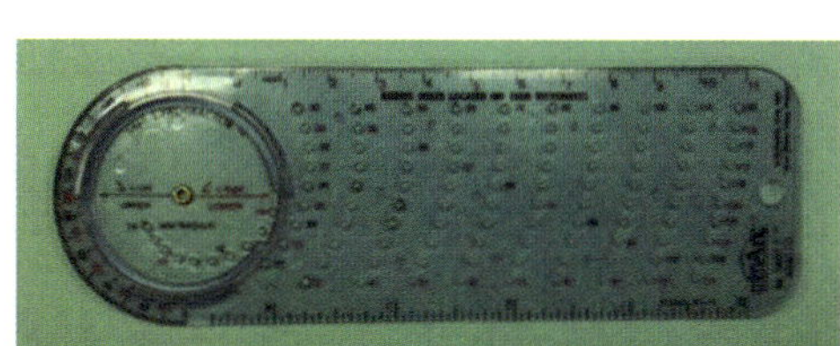

Safety compass

Pencil and String compass

All three kinds of compasses have pros and cons; use whichever seems best for students. Over the next several lessons, a compass is critical in studying the criteria that determine a unique triangle.

Exploratory Challenge (25 minutes)

Ideally, the Exploratory Challenge is done in small groups so that students can compare and discuss the constructions as they finish them. After the allotted 25 minutes, or periodically, responses can be discussed as a whole class by sharing individual work from each group. Another option is to post examples of work around the classroom at the end of the Exploratory Challenge and have a gallery walk.

Regarding the rest of the Exploratory Challenge:

- What, if anything, is challenging about the problems?
  - *Reading and following the steps correctly.*
- What can groups do to make sure that everyone proceeds through each problem correctly?
  - *Discuss each step and decide what it means before constructing it. Conversely, groups could do the step, and if there are differences, they could discuss which construction seems correct against the written instruction.*

Exploratory Challenge

Use a ruler, protractor, and compass to complete the following problems.

1. **Use your ruler to draw three segments of the following lengths: 4 cm, 7.2 cm, and 12.8 cm. Label each segment with its measurement.**

4 cm

7.2 cm

12.8 cm

Remind students how to measure angles accurately using a protractor:

1. Place the center notch of the protractor on the vertex.
2. Put the pencil point through the notch, and move the straightedge into alignment.
3. When measuring angles, it is sometimes necessary to extend the sides of the angle so that they intersect with the protractor's scale.

Refer to Grade 4 Module 4 Topic B for more information on how to instruct students to measure angles.

2. **Draw complementary angles so that one angle is $35°$. Label each angle with its measurement. Are the angles required to be adjacent?**

   The complementary angles do not need to be adjacent; the sum of the measurements of the angles needs to be $90°$.

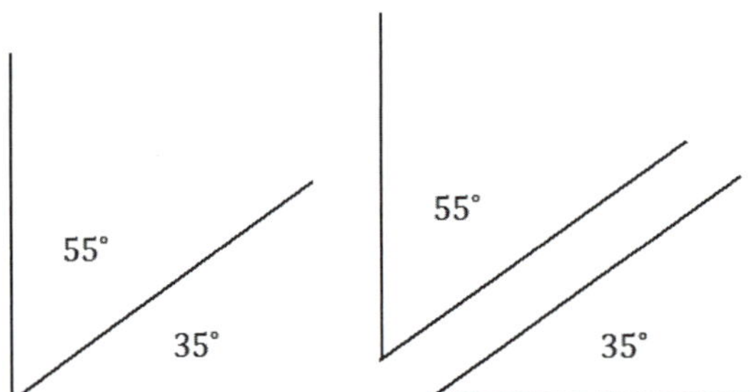

- How will you begin Exploratory Challenge Problem 3?
  - *I will draw an angle with a measurement of* $125°$ *and then extend the rays through the vertex so that the figure looks like an X. Since one angle will have a measurement of* $125°$*, the adjacent angle on the line will measure* $55°$.

3. **Draw vertical angles so that one angle is $125°$. Label each angle formed with its measurement.**

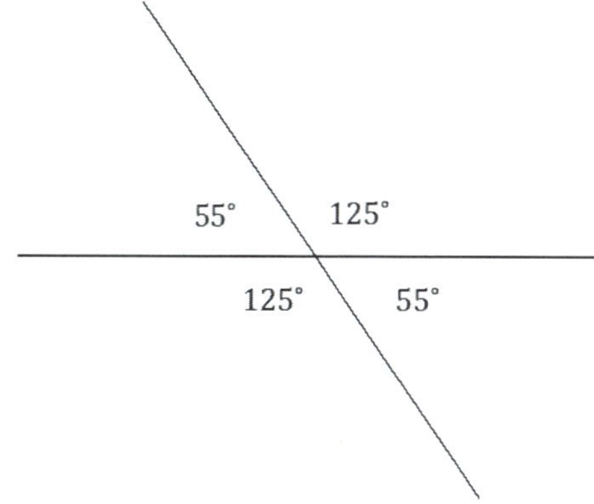

4. **Draw three distinct segments of lengths 2 cm, 4 cm, and 6 cm. Use your compass to draw three circles, each with a radius of one of the drawn segments. Label each radius with its measurement.**

   Due to space restrictions, only the two smaller circles are shown here:

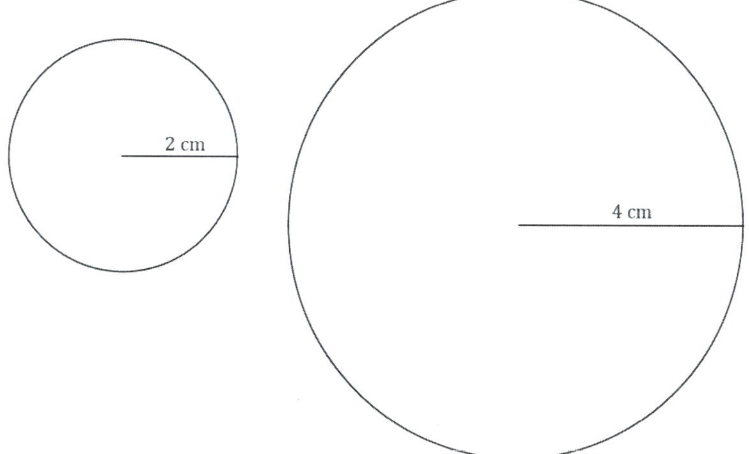

5. Draw three adjacent angles a, b, and c so that $a = 25°$, $b = 90°$, and $c = 50°$. Label each angle with its measurement.

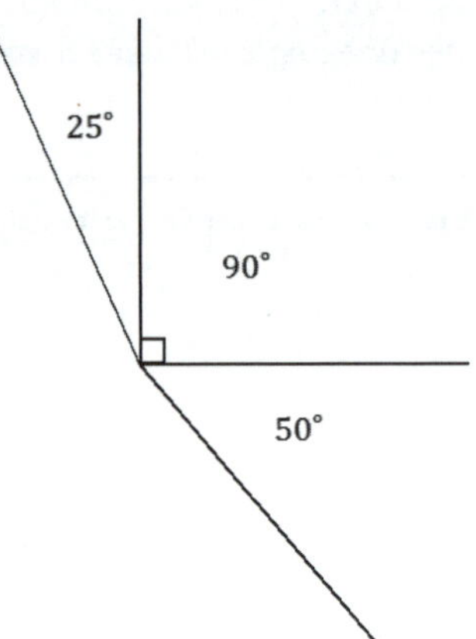

6. Draw a rectangle $ABCD$ so that $AB = CD = 8$ cm and $BC = AD = 3$ cm.

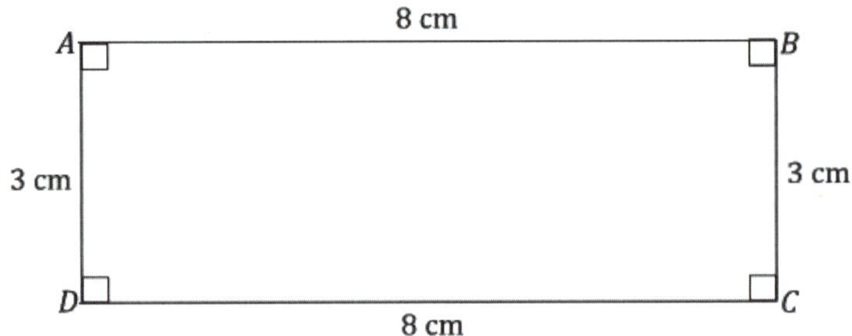

7. Draw a segment AB that is 5 cm in length. Draw a second segment that is longer than $\overline{AB}$, and label one endpoint C. Use your compass to find a point on your second segment, which will be labeled D, so that $CD = AB$.

A 5 cm B

C D

8. Draw a segment AB with a length of your choice. Use your compass to construct two circles:

   i. A circle with center A and radius AB.

   ii. A circle with center B and radius BA.

   Describe the construction in a sentence.

   Two circles with radius AB are drawn; one has its center at A, and the other has its center at B.

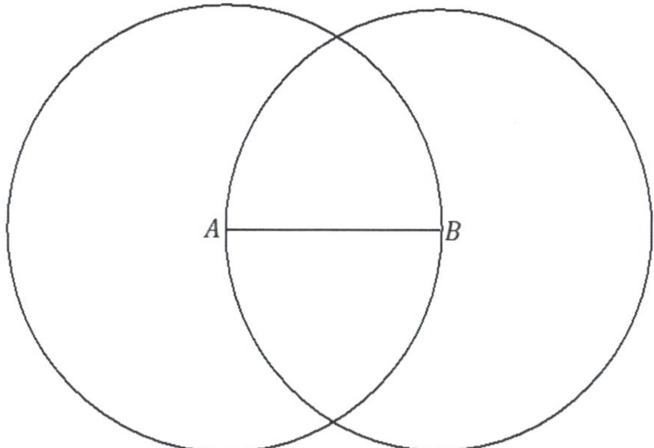

9. Draw a horizontal segment AB, 12 cm in length.

   a. Label a point O on $\overline{AB}$ that is 4 cm from B.

   b. Point O will be the vertex of an angle COB.

   c. Draw ray OC so that the ray is above $\overline{AB}$ and $\angle COB = 30°$.

   d. Draw a point P on $\overline{AB}$ that is 4 cm from A.

   e. Point P will be the vertex of an angle QPO.

   f. Draw ray PQ so that the ray is above $\overline{AB}$ and $\angle QPO = 30°$.

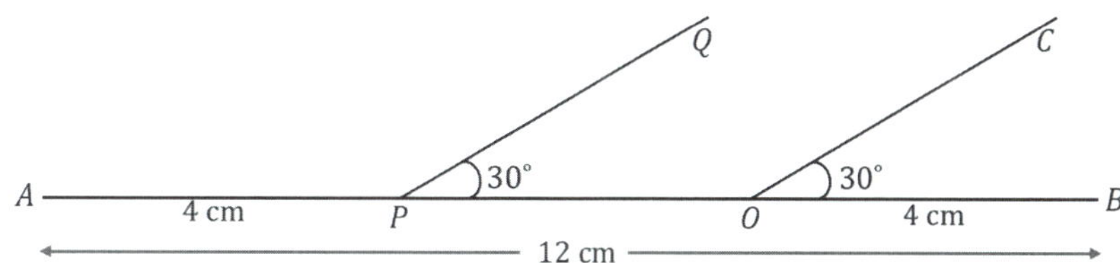

10. Draw segment AB of length 4 cm. Draw two circles that are the same size, one with center A and one with center B (i.e., do not adjust your compass in between) with a radius of a length that allows the two circles to intersect in two distinct locations. Label the points where the two circles intersect C and D. Join A and C with a segment; join B and C with a segment. Join A and D with a segment; join B and D with a segment.

    What kind of triangles are $\triangle ABC$ and $\triangle ABD$? Justify your response.

    $\triangle ABC$ and $\triangle ABD$ are identical isosceles triangles. Both circles are the same size (i.e., have the same radius). Furthermore, the point along each circle is the same distance away from the center no matter where you are on the circle; this means the distance from A to C is the same as the distance from B to C (the same follows for D). A triangle with at least two sides of equal length is an isosceles triangle.

Possible solution:

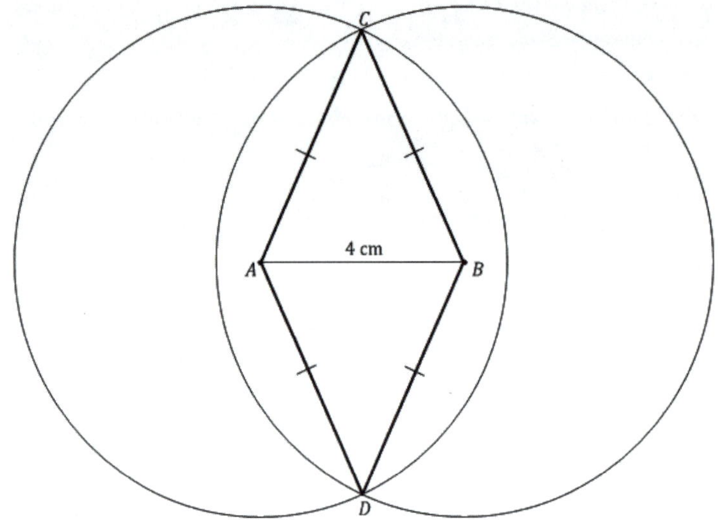

11. **Determine all possible measurements in the following triangle, and use your tools to create a copy of it.**

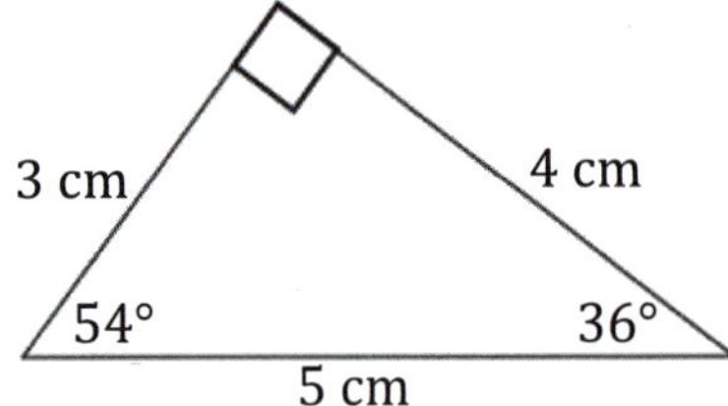

Discussion (8 minutes)

In the allotted time, review the solutions to each question as a whole group. As suggested previously, share out responses from groups, or have each group put one (or more) response up on a wall and have a gallery walk. Discuss responses to Exploratory Challenge Problems 2, 8, 9, and 10.

- Problem 2: Are the [complementary] angles required to be adjacent?
  - *No, complementary angles can be adjacent but do not have to be. The only requirement is for the sum of the measurements of the two angles to be $90°$.*
- Problem 8: Describe the construction in a sentence.
  - *Two circles with radius AB are drawn; one has its center at A, and the other has its center at B.*
- Problem 9: How would you describe the relationship between rays OC and PQ?
  - *Rays OC and PQ appear to be parallel since they both tilt or slant at the same angle to segment AB.*
- For Problem 10, emphasize that the construction requires two circles of the same size as well as circles that intersect in two locations. Problem 10: What kind of triangles are $\triangle ABC$ and $\triangle ABD$? Justify your response.
  - *$\triangle ABC$ and $\triangle ABD$ are isosceles triangles. Both circles are the same size (i.e., have the same radius). Furthermore, the point along each circle is the same distance away from the center no matter where you are on the circle. This means the distance from A to C is the same as the distance from B to C (as is the case for D). A triangle with at least two sides of equal length is an isosceles triangle.*

Closing (2 minutes)

- Three tools were used to complete the problems in the Exploratory Challenge, two of which you have already used in the last few years. What did the problems show you about the ways in which you can use a compass?
  - *A compass can be used to construct circles, to measure and mark off a segment of equal length to another segment, and to confirm the fact that the radius of the center of a circle to the circle itself remains constant no matter where you are on the circle (Problem 10).*

Lesson Summary

The compass is a tool that can be used for many purposes that include the following:

- **Constructing circles.**
- **Measuring and marking a segment of equal length to another segment.**
- **Confirming that the radius of the center of a circle to the circle itself remains constant no matter where you are on the circle.**

Exit Ticket (5 minutes)

Name ______________________________ Date ______________

Lesson 6: Drawing Geometric Shapes

Exit Ticket

1. Draw a square $PQRS$ with side length equal to 5 cm. Label the side and angle measurements.

2. Draw a segment AB, 6 cm in length. Draw a circle whose diameter is segment AB.

Exit Ticket Sample Solutions

1. Draw a square $PQRS$ with side length equal to 5 cm. Label the side and angle measurements.

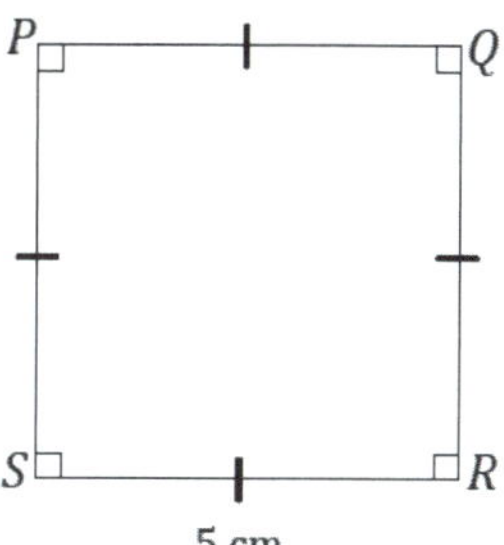

2. Draw a segment AB, 6 cm in length. Draw a circle whose diameter is segment AB.

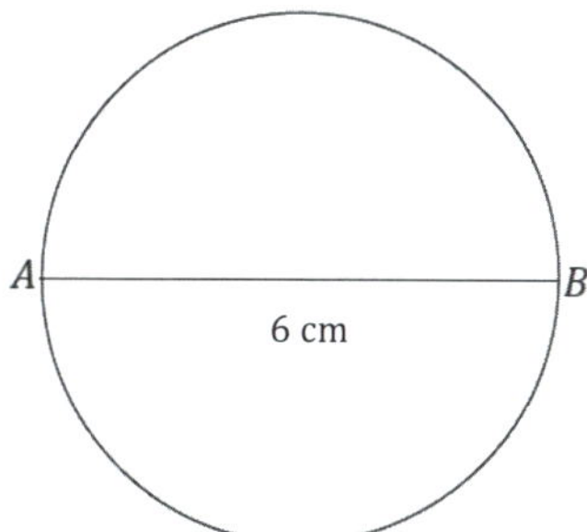

Problem Set Sample Solutions

Use a ruler, protractor, and compass to complete the following problems.

1. Draw a segment AB that is 5 cm in length and perpendicular to segment CD, which is 2 cm in length.

   One possible solution:

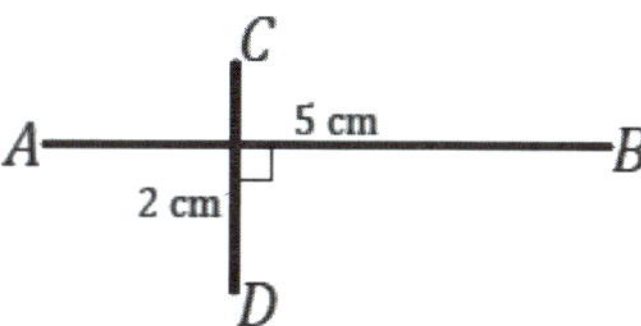

2. Draw supplementary angles so that one angle is $26°$. Label each angle with its measurement.

   Possible solutions:

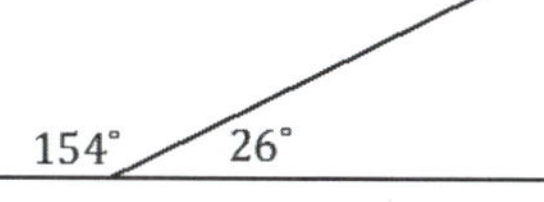

154° 26°

3. Draw $\triangle ABC$ so that $\angle B$ has a measurement of $100°$.

One possible solution:

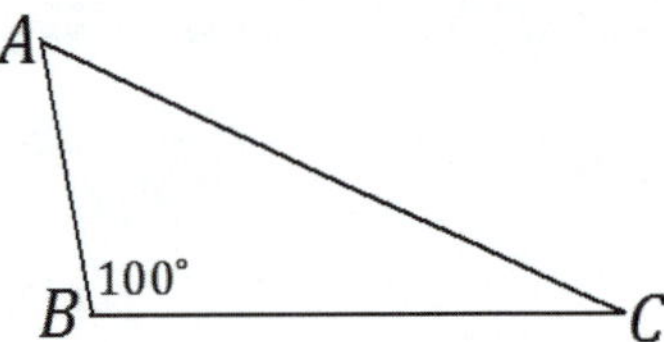

4. Draw a segment AB that is 3 cm in length. Draw a circle with center A and radius AB. Draw a second circle with diameter AB.

One possible solution:

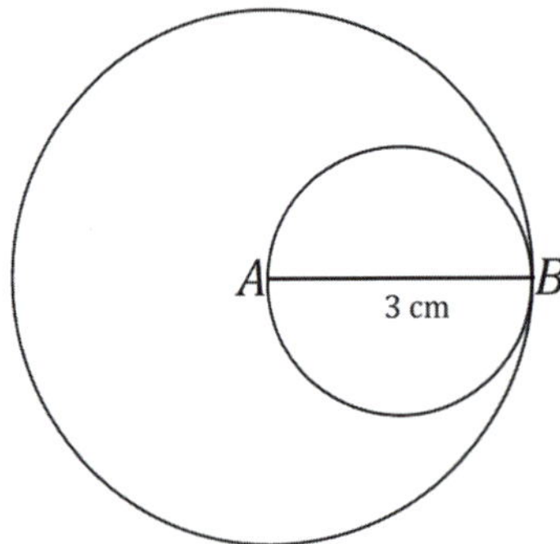

5. Draw an isosceles $\triangle ABC$. Begin by drawing $\angle A$ with a measurement of $80°$. Use the rays of $\angle A$ as the equal legs of the triangle. Choose a length of your choice for the legs, and use your compass to mark off each leg. Label each marked point with B and C. Label all angle measurements.

One possible solution:

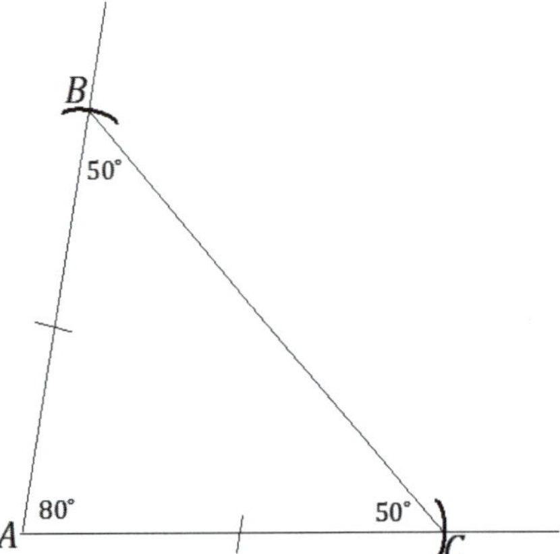

6. Draw an isosceles $\triangle DEF$. Begin by drawing a horizontal segment DE that is 6 cm in length. Use your protractor to draw $\angle D$ and $\angle E$ so that the measurements of both angles are $30°$. If the non-horizontal rays of $\angle D$ and $\angle E$ do not already cross, extend each ray until the two rays intersect. Label the point of intersection F. Label all side and angle measurements.

One possible solution:

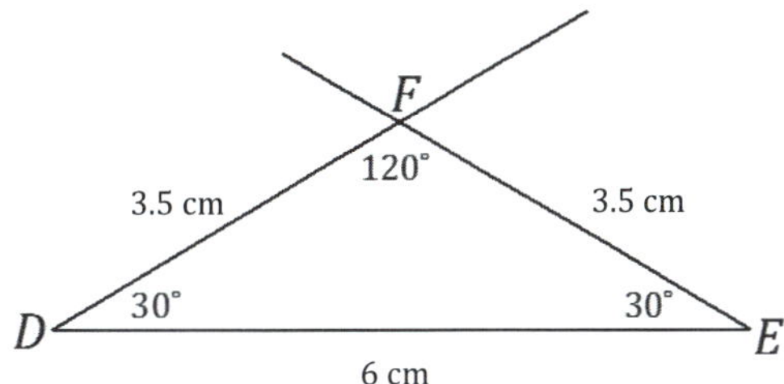

7. Draw a segment AB that is 7 cm in length. Draw a circle with center A and a circle with center B so that the circles are not the same size, but do intersect in two distinct locations. Label one of these intersections C. Join A to C and B to C to form $\triangle ABC$.

One possible solution:

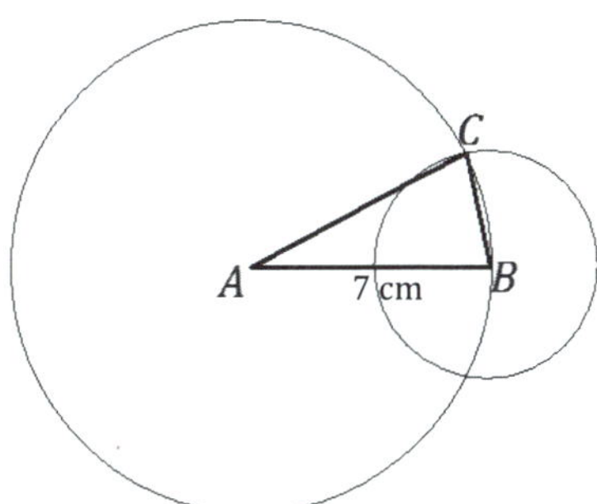

8. Draw an isosceles trapezoid $WXYZ$ with two equal base angles, $\angle W$ and $\angle X$, that each measures $110°$. Use your compass to create the two equal sides of the trapezoid. Leave arc marks as evidence of the use of your compass. Label all angle measurements. Explain how you constructed the trapezoid.

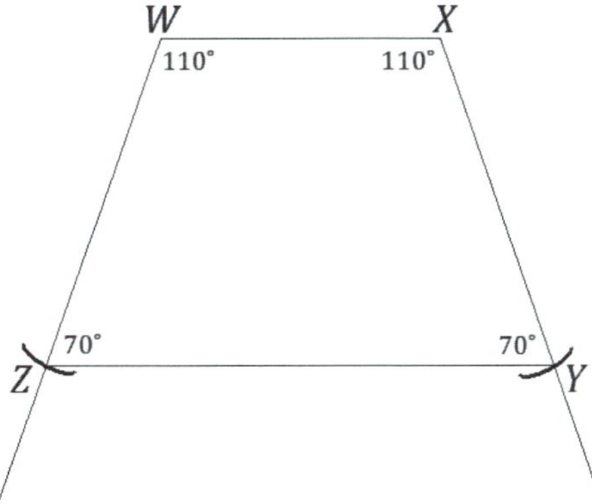

Draw segment WX. Use a protractor and segment WX to draw $\angle XWZ$ at a measurement of $110°$; do the same to draw $\angle WXY$. (Note: When drawing $\overrightarrow{WZ}$ and $\overrightarrow{XY}$, length is not specified, so students should have rays long enough so that they can use a compass to mark off lengths that are the same along each ray in the next step.) Place the point of the compass at W, adjust the compass to a desired width, and mark an arc so that it crosses $\overrightarrow{WZ}$. Label the intersection as Z. Do the same from X along $\overrightarrow{XY}$, and mark the intersection as Y. Finally, join Z and Y to form isosceles trapezoid $WXYZ$.

Lesson 7: Drawing Parallelograms

Student Outcomes

- Students use a protractor, ruler, and setsquare to draw parallelograms based on given conditions.

Lesson Notes

In Lesson 6, students drew a series of figures (e.g., complementary angles, vertical angles, circles of different radii, and isosceles triangles). Lesson 7 familiarizes students with a geometry tool known as the setsquare, and they use it to draw parallelograms under a variety of conditions.

Similar to the previous lesson, students should have experience with using tools and making comparisons between drawings done with tools and drawings done freehand. Work in this lesson embodies Mathematical Practice 5.

Classwork

Opening (5 minutes)

A setsquare is a triangle with a right angle. It can be made out of plastic or out of paper. Have students create their own setsquares out of paper or cardstock.

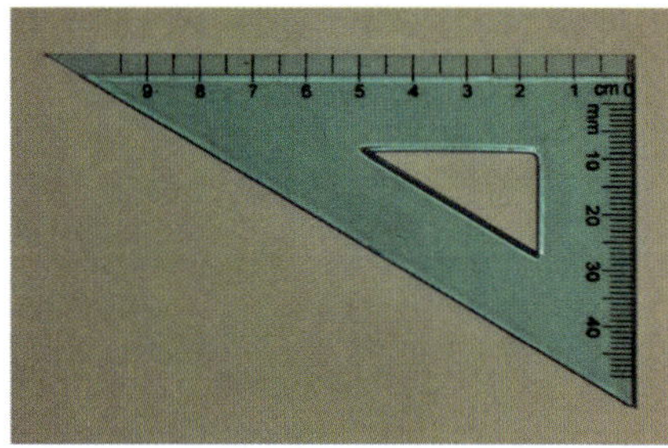

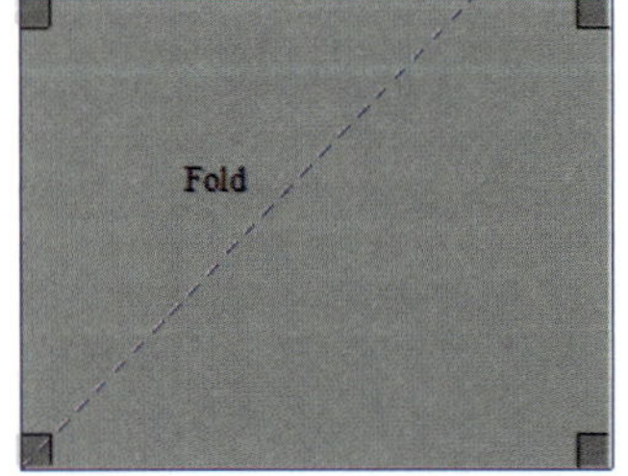

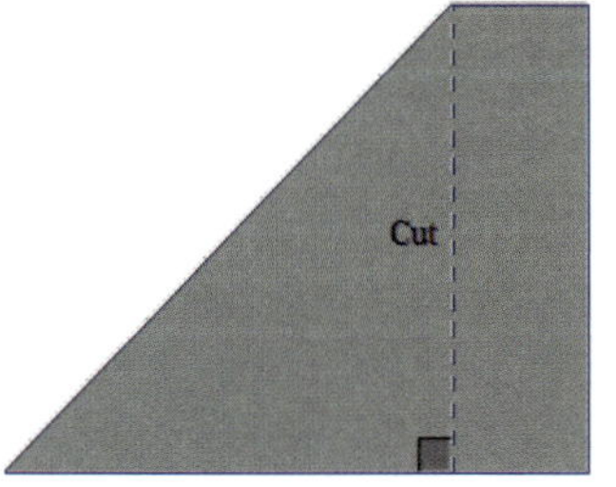

Place tape on the edges of the setsquare.

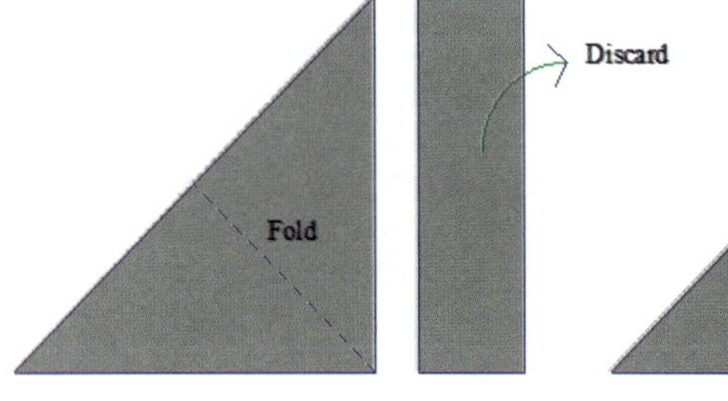

Setsquare

Have students use a setsquare and a ruler to check if lines are parallel.

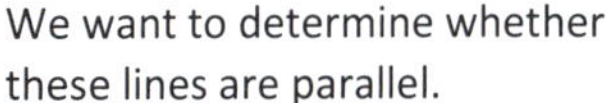

We want to determine whether these lines are parallel.

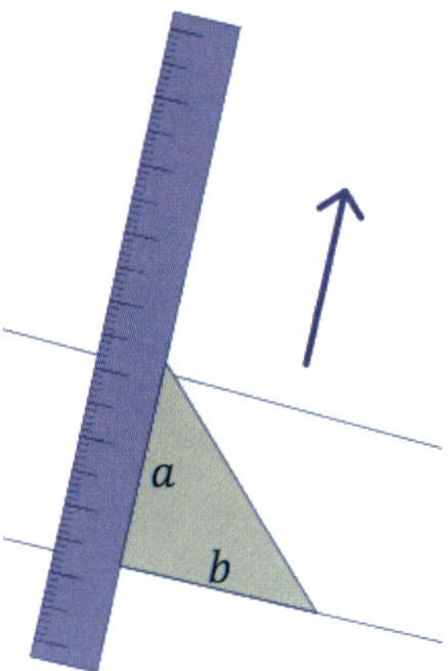

Identify the two sides of the setsquare that meet at the right angle as shown, a and b (a and b are legs of the setsquare). Place the ruler against a, and align b with one of the two lines. Notice that the ruler must be perpendicular to the line.

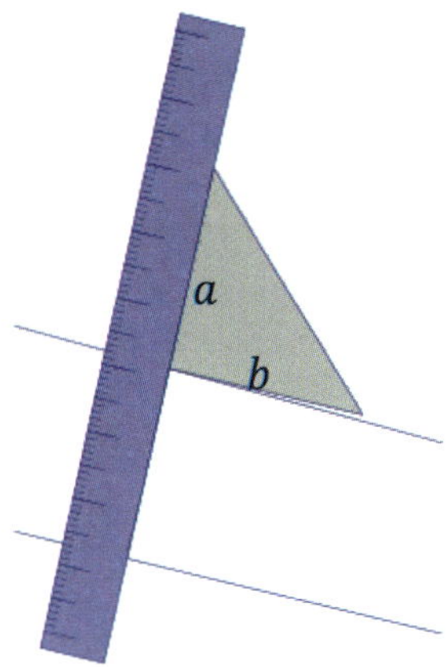

Slide the setsquare along the ruler. Try to line b up with the other line. If the edge of the setsquare aligns with the line, the two lines are parallel. If the edge of the setsquare does not align with the line, then the two lines are not parallel.

Have students use a setsquare and a ruler to draw a line parallel to $\overline{AB}$ through the point C.

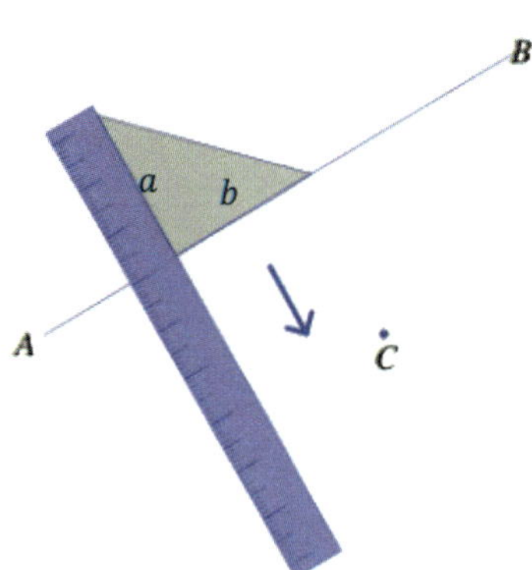

We want to draw a line through C parallel to $\overleftrightarrow{AB}$. Align b with $\overleftrightarrow{AB}$.

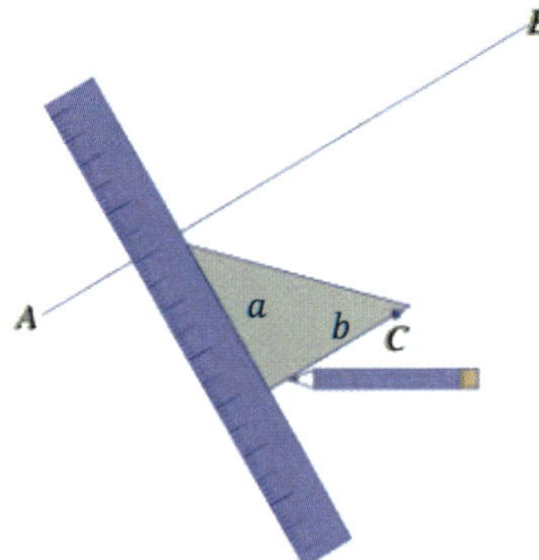

Slide the setsquare along the ruler until it is possible to draw a segment through C along b.

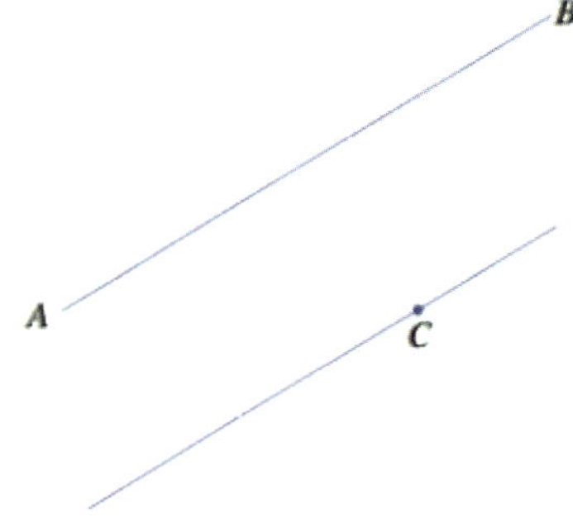

Use your ruler to extend the segment through C. The two lines are parallel.

Example 1 (5 minutes)

> *Scaffolding:*
> Consider providing a visual reminder (e.g., a chart handout or poster) naming the quadrilaterals and their key properties.

Example 1

Use what you know about drawing parallel lines with a setsquare to draw rectangle $ABCD$ with dimensions of your choice. State the steps you used to draw your rectangle, and compare those steps to those of a partner.

Possible steps: Draw $\overline{AB}$ first. Align the setsquare so that one leg aligns with $\overline{AB}$, and place the ruler against the other leg of the setsquare; mark a point X, __cm away from $\overline{AB}$. Draw a line parallel to $\overline{AB}$ through X. Realign the setsquare with $\overline{AB}$, situate the ruler so that it passes through A, and draw a segment with a length of __cm. Mark the intersection of the line through A and the parallel line to $\overline{AB}$ as D. $\overline{AD}$ is now drawn perpendicular to $\overline{AB}$. Repeat the steps to determine C.

Example 2 (7 minutes)

Example 2

Use what you know about drawing parallel lines with a setsquare to draw rectangle $ABCD$ with $AB = 3$ cm and $BC = 5$ cm. Write a plan for the steps you will take to draw $ABCD$.

Draw $\overline{AB}$ first. Align the setsquare so that one leg aligns with $\overline{AB}$, and place the ruler against the other leg of the setsquare; mark a point X 5 cm away from $\overline{AB}$. Slide the setsquare along the ruler until it aligns with X. Draw a line parallel to $\overline{AB}$ through X. To create the right angle at A, align the setsquare so that the leg of the setsquare aligns with $\overline{AB}$, situate the ruler so that the outer edge of the ruler passes through A, and draw a line through A. Mark the intersection of the line through A and the parallel line to $\overline{AB}$ as D. Repeat the steps to determine C.

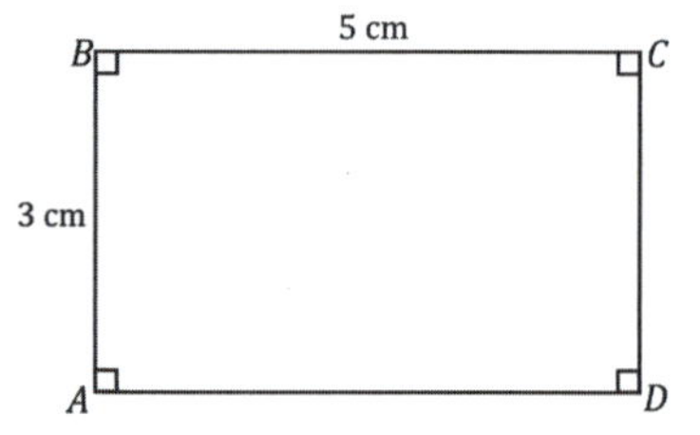

Make sure that students label all vertices, right angles, and measurements. This example also provides an opportunity to review the definition of diagonal:

- In a quadrilateral $ABCD$, $\overline{AC}$ and $\overline{BD}$ would be called the diagonals of the quadrilateral.

With respect to the solution, students will not respond (and are not expected to respond) with the level of detail found in the solution. The goal is to get them thinking about how they will use their newly acquired skills to draw a rectangle with the setsquare. Share the solution so they have exposure to the precision needed for clear instructions. Eventually, in Geometry, students will write instructions so precise that a person who doesn't know what a rectangle is would still be able to draw one by using the instructions.

Example 3 (7 minutes)

Example 3

Use a setsquare, ruler, and protractor to draw parallelogram $PQRS$ so that the measurement of $\angle P$ is $50°$, $PQ = 5$ cm, the measurement of $\angle Q$ is $130°$, and the length of the altitude to $\overline{PQ}$ is 4 cm.

Steps to draw the figure: Draw $\overline{PQ}$ first. Align the setsquare and ruler so one leg of the setsquare aligns with $\overline{PQ}$, and mark a point X 4 cm from $\overline{PQ}$. Slide the setsquare along the ruler so that one side of the setsquare passes through X, and draw a line through X; this line is parallel to $\overline{PQ}$. Using $\overline{PQ}$ as one ray of $\angle P$, draw $\angle P$ so that the measurement of $\angle P$ is $50°$ and that the ray PS intersects with the line parallel to $\overline{PQ}$ (the intersection is S). Draw $\angle Q$ so that the measurement of $\angle Q$ is $130°$; the ray QR should be drawn to intersect with the line parallel to $\overline{PQ}$ (the intersection is R).

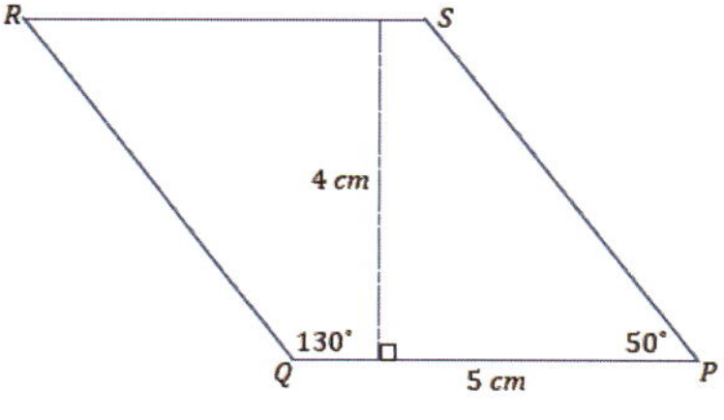

Exercise 1 (6 minutes)

Exercise 1

Use a setsquare, ruler, and protractor to draw parallelogram $DEFG$ so that the measurement of $\angle D$ is $40°$, $DE = 3$ cm, the measurement of $\angle E$ is $140°$, and the length of the altitude to $\overline{DE}$ is 5 cm.

Steps to draw the figure: Draw $\overline{DE}$ first. Align the setsquare and ruler so one leg of the setsquare aligns with $\overline{DE}$, and mark a point X 5 cm from $\overline{DE}$. Slide the setsquare along the ruler so that one side of the setsquare passes through X, and draw a line through X; this line is parallel to $\overline{DE}$. Using $\overline{DE}$ as one ray of $\angle D$, draw $\angle D$ so that the measurement of $\angle D$ is $40°$ and that the ray DG intersects with the line parallel to $\overline{DE}$ (the intersection is G). Draw $\angle E$ so that the measurement of $\angle E$ is $140°$; the ray EF should be drawn to intersect with the line parallel to $\overline{DE}$ (the intersection is F).

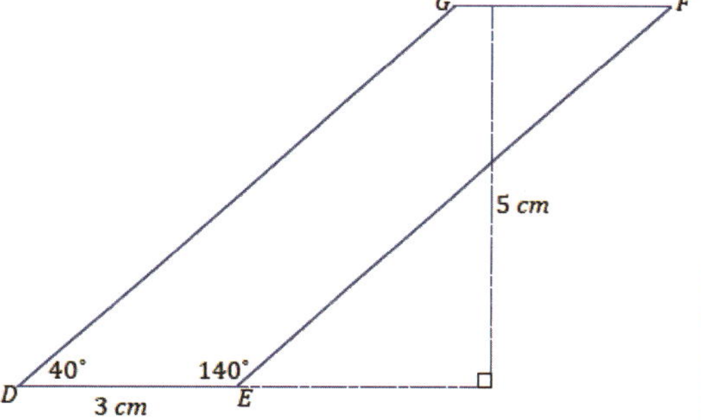

Example 4 (7 minutes)

Example 4

Use a setsquare, ruler, and protractor to draw rhombus $ABCD$ so that the measurement of $\angle A$ is $80°$, the measurement of $\angle B$ is $100°$, and each side of the rhombus measures 5 cm.

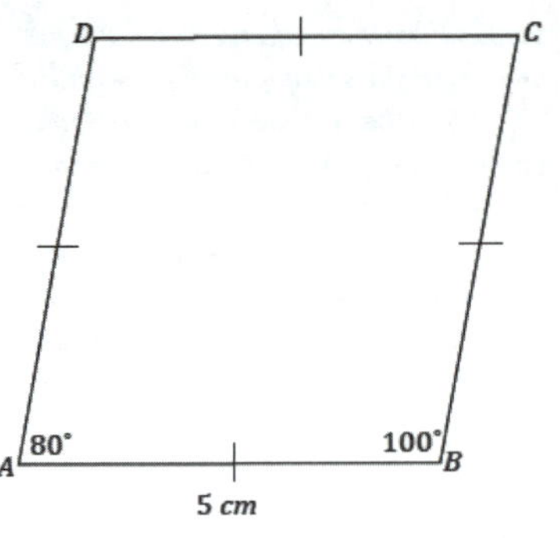

Steps to draw the figure: Draw $\overline{AB}$ first. Using $\overline{AB}$ as one ray of $\angle A$, draw $\angle A$ so that the measurement of $\angle A$ is $80°$. The other ray, or the to-be side of the rhombus, AD, should be 5 cm in length; label the endpoint of the segment as D. Align the setsquare and ruler so one leg of the setsquare aligns with $\overline{AB}$ and the edge of the ruler passes through D. Slide the setsquare along the ruler so that the edge of the setsquare passes through D, and draw a line along the edge of the setsquare. This line is parallel to $\overline{AB}$. Now align the setsquare and ruler so one leg of the setsquare aligns with $\overline{AD}$ and the edge of the ruler passes through B. Slide the setsquare along the ruler so that the edge of the setsquare passes through B, and draw a line along the edge of the setsquare. This line is parallel to $\overline{AD}$. Along this line, measure a segment 5 cm with B as one endpoint, and label the other endpoint C. Join C to D.

Closing (1 minute)

- Why are setsquares useful in drawing parallelograms?
  - *They give us a means to draw parallel lines for the sides of parallelograms.*

Lesson Summary

A protractor, ruler, and setsquare are necessary tools to construct a parallelogram. A setsquare is the tool that gives a means to draw parallel lines for the sides of a parallelogram.

Exit Ticket (7 minutes)

Name ______________________________ Date ____________

Lesson 7: Drawing Parallelograms

Exit Ticket

Use what you know about drawing parallel lines with a setsquare to draw square $ABCD$ with $AB = 5$ cm. Explain how you created your drawing.

Exit Ticket Sample Solutions

Use what you know about drawing parallel lines with a setsquare to draw square $ABCD$ with $AB = 5$ cm. Explain how you created your drawing.

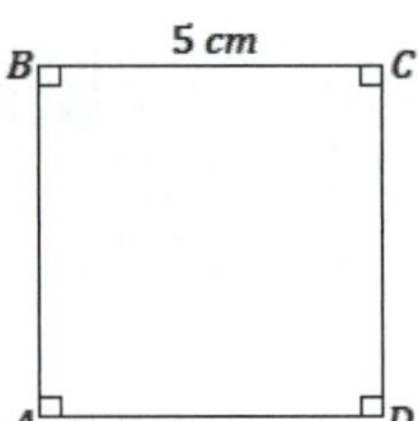

Draw $\overline{AB}$ (any side will do here) first. Align the setsquare and ruler so that one leg of the setsquare aligns with $\overline{AB}$; mark a point X 5 cm away from $\overline{AB}$. Draw a line parallel to $\overline{AB}$ through X. To create the right angle at A, align the setsquare so that the leg of the setsquare aligns with $\overline{AB}$, situate the ruler so that the outer edge of the ruler passes through A, and draw a line through A. Mark the intersection of the line through A and the parallel line to $\overline{AB}$ as D; join A and D. Repeat the steps to determine C, and join B and C.

Problem Set Sample Solutions

1. Draw rectangle $ABCD$ with $AB = 5$ cm and $BC = 7$ cm.

   Steps to draw the figure: Draw $\overline{AB}$ first. Align the setsquare so that one leg aligns with $\overline{AB}$, and place the ruler against the other leg of the setsquare; mark a point X 7 cm away from $\overline{AB}$. Draw a line parallel to $\overline{AB}$ through X. To create the right angle at A, align the setsquare so that its leg aligns with $\overline{AB}$, situate the ruler so that the outer edge of the ruler passes through A, and draw a line through A. Mark the intersection of the line through A and the parallel line to $\overline{AB}$ as D. Repeat the steps to determine C.

2. Use a setsquare, ruler, and protractor to draw parallelogram $PQRS$ so that the measurement of $\angle P$ is $65°$, $PQ = 8$ cm, the measurement of $\angle Q$ is $115°$, and the length of the altitude to $\overline{PQ}$ is 3 cm.

   Steps to draw the figure: Draw $\overline{PQ}$ first. Align the setsquare and ruler so one leg of the setsquare aligns with $\overline{PQ}$, and mark a point X 3 cm from $\overline{PQ}$. Slide the setsquare along the ruler so that one side of the setsquare passes through X, and draw a line through X; this line is parallel to $\overline{PQ}$. Using $\overline{PQ}$ as one ray of $\angle P$, draw $\angle P$ so that the measurement of $\angle P$ is $65°$ and that the ray PS intersects with the line parallel to $\overline{PQ}$ (the intersection is S). Draw $\angle Q$ so that the measurement of $\angle Q$ is $115°$; the ray QR should be drawn to intersect with the line parallel to $\overline{PQ}$ (the intersection is R).

3. Use a setsquare, ruler, and protractor to draw rhombus $ABCD$ so that the measurement of $\angle A$ is $60°$, and each side of the rhombus measures 5 cm.

   Steps to draw the figure: Draw $\overline{AB}$ first. Using $\overline{AB}$ as one ray of $\angle A$, draw $\angle A$ so that the measurement of $\angle A$ is $60°$. The other ray, or to-be side of the rhombus, AD, should be 5 cm in length; label the endpoint of the segment as D. Align the setsquare and ruler so one leg of the setsquare aligns with $\overline{AB}$ and the edge of the ruler passes through D. Slide the setsquare along the ruler so that the edge of the setsquare passes through D, and draw a line along the edge of the setsquare. This line is parallel to $\overline{AB}$. Now, align the setsquare and ruler so one leg of the setsquare aligns with $\overline{AD}$ and the edge of the ruler passes through B. Slide the setsquare along the ruler so that the edge of the setsquare passes through B, and draw a line along the edge of the setsquare. This line is parallel to $\overline{AD}$. Along this line, measure a segment 5 cm with B as one endpoint, and label the other endpoint C. Join C to D.

The following table contains partial information for parallelogram $ABCD$. Using no tools, make a sketch of the parallelogram. Then, use a ruler, protractor, and setsquare to draw an accurate picture. Finally, complete the table with the unknown lengths.

Two of the sketches are provided.

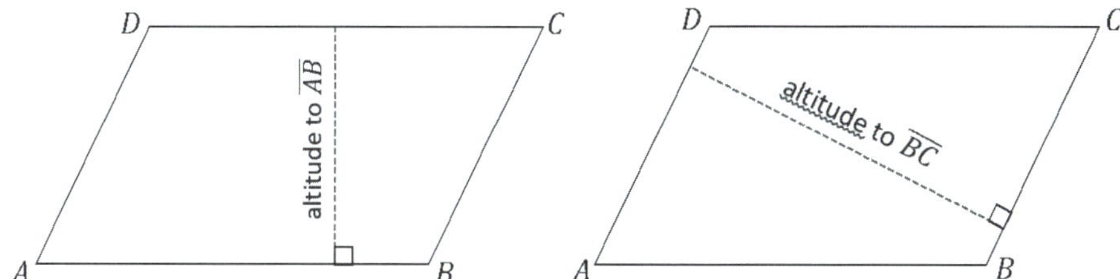

	$\angle A$	AB	Altitude to $\overline{AB}$	BC	Altitude to $\overline{BC}$
4.	$45°$	5 cm	2.8 cm	4 cm	3.5 cm
5.	$50°$	3 cm	2.3 cm	3 cm	2.3 cm
6.	$60°$	4 cm	4 cm	4.6 cm	3.5 cm

7. **Use what you know about drawing parallel lines with a setsquare to draw trapezoid $ABCD$ with parallel sides $\overline{AB}$ and $\overline{CD}$. The length of $\overline{AB}$ is 3 cm, and the length of $\overline{CD}$ is 5 cm; the height between the parallel sides is 4 cm. Write a plan for the steps you will take to draw $ABCD$.**

   Draw $\overline{AB}$ (or $\overline{CD}$) first. Align the setsquare and ruler so that one leg of the setsquare aligns with $\overline{AB}$; mark a point X 4 cm away from $\overline{AB}$. Draw a line parallel to $\overline{AB}$ through X. Once a line parallel to $\overline{AB}$ has been drawn through X, measure a portion of the line to be 3 cm, and label the endpoint on the right as C and the other endpoint D. Join D to A and C to B.

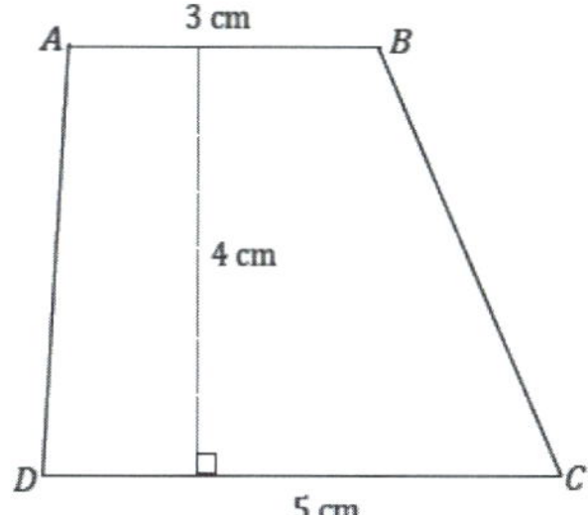

8. **Use the appropriate tools to draw rectangle $FIND$ with $FI = 5$ cm and $IN = 10$ cm.**

   Draw $\overline{FI}$ first. Align the setsquare so that one leg aligns with $\overline{FI}$, and place the ruler against the other leg of the setsquare; mark a point X 10 cm away from $\overline{FI}$. Draw a line parallel to $\overline{FI}$ through X. To create the right angle at F, align the setsquare so that its leg aligns with $\overline{FI}$, and situate the ruler so that its outer edge passes through F, and then draw a line through F. Mark the intersection of the line through F and the parallel line to $\overline{FI}$ as D. Repeat the steps to determine N.

9. **Challenge: Determine the area of the largest rectangle that will fit inside an equilateral triangle with side length 5 cm.**

   Students will quickly discover that rectangles of different dimensions can be drawn; finding the largest rectangle may take multiple efforts. The maximum possible area is 5.4 cm^2.

Supplement

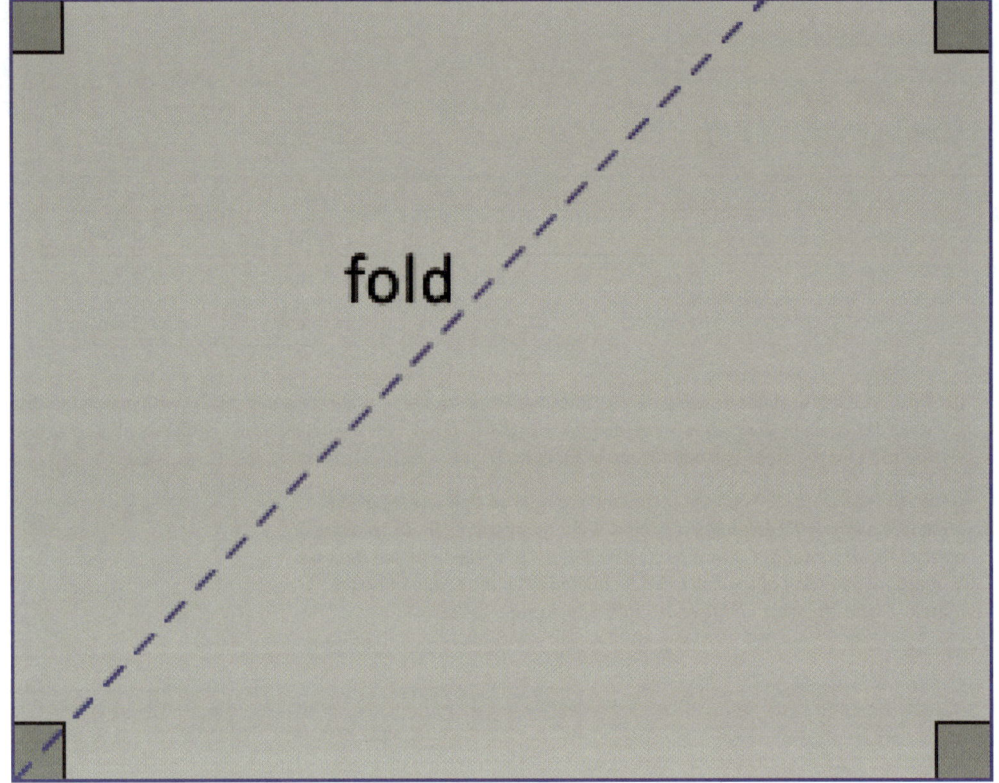

Lesson 8: Drawing Triangles

Student Outcomes

- Students draw triangles under different criteria to explore which criteria result in many, a few, or one triangle.

Lesson Notes

Students should end this lesson understanding the question that drives Lessons 8–11: What conditions (i.e., how many measurements and what arrangement of measurements) are needed to produce identical triangles? Likewise, what conditions are needed to produce a unique triangle? Understanding how a triangle is put together under given conditions helps answer this question. Students arrive at this question after drawing several triangles based on conditions that yield many triangles, one triangle, and a handful of triangles. After each drawing, students consider whether the conditions yielded identical triangles. Students continue to learn how to use their tools to draw figures under provided conditions.

Classwork

Exercises 1–2 (10 minutes)

Scaffolding:

Students may benefit from explicit modeling of the use of the protractor and ruler to make this construction. Seeing an example of the product and the process aids struggling students.

MP.3

Exercises 1–2

1. **Use your protractor and ruler to draw right triangle DEF. Label all sides and angle measurements.**
   a. **Predict how many of the right triangles drawn in class are identical to the triangle you have drawn.**

      Answers will vary; students may say that they should all be the same since the direction is to draw a right triangle.

   b. **How many of the right triangles drawn in class are identical to the triangle you drew? Were you correct in your prediction?**

      Drawings will vary; most likely few or none of the triangles in the class are identical. Ask students to reflect on why their prediction was incorrect if it was in fact incorrect.

- Why is it possible to have so many different triangles? How could we change the question so that more people could draw the same triangle? Elicit suggestions for more criteria regarding the right triangle.
  - *There are many ways to create a right triangle; there is only one piece of information to use when building a triangle. For people to have the same triangle, we would have to know more about the triangle than just its $90°$ angle.*

Take time at the close of this exercise to introduce students to prime notation.

- We use prime notation to distinguish two or more figures that are related in some way. Take, for example, two different right triangles that have equal side lengths and equal angle measures under some correspondence. If the first triangle is $\triangle DEF$ as shown, what letters should we use for the vertices of the second triangle?

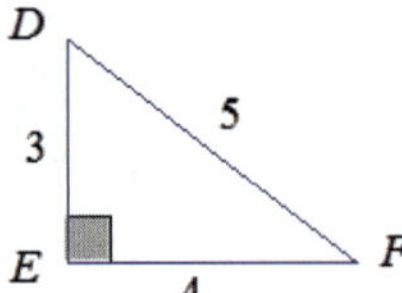

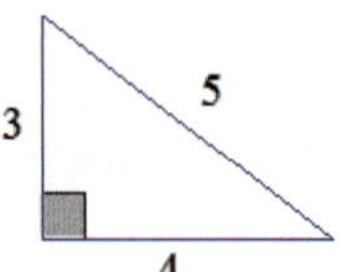

- We don't want to use D, E, or F because they have already been used, and it would be confusing to have two different points with the same name. Instead, we could use D', E', and F' (read: D prime, E prime, and F prime). This way the letters show the connections between the two triangles.

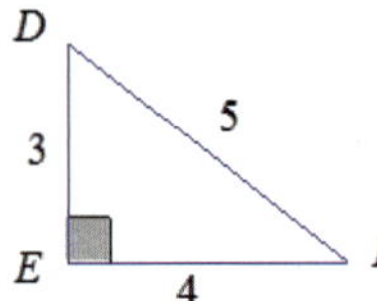

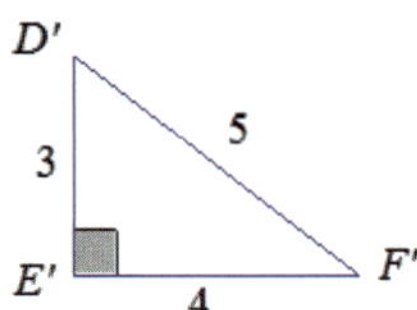

- If there were a third triangle, we could use D'', E'', and F'' (read: D double prime, E double prime, and F double prime).

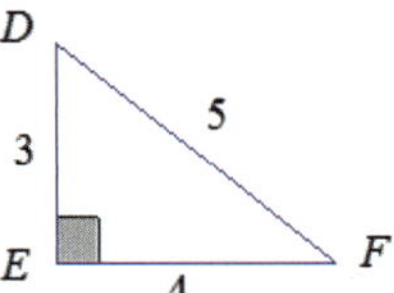

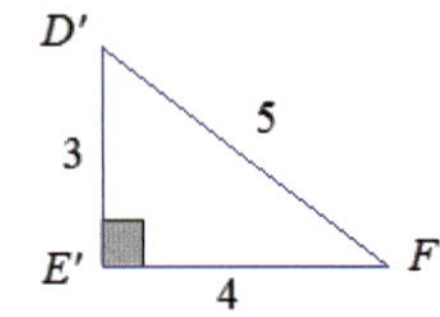

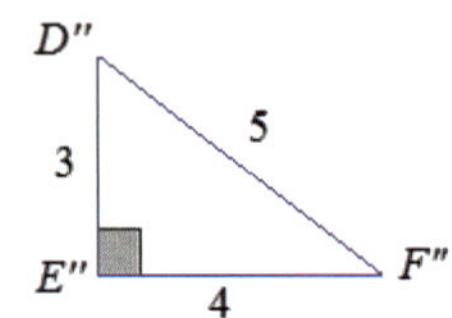

MP.5

2. Given the following three sides of $\triangle ABC$, use your compass to copy the triangle. The longest side has been copied for you already. Label the new triangle $A'B'C'$, and indicate all side and angle measurements. For a reminder of how to begin, refer to Lesson 6 Exploratory Challenge Problem 10.

A_______________B

B_______________________C

A______________________________C

A______________________________C

Students must learn how to determine the third vertex of a triangle, given three side lengths. This skill is anchored in the understanding that a circle drawn with a radius of a given segment shows every possible location of one endpoint of that segment (with the center being the other endpoint).

Depending on how challenging students find the task, the following instructions can be provided as a scaffold to the problem. Note that student drawings use prime notation, whereas the original segments do not.

i. Draw a circle with center A' and radius AB.

ii. Draw a circle with center C' and radius BC.

iii. Label the point of intersection of the two circles above $A'C'$ as B' (the intersection below $A'C'$ works as well).

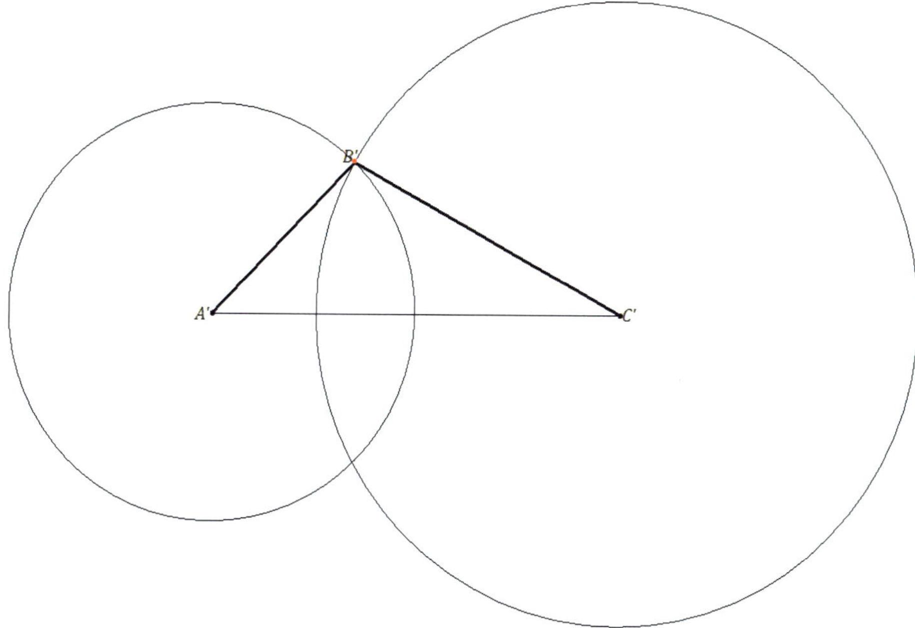

- How many of the triangles drawn in class are identical?
  - *All the drawings should be identical. With three provided side lengths, there is only one way to draw the triangle.*

Exploratory Challenge (25 minutes)

In the Exploratory Challenge, students draw a triangle given two angle measurements and the length of a side. Then, they rearrange the measurements in as many ways as possible and determine whether the triangles they drew are all identical. The goal is to conclude the lesson with the question: Which pieces and what arrangement of those pieces guarantees that the triangles drawn are identical? This question sets the stage for the next several lessons.

The Exploratory Challenge is written assuming students are using a protractor, ruler, and compass. Triangles in the Exploratory Challenge have been drawn on grid paper to facilitate the measurement process. When comparing different triangle drawings, the use of the grid provides a means to quickly assess the length of a given side. An ideal tool to have at this stage is an angle-maker, which is really a protractor, adjustable triangle, and ruler all in one. Using this tool here is fitting because it facilitates the drawing process in questions like part (b).

Exploratory Challenge

A triangle is to be drawn provided the following conditions: the measurements of two angles are 30° and 60°, and the length of a side is 10 cm. Note that where each of these measurements is positioned is not fixed.

a. How is the premise of this problem different from Exercise 2?

In that exercise, we drew a triangle with three provided lengths, while in this problem we are provided two angle measurements and one side length; therefore, the process of drawing this triangle will not require a compass at all.

MP.3

b. Given these measurements, do you think it will be possible to draw more than one triangle so that the triangles drawn will be different from each other? Or do you think attempting to draw more than one triangle with these measurements will keep producing the same triangle, just turned around or flipped about?

Responses will vary. Possible response: I think more than one triangle can be drawn because we only know the length of one side, and the lengths of the two remaining sides are still unknown. Since two side lengths are unknown, it is possible to have different side lengths and build several different triangles.

c. Based on the provided measurements, draw $\triangle ABC$ so that $\angle A = 30^\circ$, $\angle B = 60^\circ$, and $AB = 10$ cm. Describe how the 10 cm side is positioned.

The 10 cm side is between $\angle A$ and $\angle B$.

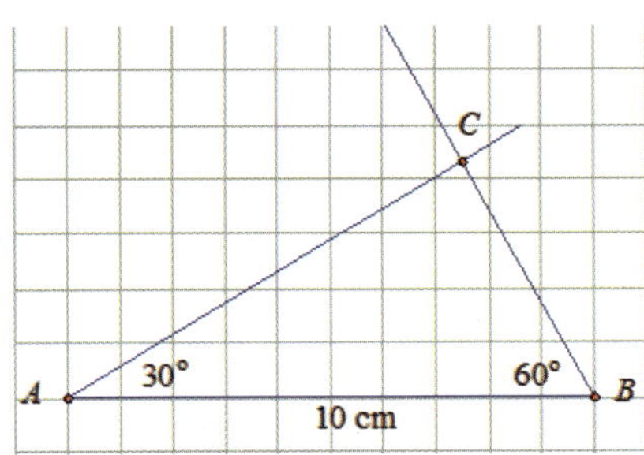

d. Now, using the same measurements, draw $\triangle A'B'C'$ so that $\angle A' = 30^\circ$, $\angle B' = 60^\circ$, and $AC = 10$ cm. Described how the 10 cm side is positioned.

The 10 cm side is opposite to $\angle B$.

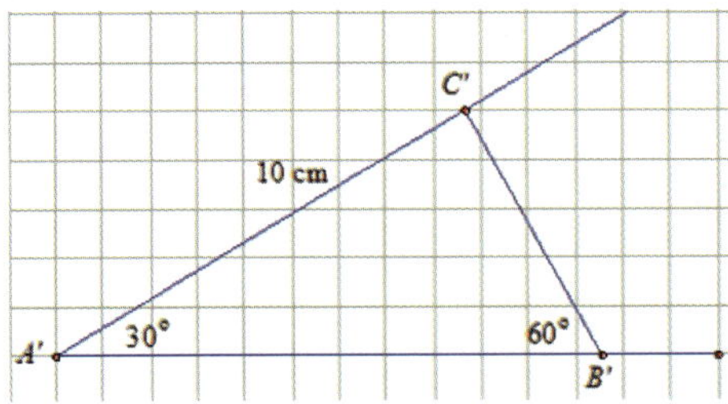

e. Lastly, again, using the same measurements, draw $\triangle A''B''C''$ so that $\angle A'' = 30°$, $\angle B'' = 60°$, and $B''C'' = 10$ cm. Describe how the 10 cm side is positioned.

The 10 cm side is opposite to $\angle A$.

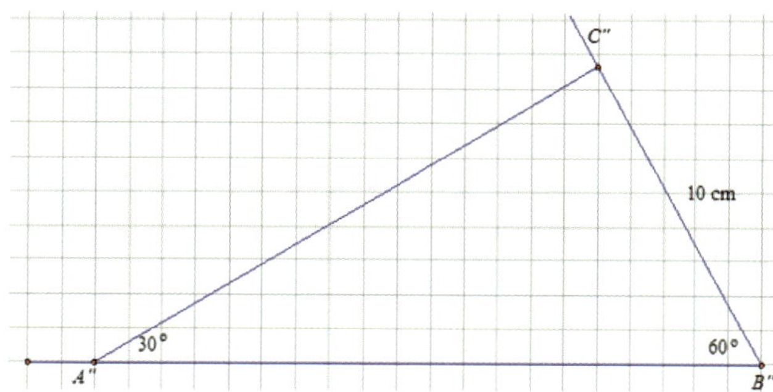

f. Are the three drawn triangles identical? Justify your response using measurements.

No. If the triangles were identical, then the $30°$ and $60°$ angles would match, and the other angles, $\angle C$, $\angle C'$, and $\angle C''$ would have to match, too. The side opposite $\angle C$ is 10 cm. The side opposite $\angle C'$ is between 11 and 12 cm. The side opposite $\angle C''$ is 20 cm. There is no correspondence to match up all the angles and all the sides; therefore, the triangles are not identical.

g. Draw $\triangle A'''B'''C'''$ so that $\angle B''' = 30°$, $\angle C''' = 60°$, and $B'''C''' = 10$ cm. Is it identical to any of the three triangles already drawn?

It is identical to the triangle in part (d).

h. Draw another triangle that meets the criteria of this challenge. Is it possible to draw any other triangles that would be different from the three drawn above?

No, it will be identical to one of the triangles above. Even though the same letters may not line up, the triangle can be rotated or flipped so that there will be some correspondence that matches up equal sides and equal angles.

Discussion (5 minutes)

- In parts (c)–(e) of the Exploratory Challenge, you were given three measurements, two angle measurements and a side length to use to draw a triangle. How many nonidentical triangles were produced under these given conditions?
  - *Three nonidentical triangles*
- If we wanted to draw more triangles, is it possible that we would draw more nonidentical triangles?
  - *We tried to produce another triangle in part (g), but we created a copy of the triangle in part (d). Any attempt at a new triangle will result in a copy of one of the triangles already drawn.*
- If the given conditions had produced just one triangle—in other words, had we attempted parts (c)–(e) and produced the same triangle, including one that was simply a rotated or flipped version of the others—then we would have produced a unique triangle.
- Provided two angle measurements and a side length, without any direction with respect to the arrangement of those measurements, we produced triangles that were nonidentical after testing different arrangements of the provided parts.

> *Scaffolding:*
> To help students keep track of the conditions that do and do not produce unique triangles, it may be helpful to track the conditions in a chart with examples. Students can add to the chart during the closing of each lesson.

- Think back to Exercises 1–2. With a single criterion, a right angle, we were able to draw many triangles. With the criteria of two angle measurements and a side length—and no instruction regarding the arrangement—we drew three different triangles.
- What conditions do you think produce a unique triangle? In other words, what given conditions yield the same triangle or identical triangles no matter how many arrangements are drawn? Are there any conditions you know for certain, without any testing, that produce a unique triangle? Encourage students to write a response to this question and share with a neighbor.
  - *Providing all six measurements of a triangle (three angle measurements and three side lengths) and their arrangement will guarantee a unique triangle.*

MP.3

- All six measurements and their arrangement will indeed guarantee a unique triangle. Is it possible to have less information than all six measurements and their respective arrangements and still produce a unique triangle?
  - *Responses will vary.*
- This question guides us in our next five lessons.

Closing (1 minute)

We have seen a variety of conditions under which triangles were drawn. Our examples showed that just because a condition is given, it does not necessarily imply that the triangle you draw will be identical to another person's drawing given those same conditions. We now want to determine exactly what conditions produce identical triangles.

Have students record a table like the following in their notebooks to keep track of the criteria that determine a unique triangle.

What Criteria Produce Unique Triangles?

Criteria	Example
Three angle measurements and three side lengths	10 cm, 10 cm, 17.4 cm; 120°, 30°, 30° There is only one triangle with side lengths 10 cm, 10 cm, and 17.4 cm, with angles $30°$, $30°$, and $120°$ as arranged above.

Lesson Summary

The following conditions produce identical triangles:

What Criteria Produce Unique Triangles?

Criteria	Example

Exit Ticket (4 minutes)

Name ____ Date ____________________

Lesson 8: Drawing Triangles

Exit Ticket

1. A student is given the following three side lengths of a triangle to use to draw a triangle.

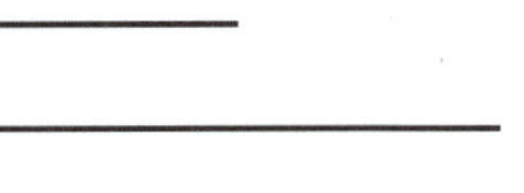

The student uses the longest of the three segments as side $\overline{AB}$ of triangle $\triangle ABC$. Explain what the student is doing with the two shorter lengths in the work below. Then, complete the drawing of the triangle.

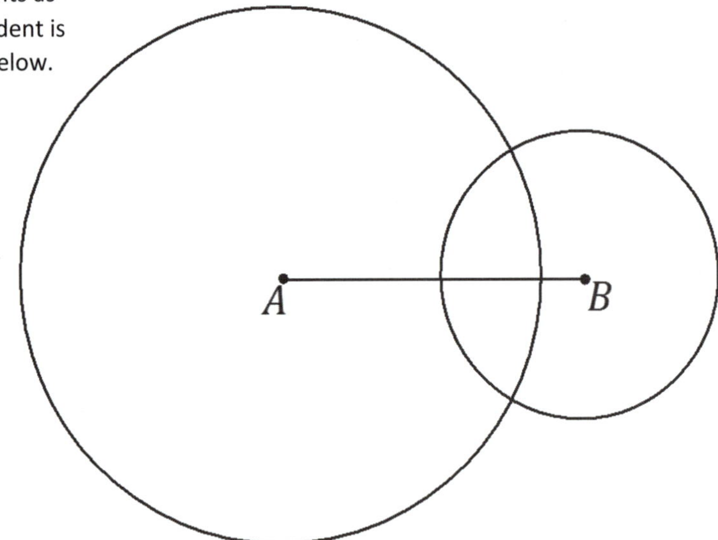

2. Explain why the three triangles constructed in parts (c), (d), and (e) of the Exploratory Challenge were nonidentical.

EUREKA MATH™

Exit Ticket Sample Solutions

1. A student is given the following three side lengths of a triangle to use to draw a triangle.

   The student uses the longest of the three segments as side $\overline{AB}$ of $\triangle ABC$. Explain what the student is doing with the two shorter lengths in the work below. Then, complete the drawing of the triangle.

   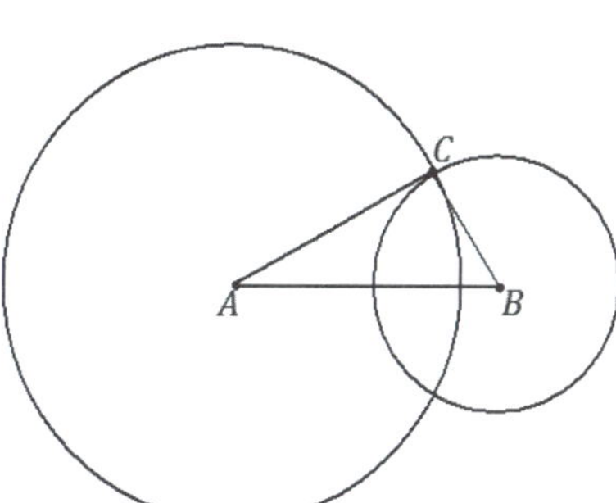

   The student drew a circle with center A and a radius equal in length to the medium segment and a circle with center B and a radius equal in length to the smallest segment. The points of the circle A are all a distance equal to the medium segment from point A, and the points of the circle B are all a distance equal to the smallest segment from point B. The point where the two circles intersect indicates where both segments would meet when drawn from A and B, respectively.

2. Explain why the three triangles constructed in parts (c), (d), and (e) of the Exploratory Challenge were nonidentical.

   They were nonidentical because the two angles and one side length could be arranged in different ways that affected the structure of the triangle. The different arrangements resulted in differences in angle measurements and side lengths in the remaining parts.

Problem Set Sample Solutions

1. Draw three different acute triangles XYZ, $X'Y'Z'$, and $X''Y''Z''$ so that one angle in each triangle is $45°$. Label all sides and angle measurements. Why are your triangles not identical?

   Drawings will vary; the angle measurements are not equal from triangle to triangle, so there is no correspondence that will match equal angles to equal angles.

2. Draw three different equilateral triangles ABC, $A'B'C'$, and $A''B''C''$. A side length of $\triangle ABC$ is 3 cm. A side length of $\triangle A'B'C'$ is 5 cm. A side length of $\triangle A''B''C''$ is 7 cm. Label all sides and angle measurements. Why are your triangles not identical?

   The location of vertices may vary; all angle measurements are $60°$. Though there is a correspondence that will match equal angles to equal angles, there is no correspondence that will match equal sides to equal sides.

   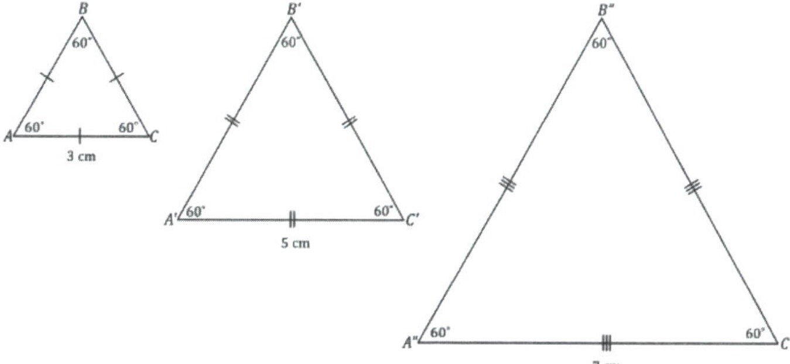

3. Draw as many isosceles triangles that satisfy the following conditions: one angle measures $110°$, and one side measures 6 cm. Label all angle and side measurements. How many triangles can be drawn under these conditions?

Two triangles

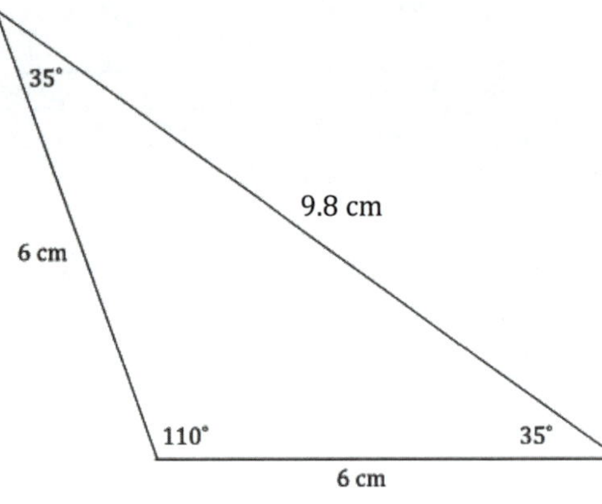

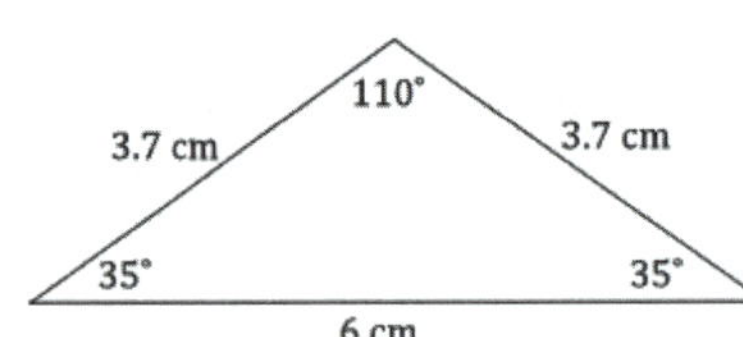

4. Draw three nonidentical triangles so that two angles measure $50°$ and $60°$ and one side measures 5 cm.

   a. Why are the triangles not identical?

   Though there is a correspondence that will match equal angles to equal angles, there is no correspondence that will match equal sides to equal sides.

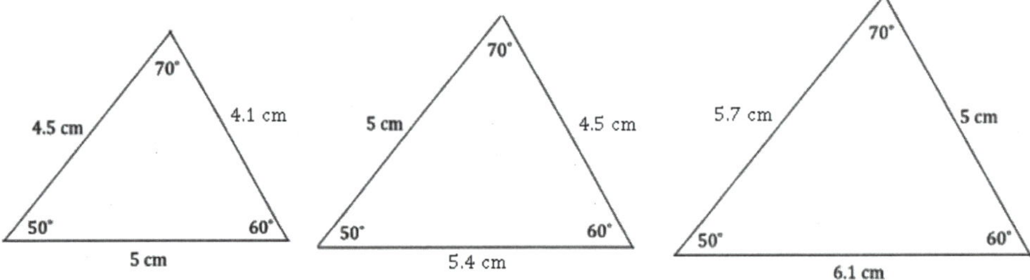

   b. Based on the diagrams you drew for part (a) and for Problem 2, what can you generalize about the criterion of three given angles in a triangle? Does this criterion determine a unique triangle?

   No, it is possible to draw nonidentical triangles that all have the same three angle measurements but have different corresponding side lengths.

Lesson 9: Conditions for a Unique Triangle—Three Sides and Two Sides and the Included Angle

Student Outcomes

- Students understand that two triangles are identical if all corresponding sides are equal under some correspondence; three side lengths of a triangle determine a unique triangle.
- Students understand that two triangles are identical if two corresponding sides and the included angle are equal under some correspondence; two sides and an included angle of a triangle determine a unique triangle.

Lesson Notes

Students finished Lesson 8 with the driving question: What conditions produce identical triangles? More specifically, given a few measurements of the sides and angles of a known triangle, but not necessarily given the relationship of those sides and angles, is it possible to produce a triangle identical to the original triangle? This question can be rephrased as, "Which conditions yield a unique triangle?" If several attempts were made to draw triangles under the provided conditions, would it be possible to draw several nonidentical triangles? In Lesson 9, students draw all variations of a triangle with all three side lengths provided. They also draw all variations of a triangle with two side lengths and the included angle provided. They conclude that drawing a triangle under either of these conditions always yields a unique triangle.

Classwork

Opening (5 minutes)

Students have learned that triangles are identical if there is a correspondence between the triangles that matches sides of equal lengths and matches angles of equal measurement. What conditions on a triangle always produce identical triangles? In other words, what conditions on a triangle determine a unique triangle?

- Given a triangle, we consider conditions on the triangle such as the measurements of angles, the measurements of sides, and the relationship between those angles and sides.
- If we measure all of the angles and sides and give all the relationships between angles and sides, then any other triangle satisfying the same conditions will be identical to our given triangle.
- If we give too *few* conditions on a triangle, such as the length of one side and the measurement of one angle, then there will be many nonidentical triangles that satisfy the conditions.
- Sometimes just a few specific conditions on a triangle make it so that every triangle satisfying those conditions is identical to the given triangle. In this case, we say the conditions on a triangle determine a *unique triangle*; that is, all triangles created using those conditions will be identical.

Exploratory Challenge (25 minutes)

Students draw triangles under two different conditions. Exploratory Challenge Problems 1 and 2 are examples designed to illustrate the three sides condition; Exploratory Challenge Problems 3 and 4 are examples designed to illustrate the two sides and included angle condition. In all four cases (under two kinds of conditions), students see that the conditions always yield a unique triangle. Once students have read the instructions, ask them to record their predictions about how many different triangles can be generated under each set of conditions.

Scaffolding:

Refer students to Lesson 8 Exercise 2, for additional support. Additionally, it may be helpful to provide students with manipulatives (e.g., straws) that model three lengths with which to build the triangle.

MP.5

Exploratory Challenge

1. **A triangle XYZ exists with side lengths of the segments below. Draw $\triangle X'Y'Z'$ with the same side lengths as $\triangle XYZ$. Use your compass to determine the sides of $\triangle X'Y'Z'$. Use your ruler to measure side lengths. Leave all construction marks as evidence of your work, and label all side and angle measurements.**

   Under what condition is $\triangle X'Y'Z'$ drawn? Compare the triangle you drew to two of your peers' triangles. Are the triangles identical? Did the condition determine a unique triangle? Use your construction to explain why. Do the results differ from your predictions?

   The condition on $\triangle X'Y'Z'$ is the three side lengths. All of the triangles are identical; the condition determined a unique triangle. After drawing the longest side length, I used the compass to locate the third vertex of the triangle by drawing two circles, one with a radius of the smallest side length and the other with a radius of the medium side length. Each circle was centered at one end of the longest side length. Two possible locations were determined by the intersections of the circles, but both determined the same triangle. One is just a flipped version of the other. The three sides condition determined a unique triangle.

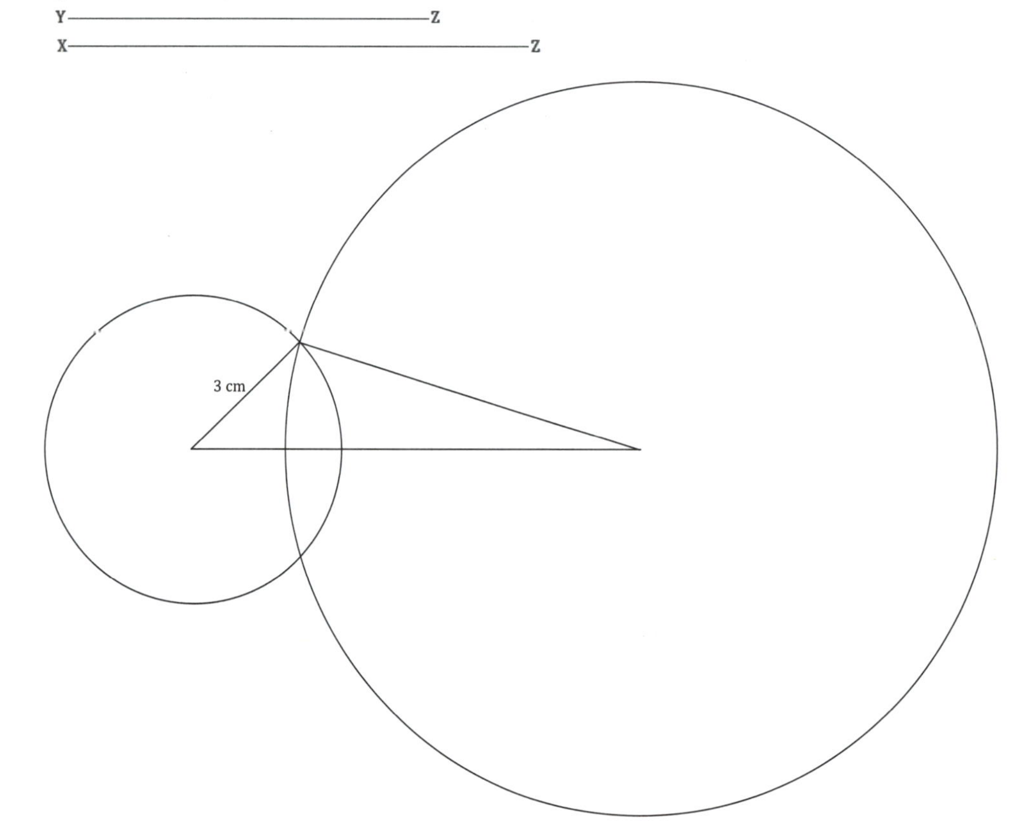

MP.5

2. $\triangle ABC$ is located below. Copy the sides of the triangle to create $\triangle A'B'C'$. Use your compass to determine the sides of $\triangle A'B'C'$. Use your ruler to measure side lengths. Leave all construction marks as evidence of your work, and label all side and angle measurements.

   Under what condition is $\triangle A'B'C'$ drawn? Compare the triangle you drew to two of your peers' triangles. Are the triangles identical? Did the condition determine a unique triangle? Use your construction to explain why.

   The condition on $\triangle A'B'C'$ is the three side lengths. All of the triangles are identical; the condition determined a unique triangle. After drawing the longest side length, I used the compass to locate the third vertex of the triangle by drawing two circles, one with a radius of the smallest side length and the other with a radius of the medium side length. Each circle was centered at one end of the longest side length. Two possible locations were determined by the intersections of the circles, but both determined the same triangle. One is just a flipped version of the other. The three sides condition determined a unique triangle.

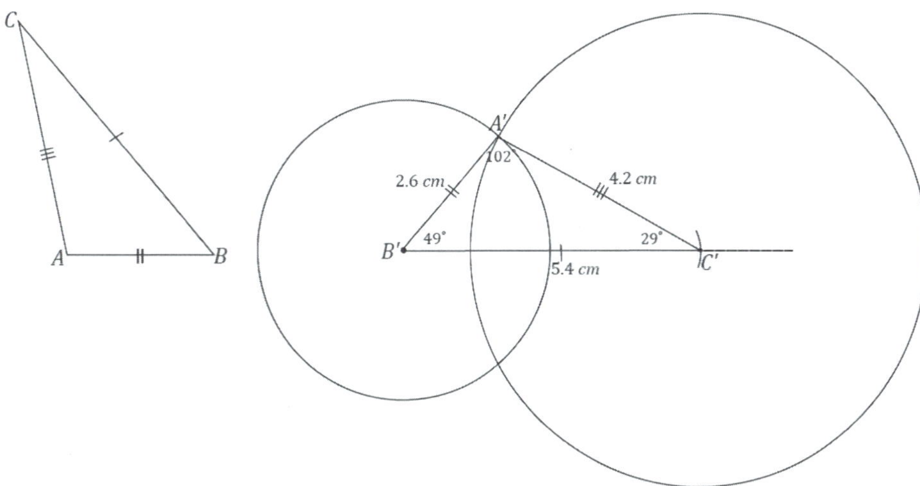

3. A triangle DEF has an angle of $40°$ adjacent to side lengths of 4 cm and 7 cm. Construct $\triangle D'E'F'$ with side lengths $D'E' = 4$ cm, $D'F' = 7$ cm, and included angle $\angle D' = 40°$. Use your compass to draw the sides of $\triangle D'E'F'$. Use your ruler to measure side lengths. Leave all construction marks as evidence of your work, and label all side and angle measurements.

   Under what condition is $\triangle D'E'F'$ drawn? Compare the triangle you drew to two of your peers' triangles. Did the condition determine a unique triangle? Use your construction to explain why.

   The condition on $\triangle D'E'F'$ is two side lengths and the included angle measurement. All of the triangles are identical; the condition determined a unique triangle. Once the $40°$ angle is drawn and the 4 cm and 7 cm side lengths are marked off on the rays of the angle, there is only one place the third side of the triangle can be. Therefore, all triangles drawn under this condition will be identical. Switching the 4 cm and 7 cm sides also gives a triangle satisfying the conditions, but it is just a flipped version of the other.

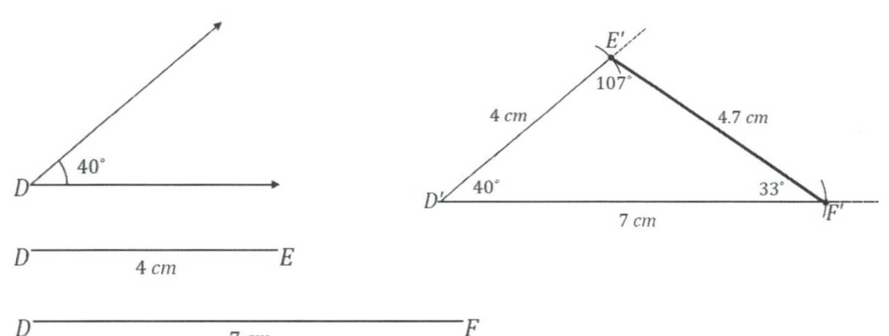

Scaffolding:

Consider providing students with manipulatives (e.g., paperclips for angles) with which to build the triangle.

MP.5

4. $\triangle XYZ$ has side lengths $XY = 2.5$ cm, $XZ = 4$ cm, and $\angle X = 120°$. Draw $\triangle X'Y'Z'$ under the same conditions. Use your compass and protractor to draw the sides of $\triangle X'Y'Z'$. Use your ruler to measure side lengths. Leave all construction marks as evidence of your work, and label all side and angle measurements.

   Under what condition is $\triangle X'Y'Z'$ drawn? Compare the triangle you drew to two of your peers' triangles. Are the triangles identical? Did the condition determine a unique triangle? Use your construction to explain why.

   The condition on $\triangle X'Y'Z'$ is two side lengths and the included angle measurement. The triangle is identical to other triangles drawn under this condition; the conditions produced a unique triangle. Once the $120°$ angle is drawn and the 2.5 cm and 7 cm side lengths are marked off on the rays of the angle, there is only one place the third side of the triangle can be. Therefore, all triangles drawn under these conditions will be identical. Switching the 2.5 cm and 7 cm sides also gives a triangle satisfying the conditions, but it is just a flipped version of the other.

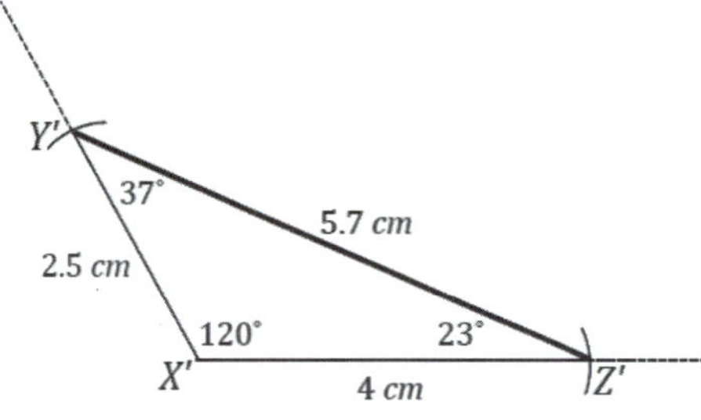

Discussion (10 minutes)

Review responses as a whole group either by sharing out responses from each group or by doing a gallery walk. Consider asking students to write a reflection on the conclusions they reached, either before or after the discussion.

In Lesson 8, students discovered that, depending on the condition provided, it is possible to produce many nonidentical triangles, a few nonidentical triangles, and, sometimes, identical triangles. The question posed at the close of the lesson asked what kinds of conditions produce identical triangles; in other words, determine a unique triangle. The examples in the Exploratory Challenge demonstrate how the three sides condition and the two sides and included angle condition always determine a unique triangle.

- One of the conditions we saw in Lesson 8 provided two angles and a side, by which a maximum of three nonidentical triangles could be drawn. Today, we saw that two sides and an included angle determine a single, unique triangle. What differences exist between these two sets of conditions?
  - *The condition from Lesson 8, two angles and a side, involves different parts of a triangle from the condition in Lesson 9, two sides and an angle. Furthermore, the conditions in Lesson 9 also have a specific arrangement. The angle is specified to be between the sides, while there was no specification for the arrangement of the parts in the condition from Lesson 8.*
- Does the arrangement of the parts play a role in determining whether provided conditions determine a unique triangle?
  - *It seems like it might, but we will have to test out other pieces and other arrangements to be sure.*

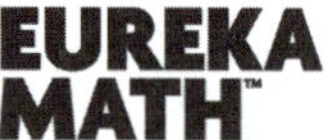

Closing (1 minute)

By drawing triangles under the three sides condition and the two sides and an included angle condition, we saw that there is only one way to draw triangles under each of the conditions, which determines a unique triangle.

The term *diagonal* is used for several Problem Set questions. Alert students to expect this and review the definition provided in the Lesson Summary.

> **Lesson Summary**
>
> **The following conditions determine a unique triangle:**
>
> - **Three sides.**
> - **Two sides and an included angle.**

Exit Ticket (4 minutes)

Name ______________________________ Date ______________

Lesson 9: Conditions for a Unique Triangle—Three Sides and Two Sides and the Included Angle

Exit Ticket

Choose either the three sides condition or the two sides and included angle condition, and explain why the condition determines a unique triangle.

Exit Ticket Sample Solutions

Choose either the three sides condition or the two sides and included angle condition, and explain why the condition determines a unique triangle.

In drawing a triangle with three provided side lengths, there is only one way to draw the triangle. After drawing one length, use the other two lengths to draw circles with the lengths as the respective radii of each circle, centered at either end of the segment drawn first. Regardless of which order of segments is used, there is only one unique triangle that can be drawn.

In drawing a triangle with two side lengths and included angle provided, there is only one way to draw the triangle. After drawing the angle and marking off the two side lengths on the rays of the angle, there is only one possible place to position the third side of the triangle, which also determines the two remaining angle measures of the triangle. Therefore, the two sides and included angle condition determines a unique triangle.

Problem Set Sample Solutions

1. A triangle with side lengths 3 cm, 4 cm, and 5 cm exists. Use your compass and ruler to draw a triangle with the same side lengths. Leave all construction marks as evidence of your work, and label all side and angle measurements.

   Under what condition is the triangle drawn? Compare the triangle you drew to two of your peers' triangles. Are the triangles identical? Did the condition determine a unique triangle? Use your construction to explain why.

   The triangles are identical; the three sides condition determined a unique triangle. After drawing the longest side length, I used the compass to locate the third vertex of the triangle by drawing two circles, one with a radius of the smallest side length and the other with a radius of the medium side length. Each circle was centered at one end of the longest side length. Two possible locations were determined by the intersections of the circles, but both determined the same triangle; one is just a flipped version of the other. The three sides condition determined a unique triangle.

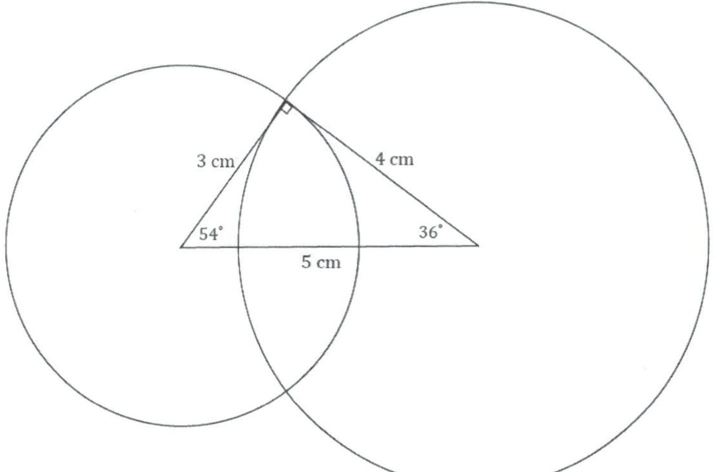

2. Draw triangles under the conditions described below.

   a. A triangle has side lengths 5 cm and 6 cm. Draw two nonidentical triangles that satisfy these conditions. Explain why your triangles are not identical.

      Solutions will vary; check to see that the conditions are satisfied in each triangle. The triangles cannot be identical because there is no correspondence that will match equal corresponding sides and equal angles between the triangles.

   b. A triangle has a side length of 7 cm opposite a $45°$ angle. Draw two nonidentical triangles that satisfy these conditions. Explain why your triangles are not identical.

      Solutions will vary; check to see that the conditions are satisfied in each triangle. The triangles cannot be identical because there is no correspondence that will match equal corresponding sides and equal angles between the triangles.

3. Diagonal $\overline{BD}$ is drawn in square $ABCD$. Describe what condition(s) can be used to justify that $\triangle ABD$ is identical to $\triangle CBD$. What can you say about the measures of $\angle ABD$ and $\angle CBD$? Support your answers with a diagram and explanation of the correspondence(s) that exists.

   Two possible conditions can be used to justify that $\triangle ABD$ is identical to $\triangle CBD$:

   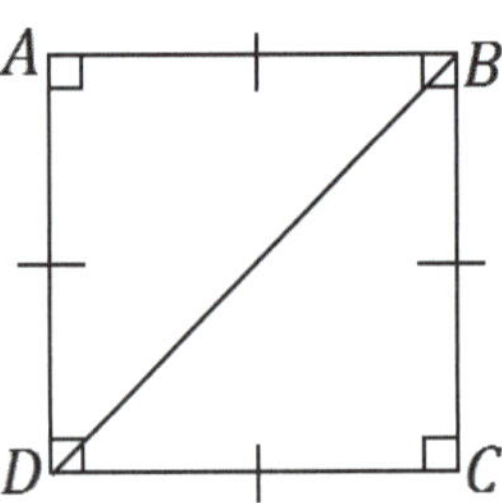

   $\triangle ABD$ is identical to $\triangle CBD$ by the two sides and included angle condition. Since all four sides of a square are equal in length, $AB = CB$ and $AD = CD$. All four angles in a square are right angles; therefore, they are equal in measurement: $\angle A = \angle C$. The two sides and included angle condition is satisfied by the same measurements in both triangles. Since the two sides and included angle condition determine a unique triangle, $\triangle ABD$ must be identical to $\triangle CBD$. The correspondence $\triangle ABD \leftrightarrow \triangle CBD$ matches corresponding equal sides and corresponding angles. It matches $\angle ABD$ with $\angle CBD$, so the two angles have equal measure and angle sum of $90°$; therefore, each angle measures $45°$.

   $\triangle ABD$ is identical to $\triangle CBD$ by the three sides condition. Again, all four sides of the square are equal in length; therefore, $AB = CB$, and $AD = CD$. $\overline{BD}$ is a side to both $\triangle ABD$ and $\triangle CBD$, and $BD = BD$. The three sides condition is satisfied by the same measurements in both triangles. Since the three sides condition determines a unique triangle, $\triangle ABD$ must be identical to $\triangle CBD$. The correspondence $\triangle ABD \leftrightarrow \triangle CBD$ matches equal corresponding sides and equal corresponding angles. It matches $\angle ABD$ with $\angle CBD$, so the two angles have equal measure and angle sum of $90°$; therefore, each angle measures $45°$.

4. Diagonals $\overline{BD}$ and $\overline{AC}$ are drawn in square $ABCD$. Show that $\triangle ABC$ is identical to $\triangle BAD$, and then use this information to show that the diagonals are equal in length.

   Use the two sides and included angle condition to show $\triangle ABC$ is identical to $\triangle BAD$; then, use the correspondence $\triangle ABC \leftrightarrow \triangle BAD$ to conclude $AC = BD$.

   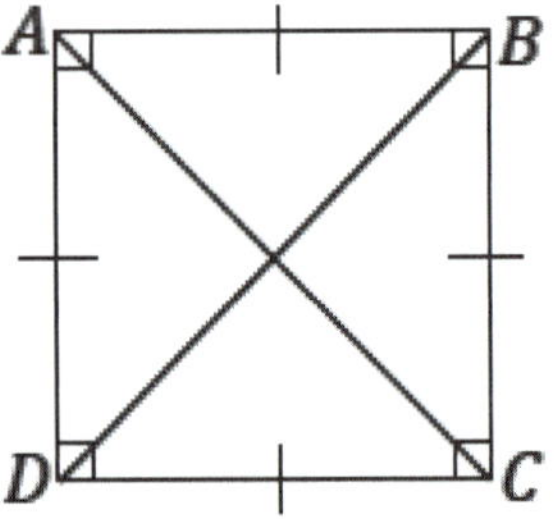

   $\triangle ABC$ is identical to $\triangle BAD$ by the two sides and included angle condition. Since $\overline{AB}$ and $\overline{BA}$ determine the same line segment, $AB = BA$. Since all four sides of a square are equal in length, then $BC = AD$. All four angles in a square are right angles and are equal in measurement; therefore, $\angle B = \angle A$. The two sides and included angle condition is satisfied by the same measurements in both triangles. Since the two sides and included angle condition determines a unique triangle, $\triangle ABC$ must be identical to $\triangle BAD$. The correspondence $\triangle ABC \leftrightarrow \triangle BAD$ matches corresponding equal sides and corresponding equal angles. It matches the diagonals $\overline{AC}$ and $\overline{BD}$. Therefore, $AC = BD$.

5. Diagonal $\overline{QS}$ is drawn in rhombus $PQRS$. Describe the condition(s) that can be used to justify that $\triangle PQS$ is identical to $\triangle RQS$. Can you conclude that the measures of $\angle PQS$ and $\angle RQS$ are the same? Support your answer with a diagram and explanation of the correspondence(s) that exists.

$\triangle PQS$ is identical to $\triangle RQS$ by the three sides condition. All four sides of a rhombus are equal in length; therefore, $PQ = RQ$ and $PS = RS$. $\overline{QS}$ is a side to both $\triangle PQS$ and $\triangle RQS$, and $QS = QS$. The three sides condition is satisfied by the same measurements in both triangles. Since the three sides condition determines a unique triangle, $\triangle PQS$ must be identical to $\triangle RQS$. The correspondence $\triangle PQS \leftrightarrow \triangle RQS$ matches equal corresponding sides and equal corresponding angles. The correspondence matches $\angle PQS$ and $\angle RQS$; therefore, they must have the same measure.

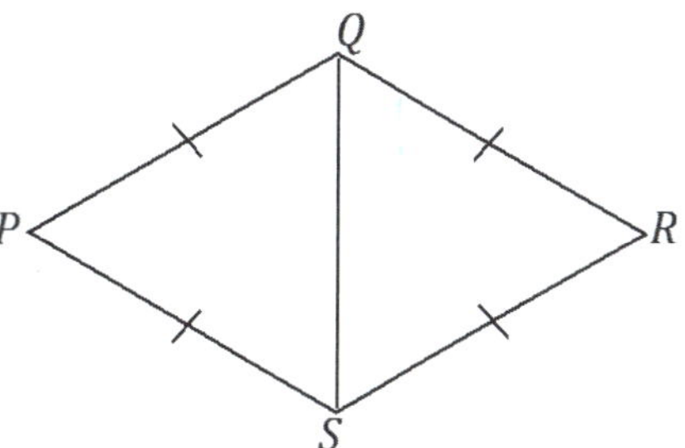

6. Diagonals $\overline{QS}$ and $\overline{PR}$ are drawn in rhombus $PQRS$ and meet at point T. Describe the condition(s) that can be used to justify that $\triangle PQT$ is identical to $\triangle RQT$. Can you conclude that the line segments PR and QS are perpendicular to each other? Support your answers with a diagram and explanation of the correspondence(s) that exists.

$\triangle PQT$ is identical to $\triangle RQT$ by the two sides and included angle condition. All four sides of a rhombus are equal in length; therefore, $PQ = RQ$. $\overline{QT}$ is a side to both $\triangle PQT$ and $\triangle RQT$, and $QT = QT$. Since T lies on segment QS, then $\angle PQT = \angle PQS$ and $\angle RQT = \angle RQS$. By Problem 5, $\angle PQT = \angle RQT$, and the two sides and included angle condition is satisfied by the same measurements in both triangles. Since the two sides and included angle condition determines a unique triangle, then $\triangle PQT$ must be identical to $\triangle RQT$. The correspondence $\triangle PQT \leftrightarrow \triangle RQT$ matches equal corresponding sides and equal corresponding angles. The correspondence matches $\angle PTQ$ and $\angle RTQ$; therefore, they must have the same measure. The angle sum of $\angle PTQ$ and $\angle RTQ$ is $180°$; therefore, each angle is $90°$, and the diagonals are perpendicular to each other.

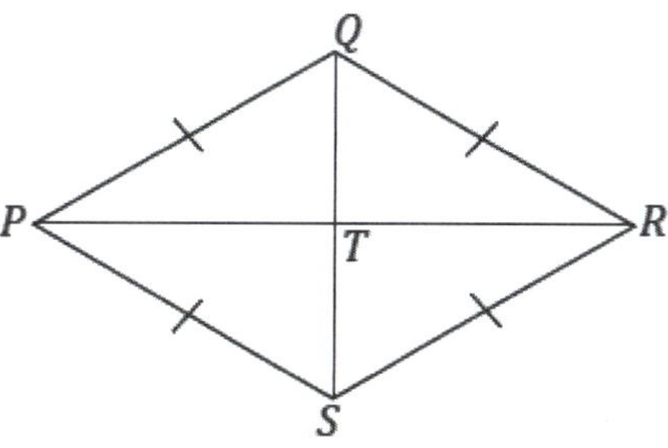

Lesson 10: Conditions for a Unique Triangle—Two Angles and a Given Side

Student Outcomes

- Students understand that two triangles are identical if two pairs of corresponding angles and one pair of corresponding sides are equal under some correspondence; two angle measurements and a given side length of a triangle determine a unique triangle.
- Students understand that the two angles and any side condition can be separated into two conditions: (1) the two angles and included side condition and (2) the two angles and the side opposite a given angle condition.

Lesson Notes

In Lesson 9, students learned two conditions that determine unique triangles: the three sides condition and the two sides and included angle condition. Drawing several examples of triangles under these conditions demonstrated that there was only one possible configuration of a triangle to be drawn. In Lesson 10, students add the two angles and one given side condition to the list of conditions that determine a unique triangle. Since this condition exists in two possible arrangements, the two angles and included side condition and the two angles and the side opposite a given angle condition, it is considered to be two conditions. Drawing a triangle under the two angles and the side opposite a given angle condition requires a step beyond other drawings because the angle opposite the given side needs to be moved around to correctly establish its location. This can be done by drawing the angle opposite the given side on a piece of patty paper, parchment paper, or regular paper. By the close of the lesson, students have a total of four conditions that determine a unique triangle.

Materials

Patty paper or parchment paper

Classwork

Opening (5 minutes)

- In Lesson 8, we explored drawing triangles under the condition that two angles and a side length were provided.
- The arrangement of these parts was not specified, and a total of three nonidentical triangles were drawn.
- In this lesson, we explore what happens when this condition is modified to take arrangement into consideration.
- Instead of drawing triangles given two angle measurements and a side length, we will draw triangles under the condition that two angles and the *included* side are provided and under the condition that two angles and the side *opposite* a given angle are provided.

Exploratory Challenge (25 minutes)

MP.5

Exploratory Challenge

1. A triangle XYZ has angle measures $\angle X = 30°$ and $\angle Y = 50°$ and included side $XY = 6$ cm. Draw $\triangle X'Y'Z'$ under the same condition as $\triangle XYZ$. Leave all construction marks as evidence of your work, and label all side and angle measurements.

   Under what condition is $\triangle X'Y'Z'$ drawn? Compare the triangle you drew to two of your peers' triangles. Are the triangles identical? Did the condition determine a unique triangle? Use your construction to explain why.

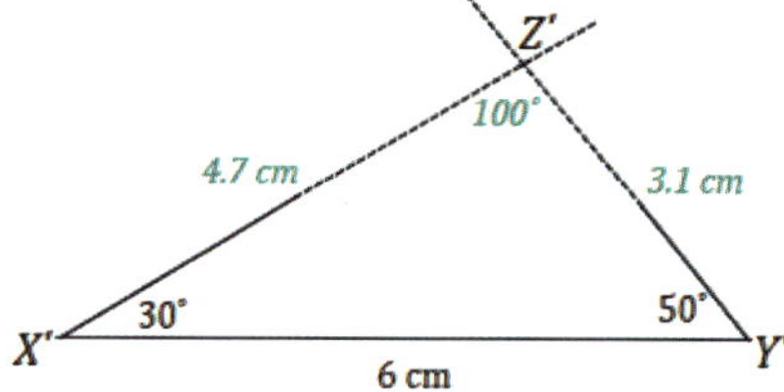

   The condition on $\triangle X'Y'Z'$ is the two angles and included side condition. All of the triangles are identical; the condition determined a unique triangle. After drawing the included side length, I used the protractor to draw the provided angle measurements at either endpoint of the included side $\overline{X'Y'}$. Since these two angle measurements are fixed, the two remaining side lengths will intersect in one location, which is the third vertex of the triangle, Z'. There is no other way to draw this triangle; therefore, the condition determines a unique triangle.

2. A triangle RST has angle measures $\angle S = 90°$ and $\angle T = 45°$ and included side $ST = 7$ cm. Draw $\triangle R'S'T'$ under the same condition. Leave all construction marks as evidence of your work, and label all side and angle measurements.

   Under what condition is $\triangle R'S'T'$ drawn? Compare the triangle you drew to two of your peers' triangles. Are the triangles identical? Did the condition determine a unique triangle? Use your construction to explain why.

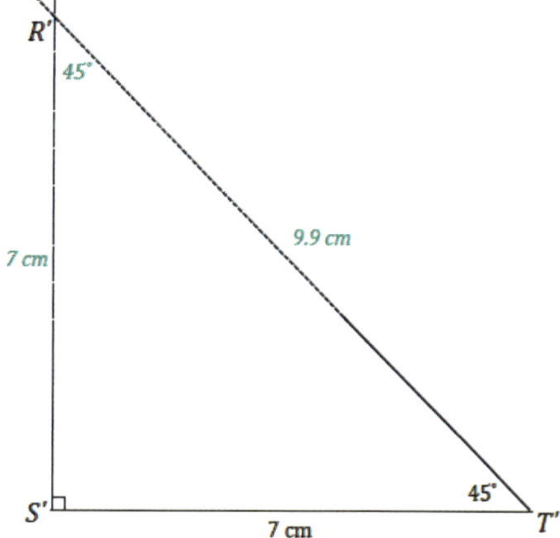

   The condition on $\triangle R'S'T'$ is the two angles and included side condition. All of the triangles are identical; the condition determined a unique triangle. After drawing the included side length, I used the protractor to draw the provided angle measurements at either endpoint of the included side $\overline{S'T'}$. The intersection of the sides of the angle is the third vertex of the triangle, R'. There is no other way to draw this triangle; therefore, the condition determines a unique triangle.

> 3. **A triangle JKL has angle measures $\angle J = 60°$ and $\angle L = 25°$ and side $KL = 5$ cm. Draw $\triangle J'K'L'$ under the same condition. Leave all construction marks as evidence of your work, and label all side and angle measurements.**
>
> **Under what condition is $\triangle J'K'L'$ drawn? Compare the triangle you drew to two of your peers' triangles. Are the triangles identical? Did the condition determine a unique triangle? Use your construction to explain why.**

Students quickly realize that this drawing has an element different from other drawings they have done.

- Do you notice anything about this drawing that is different from other drawings you have done?
  - *This drawing has parts that are not adjacent to each other, so it is harder to tell how to put the triangle together.*
- What parts of the triangle are provided, and what are their relationships?
  - *There is a side and one angle adjacent to the side, and there is one angle opposite the side.*
- Which part of this drawing can be drawn without much difficulty?
  - *The side and the angle adjacent to the side can be drawn without much difficulty.*
- How can the angle opposite the side be correctly positioned?
  - *Responses will vary. Elicit the idea of a* floating *angle, or an angle drawn on a separate piece of paper that can be moved around.*

Provide students with patty paper, parchment paper, or even small slips of regular paper, and ask them to continue the drawing by putting the angle opposite the side on this slip of paper.

Students have to line up one ray of the angle on patty paper (the angle opposite the given side) with one ray of the angle adjacent to the given side. They move the angle around until the free ray of the angle on patty paper meets the endpoint of the given segment.

1. Draw the parts that are adjacent.

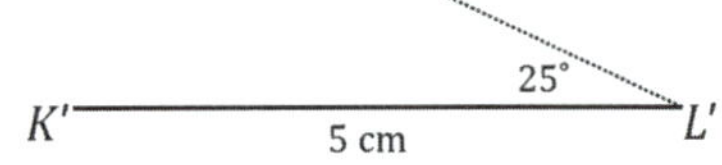

2. Draw the $60°$ angle on a piece of patty paper, and then try to align the rays of the angle so that they coincide with point K' and the non-horizontal side of $\angle L'$.

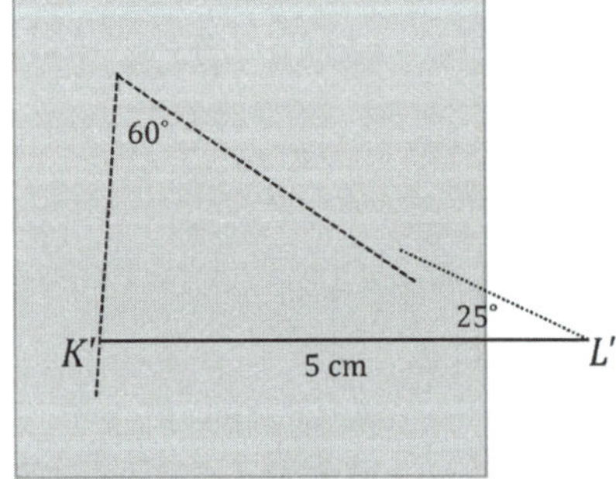

3. Move the $60°$ angle (the patty paper) around until it is in place.

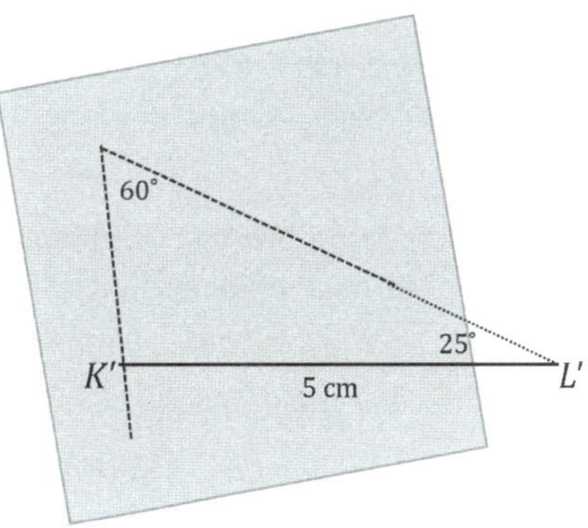

4. Once the angle is in place, the rest of the measurements can be determined.

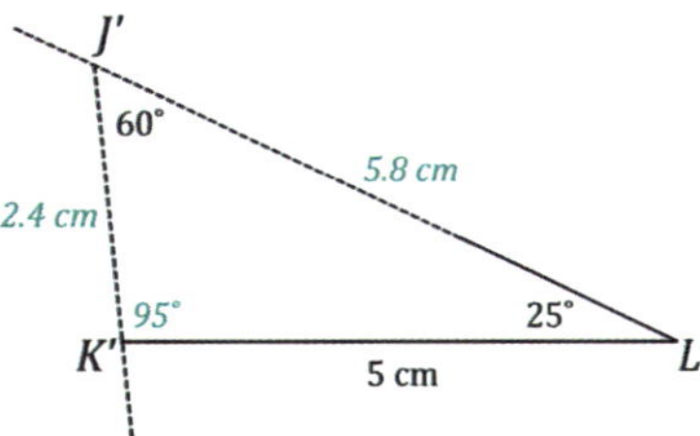

MP.5

The condition on $\triangle J'K'L'$ is the two angles and the side opposite a given angle condition. All of the triangles are identical; the condition determined a unique triangle. After drawing the given side length, I used the protractor to draw $\angle L'$ adjacent to $\overline{K'L'}$. I drew the angle opposite the given side, $\angle J'$, on a slip of paper and lined up one ray of the angle on patty paper with one ray of the angle adjacent to the given side. I moved the angle on patty paper along the coinciding rays until the free ray just met the endpoint of $\overline{K'L'}$. There is no other way to draw this triangle; therefore, the condition determines a unique triangle.

4. **A triangle ABC has angle measures $\angle C = 35°$ and $\angle B = 105°$ and side $AC = 7$ cm. Draw $\triangle A'B'C'$ under the same condition. Leave all construction marks as evidence of your work, and label all side and angle measurements.**

   Under what condition is $\triangle A'B'C'$ drawn? Compare the triangle you drew to two of your peers' triangles. Are the triangles identical? Did the condition determine a unique triangle? Use your construction to explain why.

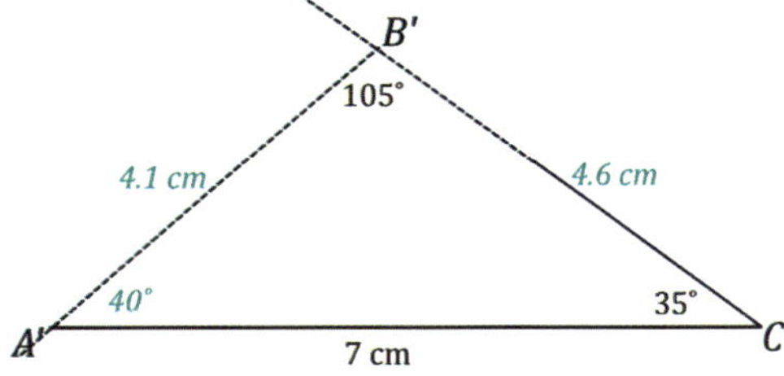

The condition on $\triangle A'B'C'$ is the two angles and the side opposite a given angle condition. All of the triangles are identical; the condition determined a unique triangle. After drawing the given side length, I used the protractor to draw $\angle C'$ adjacent to $\overline{A'C'}$. I drew the angle opposite the given side, $\angle B'$, on a slip of paper and lined up one ray of the angle on patty paper with one ray of the angle adjacent to the given side. I moved the angle on patty paper along the coinciding rays until the free ray just met the endpoint of $\overline{A'C'}$. There is no other way to draw this triangle; therefore, the condition determines a unique triangle.

Discussion (8 minutes)

Review responses to the Exploratory Challenge as a whole group, either by sharing out responses from each group or by doing a gallery walk. Consider asking students to write a reflection about what they learned from the Exploratory Challenge.

- Today we saw two conditions that determine unique triangles. What is the complete list of conditions that determine unique triangles?
  - *Three sides condition*
  - *Two sides and included angle condition*
  - *Two angles and included side condition*
  - *Two angles and the side opposite a given angle condition*

Closing (2 minutes)

The two angles and any side condition determines a unique triangle. Since the condition has two different arrangements, we separate it into two conditions: the two angles and included side condition and the two angles and the side opposite a given angle condition.

When drawing a triangle under the two angles and the side opposite a given angle condition, the angle opposite the given segment must be drawn on separate paper in order to locate the position of the third vertex.

> Lesson Summary
>
> **The following conditions determine a unique triangle:**
>
> - **Three sides.**
> - **Two sides and included angle.**
> - **Two angles and the included side.**
> - **Two angles and the side opposite.**

Exit Ticket (5 minutes)

Name ______________________________ Date ______________

Lesson 10: Conditions for a Unique Triangle—Two Angles and a Given Side

Exit Ticket

1. $\triangle ABC$ has angle measures $\angle A = 50°$ and $\angle C = 90°$ and side $AB = 5.5$ cm. Draw $\triangle A'B'C'$ under the same condition. Under what condition is $\triangle A'B'C'$ drawn? Use your construction to explain why $\triangle A'B'C'$ is or is not identical to $\triangle ABC$.

2. $\triangle PQR$ has angle measures $\angle Q = 25°$ and $\angle R = 40°$ and included side $QR = 6.5$ cm. Draw $\triangle P'Q'R'$ under the same condition. Under what condition is $\triangle P'Q'R'$ drawn? Use your construction to explain why $\triangle P'Q'R'$ is or is not identical to $\triangle PQR$.

Exit Ticket Sample Solutions

1. $\triangle ABC$ has angle measures $\angle A = 50°$ and $\angle C = 90°$ and side $AB = 5.5$ cm. Draw $\triangle A'B'C'$ under the same condition. Under what condition is $\triangle A'B'C'$ drawn? Use your construction to explain why $\triangle A'B'C'$ is or is not identical to $\triangle ABC$.

   The condition on $\triangle A'B'C'$ is the two angles and the side opposite a given angle condition. $\triangle A'B'C'$ is identical to $\triangle ABC$. After drawing the given side length, I used the protractor to draw $\angle A'$ adjacent to $\overline{A'B'}$. I drew the angle opposite the given side, $\angle B'$, on a slip of paper and lined up one ray of the angle on patty paper with one ray of the angle adjacent to the given side. I moved the angle on patty paper along the coinciding rays until the free ray just met the endpoint of $\overline{A'B'}$. There is no other way to draw this triangle; therefore, $\triangle A'B'C'$ must be identical to $\triangle ABC$.

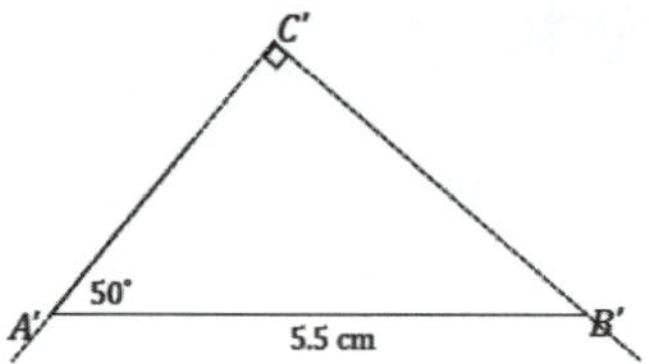

2. $\triangle PQR$ has angle measures $\angle Q = 25°$ and $\angle R = 40°$ and included side $QR = 6.5$ cm. Draw $\triangle P'Q'R'$ under the same condition. Under what condition is $\triangle P'Q'R'$ drawn? Use your construction to explain why $\triangle P'Q'R'$ is or is not identical to $\triangle PQR$.

   The condition on $\triangle P'Q'R'$ is the two angles and included side condition. $\triangle P'Q'R'$ is identical to $\triangle PQR$. After drawing the given side length, I used the protractor to draw $\angle Q'$ adjacent to $\overline{Q'R'}$. After drawing the included side length, I used the protractor to draw the provided angle measurements at either endpoint of the included side $\overline{Q'R'}$. Since these two angle measurements are fixed, the two remaining side lengths will intersect in one location, which is the third vertex of the triangle, P'. There is no other way to draw this triangle; therefore, $\triangle P'Q'R'$ must be identical to $\triangle PQR$.

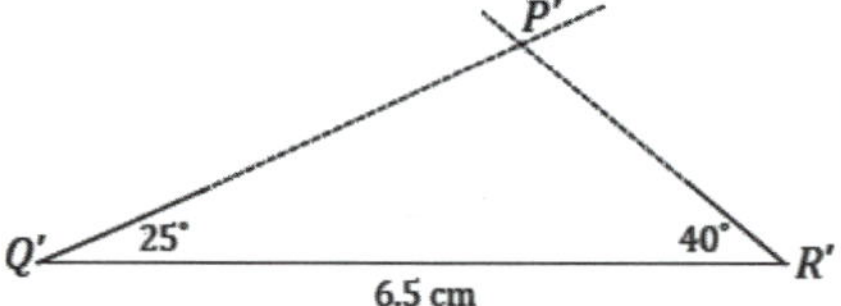

Problem Set Sample Solutions

1. In $\triangle FGH$, $\angle F = 42°$ and $\angle H = 70°$. $FH = 6$ cm. Draw $\triangle F'G'H'$ under the same condition as $\triangle FGH$. Leave all construction marks as evidence of your work, and label all side and angle measurements.

   What can you conclude about $\triangle FGH$ and $\triangle F'G'H'$? Justify your response.

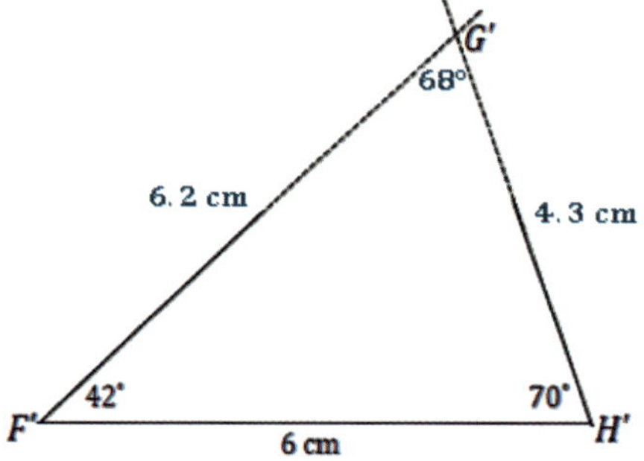

$\triangle FGH$ and $\triangle F'G'H'$ are identical triangles by the two angles and included side condition. Since both triangles are drawn under the same condition, and the two angles and included side condition determines a unique triangle, both triangles determine the same unique triangle. Therefore, they are identical.

2. In $\triangle WXY$, $\angle Y = 57°$ and $\angle W = 103°$. $YX = 6.5$ cm. Draw $\triangle W'X'Y'$ under the same condition as $\triangle WXY$. Leave all construction marks as evidence of your work, and label all side and angle measurements.

What can you conclude about $\triangle WXY$ and $\triangle W'X'Y'$? Justify your response.

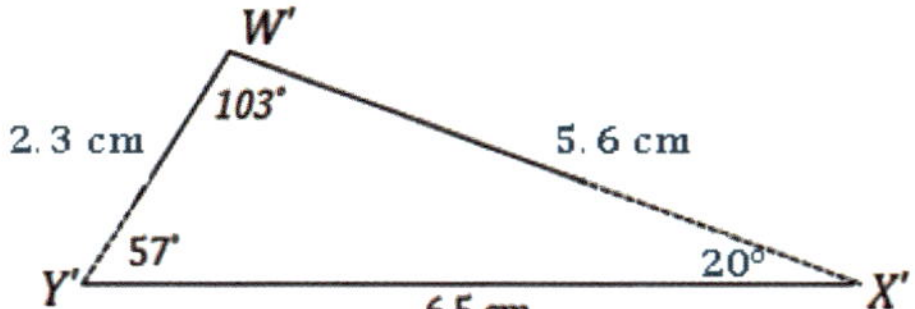

$\triangle WXY$ and $\triangle W'X'Y'$ are identical triangles by the two angles and the side opposite a given angle condition. Since both triangles are drawn under the same condition, and the two angles and the side opposite a given angle condition determines a unique triangle, both triangles determine the same unique triangle. Therefore, they are identical.

3. Points A, Z, and E are collinear, and $\angle B = \angle D$. What can be concluded about $\triangle ABZ$ and $\triangle EDZ$? Justify your answer.

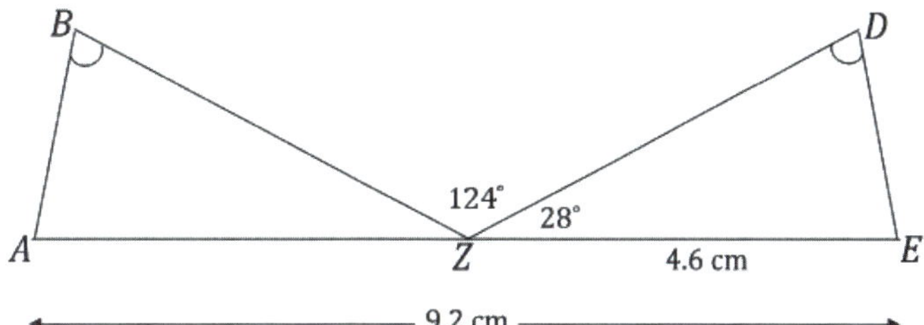

$\triangle ABZ$ and $\triangle EDZ$ are identical by the two angles and the side opposite a given angle condition. Since segments add, and AE is 9.2 cm and ZE is 4.6 cm, AZ must be 4.6 cm. Since angles on a line sum to $180°$, $\angle BZD = 124°$, and $\angle DZE = 28°$, then $\angle AZB = 28°$. From the diagram, we can see that $\angle B = \angle D$. The same measurements in both triangles satisfy the two angles and the side opposite a given angle condition, which means they both determine the same unique triangle; thus, they are identical.

4. Draw $\triangle ABC$ so that $\angle A$ has a measurement of $60°$, $\angle B$ has a measurement of $60°$, and $\overline{AB}$ has a length of 8 cm. What are the lengths of the other sides?

Both of the other side lengths are 8 cm.

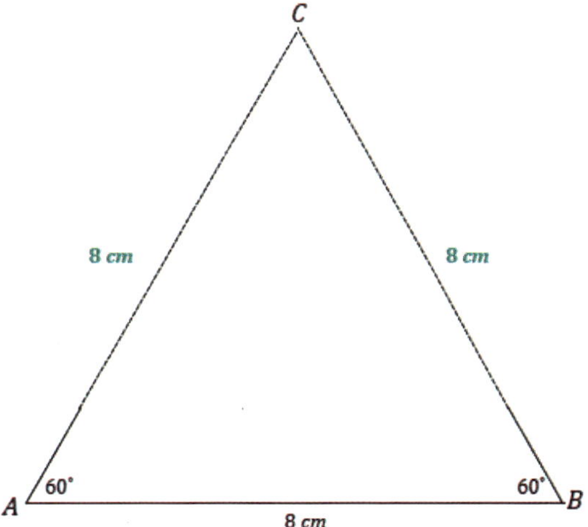

5. Draw $\triangle ABC$ so that $\angle A$ has a measurement of $30°$, $\angle B$ has a measurement of $60°$, and $\overline{BC}$ has a length of 5 cm. What is the length of the longest side?

The longest side has a length of 10 cm.

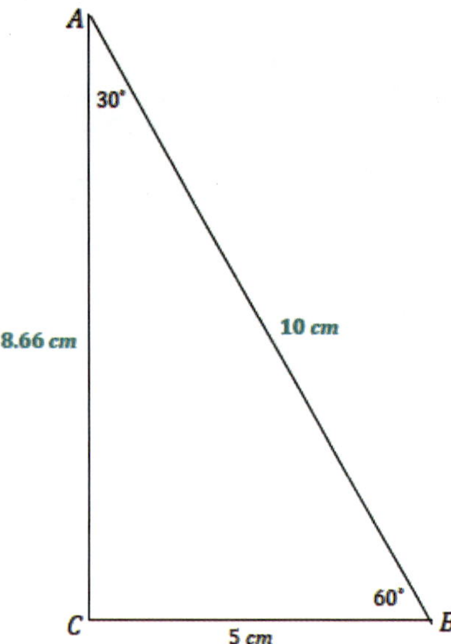

Lesson 11: Conditions on Measurements That Determine a Triangle

Student Outcomes

- Students understand that three given lengths determine a triangle, provided the largest length is less than the sum of the other two lengths; otherwise, no triangle can be formed.
- Students understand that if two side lengths of a triangle are given, then the third side length must be between the difference and the sum of the first two side lengths.
- Students understand that two angle measurements determine many triangles, provided the angle sum is less than $180°$; otherwise, no triangle can be formed.

Materials

Patty paper or parchment paper (in case dimensions of patty paper are too small)

Lesson Notes

Lesson 11 explores side-length requirements and angle requirements that determine a triangle. Students reason through three cases in the exploration regarding side-length requirements and conclude that any two side lengths must sum to be greater than the third side length. In the exploration regarding angle requirements, students observe the resulting figures in three cases and conclude that the angle sum of two angles in a triangle must be less than $180°$. Additionally, they observe that three angle measurements do not determine a unique triangle and that it is possible to draw scale drawings of a triangle with given angle measurements. Students are able to articulate the result of each case in the explorations.

Scaffolding:

An alternate activity that may increase accessibility is providing students with pieces of dry pasta (or other manipulatives) and rulers and then giving the task: "Build as many triangles as you can. Record the lengths of the three sides." Once students have constructed many triangles, ask, "What do you notice about the lengths of the sides?" Allow written and spoken responses. If necessary, ask students to construct triangles with particular dimensions (such as 3 cm, 4 cm, 1 cm; 3 cm, 4 cm, 5 cm; 3 cm, 4 cm, 8 cm) to further illustrate the concept.

Classwork

Exploratory Challenge 1 (8 minutes)

In pairs, students explore the length requirements to form a triangle.

Exploratory Challenge 1

a. **Can any three side lengths form a triangle? Why or why not?**

Possible response: Yes, because a triangle is made up of three side lengths; therefore, any three sides can be put together to form a triangle.

b. Draw a triangle according to these instructions:

- ✓ Draw segment AB of length 10 cm in your notebook.
- ✓ Draw segment BC of length 5 cm on one piece of patty paper.
- ✓ Draw segment AC of length 3 cm on the other piece of patty paper.
- ✓ Line up the appropriate endpoint on each piece of patty paper with the matching endpoint on segment AB.
- ✓ Use your pencil point to hold each patty paper in place, and adjust the paper to form $\triangle ABC$.

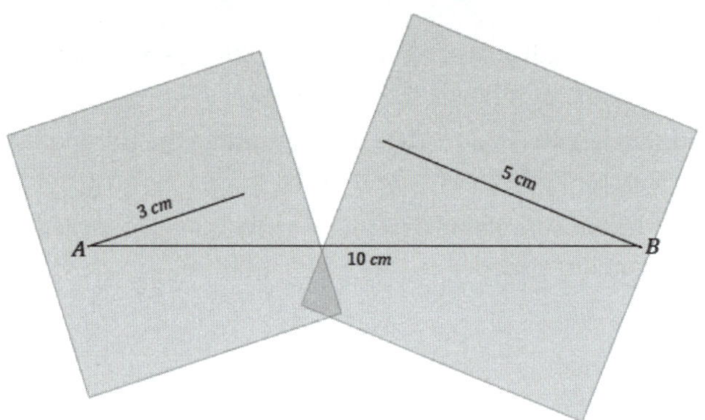

Scaffolding:

Allow students who feel comfortable with a compass to use one instead of the patty paper. The compass is adjusted to 5 cm (AC), and a circle is drawn with a center at A; similar steps are done for side $\overline{AB}$.

c. What do you notice?

$\triangle ABC$ cannot be formed because $\overline{AC}$ and $\overline{BC}$ do not meet.

d. What must be true about the sum of the lengths of $\overline{AC}$ and $\overline{BC}$ if the two segments were to just meet? Use your patty paper to verify your answer.

For $\overline{AC}$ and $\overline{BC}$ to just meet, the sum of their lengths must be equal to 10 cm.

e. Based on your conclusion for part (d), what if $AC = 3$ cm as you originally had, but $BC = 10$ cm. Could you form $\triangle ABC$?

$\triangle ABC$ can be formed because $\overline{AC}$ and $\overline{BC}$ can meet at an angle and still be anchored at A and B.

f. What must be true about the sum of the lengths of $\overline{AC}$ and $\overline{BC}$ if the two segments were to meet and form a triangle?

For $\overline{AC}$ and $\overline{BC}$ to just meet and form a triangle, the sum of their lengths must be greater than 10 cm.

Discussion (7 minutes)

MP.7

- Were you able to form $\triangle ABC$? Why or why not? Did the exercise confirm your prediction?
  - *We could not form $\triangle ABC$ because sides $\overline{BC}$ and $\overline{AC}$ are too short to meet.*
- What would the sum of the lengths of $\overline{BC}$ and $\overline{AC}$ have to be to just meet? Describe one possible set of lengths for $\overline{BC}$ and $\overline{AC}$. Would these lengths form $\triangle ABC$? Explain why or why not.
  - *The lengths would have to sum to 10 cm to just meet (e.g., $AC = 3$ cm and $BC = 7$ cm). Because the segments are anchored to either endpoint of $\overline{AB}$, the segments form a straight line or coincide with $\overline{AB}$. Therefore, $\triangle ABC$ cannot be formed since A, B, and C are collinear.*

MP.7

- If a triangle cannot be formed when the two smaller segments are too short to meet or just meet, what must be true about the sum of the lengths of the two smaller segments compared to the longest length? Explain your answer using possible measurements.
  - *The sum of the two smaller lengths must be greater than the longest length so that the vertices will not be collinear; if the sum of the two smaller lengths is greater than the longest length while anchored at either endpoint of $\overline{AB}$, the only way they can meet is if A, B, and C are not collinear. One possible set of lengths is $AC = 4$ cm and $BC = 7$ cm.*

Help students recognize this fundamental inequality by illustrating it with an image of walking between two points, A and B.

Observe the two pathways to get from A to B. Pathway 1 is a straight path from A to B. Pathway 2 requires you to walk through a point that does not lie on the straight path. Clearly, the total distance when walking through C is greater than the distance of walking the straight path. This idea can be visualized from A, B, or C. Hence, the length of any one side of a triangle is always less than the sum of the lengths of the remaining sides.

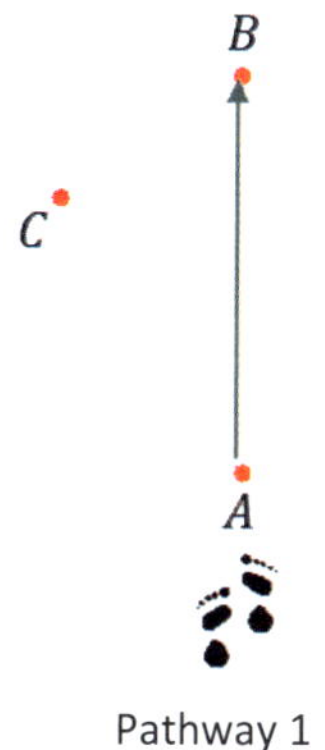

Pathway 1

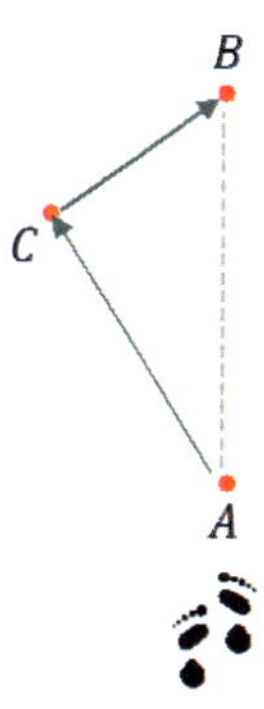

Pathway 2

- Given two side lengths of a triangle, the third side length must be between the difference of the two sides and the sum of the two sides. For example, if a triangle has two sides with lengths 2 cm and 5 cm, then the third side length must be between the difference (5 cm − 2 cm) and the sum (2 cm + 5 cm). Explanation: Let x be the length of the third side in centimeters. If $x \leq 5$, then the largest side length is 5 cm, and $2 + x > 5$, or $x > 5 - 2$. If $x \geq 5$, then x cm is the longest side length and $5 + 2 > x$. So, $5 - 2 < x < 2 + 5$.

Exercise 1 (4 minutes)

In this exercise, students must consider the length of the last side from two perspectives: one where the last side is not the longest side and one where the last side is the longest side. With these two considerations, the third side is a range of lengths, all of which satisfy the condition that the longest side length is less than the sum of the other two side lengths.

> **Exercise 1**
>
> **Two sides of $\triangle DEF$ have lengths of 5 cm and 8 cm. What are all the possible whole number lengths for the remaining side?**
>
> ***The possible whole-number side lengths in centimeters are 4, 5, 6, 7, 8, 9, 10, 11, and 12.***

Exploratory Challenge 2 (8 minutes)

Students explore the angle measurement requirements to form a triangle in pairs. Encourage students to document their exploration carefully, even if their results are not what they expected.

Exploratory Challenge 2

a. **Which of the following conditions determine a triangle? Follow the instructions to try to draw $\triangle ABC$. Segment AB has been drawn for you as a starting point in each case.**

i. **Choose measurements of $\angle A$ and $\angle B$ for $\triangle ABC$ so that the sum of measurements is greater than $180°$. Label your diagram.**

Your chosen angle measurements: $\angle A = 70°$ $\angle B = 140°$

Were you able to form a triangle? Why or why not?

Selected angle measurements and the corresponding diagram indicate one possible response.

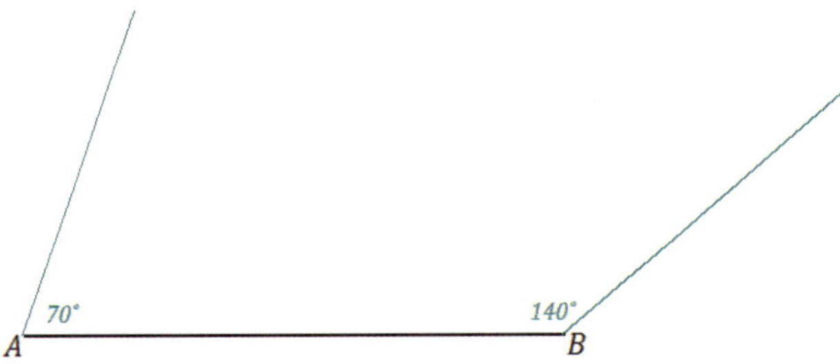

We were not able to form a triangle because the non-horizontal ray of $\angle A$ and the non-horizontal ray of $\angle B$ do not intersect.

ii. **Choose measurements of $\angle A$ and $\angle B$ for $\triangle ABC$ so that the measurement of $\angle A$ is supplementary to the measurement of $\angle B$. Label your diagram.**

Your chosen angle measurements: $\angle A = 40°$ $\angle B = 140°$

Were you able to form a triangle? Why or why not?

Selected angle measurements and the corresponding diagram indicate one possible response.

We were not able to form a triangle because the non-horizontal ray of $\angle A$ and the non-horizontal ray of $\angle B$ do not intersect; the non-horizontal rays look parallel.

iii. Choose measurements of $\angle A$ and $\angle B$ for $\triangle ABC$ so that the sum of measurements is less than $180°$. Label your diagram.

Your chosen angle measurements: $\angle A = 40°$ $\angle B = 100°$

Were you able to form a triangle? Why or why not?

Angle measurements and the corresponding diagram indicate one possible response.

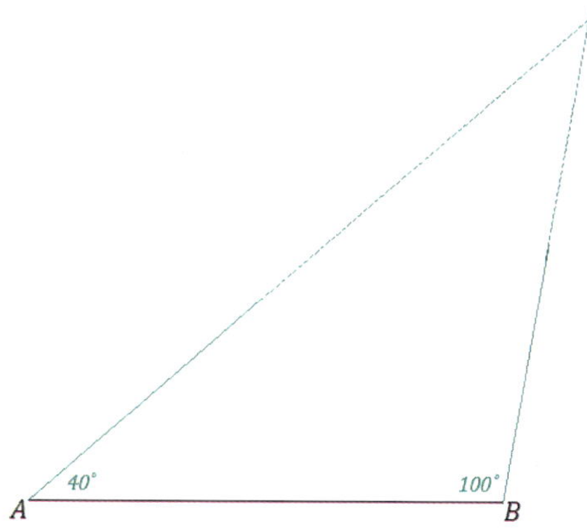

We were able to form a triangle because the non-horizontal ray of $\angle A$ and the non-horizontal ray of $\angle B$ intersect.

> *Scaffolding:*
>
> Teachers may want to demonstrate the informal proof and illustrate that the angle sum of a triangle is $180°$. Do this by tearing off the corners of a triangle and forming a straight line by placing the three angles adjacent to each other. Note that this is an extension and is not formally discussed until Grade 8.

b. Which condition must be true regarding angle measurements in order to determine a triangle?

The sum of two angle measurements of a triangle must be less than $180°$.

c. Measure and label the formed triangle in part (a) with all three side lengths and the angle measurement for $\angle C$. Now, use a protractor, ruler, and compass to draw $\triangle A'B'C'$ with the same angle measurements but side lengths that are half as long.

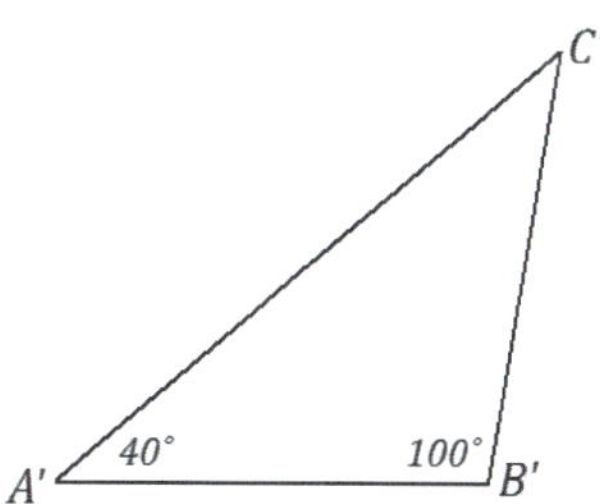

Students should begin by drawing any one side length at a length half as much as the corresponding side in $\triangle ABC$ and then drawing angles at each end of this line segment. Students should recognize that $\triangle A'B'C'$ is a scale drawing of $\triangle ABC$. Ask students to mark all length measurements as a means of verifying that they are indeed half as long as the corresponding sides of the original triangle.

d. Do the three angle measurements of a triangle determine a unique triangle? Why or why not?

Three angles do not determine a unique triangle. For a given triangle with three provided angle measurements, another triangle can be drawn with the same angle measurements but with side lengths proportional to those side lengths of the original triangle.

G7-M6-TE-B6-1.3.0-10.2015

Discussion (7 minutes)

- Why couldn't $\triangle ABC$ be formed in case (i), when the sum of the measurements of $\angle A$ and $\angle B$ were greater than $180°$?
  - *The non-horizontal rays will not intersect due to the angle they form with $\overline{AB}$.*
- Why couldn't $\triangle ABC$ be formed in case (ii), when the sum of the measurements of $\angle A$ and $\angle B$ were supplementary?
  - *The non-horizontal rays will not intersect. The lines look parallel; if they are extended, they seem to be the same distance apart from each other at any given point.*

Confirm that the two non-horizontal rays in this case are, in fact, parallel and two supplementary angle measurements in position to be two angles of a triangle will always yield parallel lines.

- What conclusion can we draw about any two angle measurements of a triangle, with respect to determining a triangle?
  - *The sum of any two angles of a triangle must be less than $180°$ in order to form the triangle.*
- Do the three angle measurements of a triangle guarantee a unique triangle?
  - *No, we drew a triangle that had the same angle measurements as our triangle in case (iii) but with side lengths that were half the length of the original triangle.*

Remind students of their work with scale drawings. Triangles that are enlargements or reductions of an original triangle all have equal corresponding angle measurements but have side lengths that are proportional.

Exercise 2 (4 minutes)

Exercise 2

Which of the following sets of angle measurements determines a triangle?

a. $30°, 120°$ ***Determines a triangle***

b. $125°, 55°$ ***Does not determine a triangle***

c. $105°, 80°$ ***Does not determine a triangle***

d. $90°, 89°$ ***Determines a triangle***

e. $91°, 89°$ ***Does not determine a triangle***

Choose one example from above that does determine a triangle and one that does not. For each, explain why it does or does not determine a triangle using words and a diagram.

Possible response:

The angle measurements in part (a) determine a triangle because the non-horizontal rays of the $30°$ *angle and the* $120°$ *angle will intersect to form a triangle.*

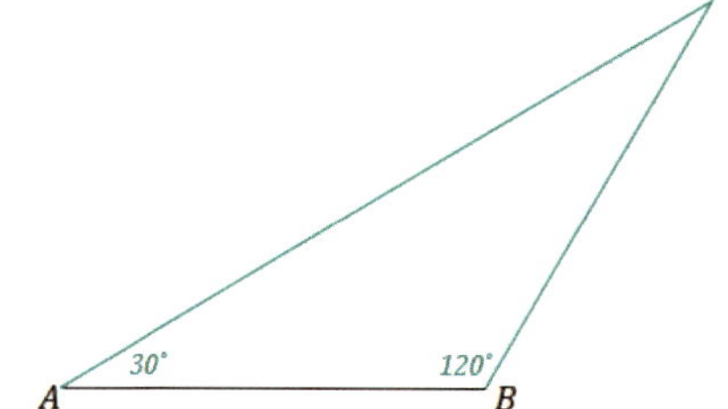

The angle measurements in part (c) do not determine a triangle because the non-horizontal rays of the $105°$ *angle and the* $80°$ *angle will not intersect to form a triangle.*

Closing (2 minutes)

Lesson Summary

- Three lengths determine a triangle provided the largest length is less than the sum of the other two lengths.
- Two angle measurements determine a triangle provided the sum of the two angle measurements is less than $180°$.
- Three given angle measurements do not determine a unique triangle.
- Scale drawings of a triangle have equal corresponding angle measurements, but corresponding side lengths are proportional.

Exit Ticket (5 minutes)

Name ______________________________ Date ______________

Lesson 11: Conditions on Measurements That Determine a Triangle

Exit Ticket

1. What is the maximum and minimum whole number side length for $\triangle XYZ$ with given side lengths of 3 cm and 5 cm? Please explain why.

2. Jill has not yet studied the angle measurement requirements to form a triangle. She begins to draw side $\overline{AB}$ of $\triangle ABC$ and considers the following angle measurements for $\angle A$ and $\angle B$. Describe the drawing that results from each set.

A ———————————— B

   a. $45°$ and $135°$

   b. $45°$ and $45°$

   c. $45°$ and $145°$

Exit Ticket Sample Solutions

1. What is the minimum and maximum whole number side length for $\triangle XYZ$ with given side lengths of 3 cm and 5 cm? Please explain why.

   Minimum: 3 cm. Maximum: 7 cm. Values above this maximum and below this minimum will not satisfy the condition that the longest side length is less than the sum of the other two side lengths.

2. Jill has not yet studied the angle measurement requirements to form a triangle. She begins to draw side $\overline{AB}$ as a horizontal segment of $\triangle ABC$ and considers the following angle measurements for $\angle A$ and $\angle B$. Describe the non-horizontal rays in the drawing that results from each set.

   A ———————— B

   a. $45°$ and $135°$

   The non-horizontal rays of $\angle A$ and $\angle B$ will not intersect to form a triangle; the rays will be parallel to each other.

   b. $45°$ and $45°$

   The non-horizontal rays of $\angle A$ and $\angle B$ will intersect to form a triangle.

   c. $45°$ and $145°$

   The non-horizontal rays of $\angle A$ and $\angle B$ will not intersect to form a triangle.

Problem Set Sample Solutions

Scaffolding:

Lessons 7 and 8 demonstrate how to use a compass for questions such as Problem 1.

1. Decide whether each set of three given lengths determines a triangle. For any set of lengths that does determine a triangle, use a ruler and compass to draw the triangle. Label all side lengths. For sets of lengths that do not determine a triangle, write "Does not determine a triangle," and justify your response.

   a. 3 cm, 4 cm, 5 cm

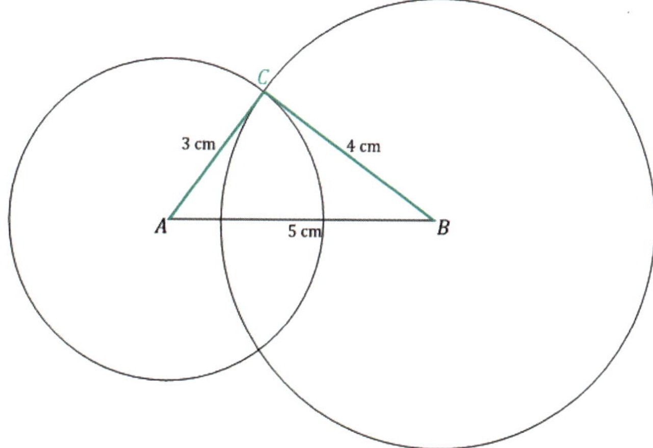

b. 1 cm, 4 cm, 5 cm

Does not determine a triangle. The shorter lengths are too short to form a triangle. They will only form a segment equal to the length of the longest side.

c. 1 cm, 5 cm, 5 cm

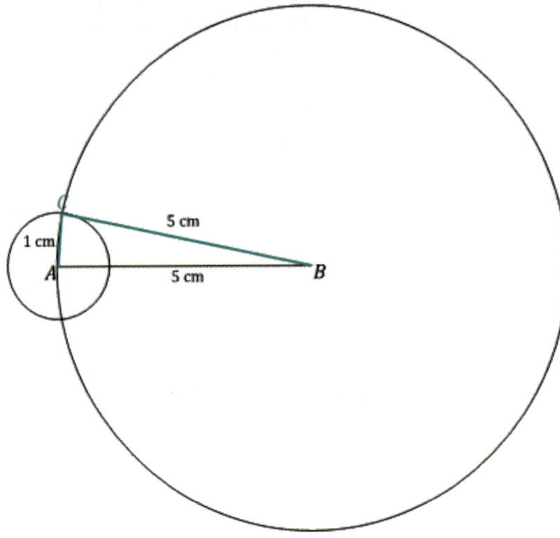

d. 8 cm, 3 cm, 4 cm

Does not determine a triangle. The shorter lengths are too short to form a triangle. They will only form a segment that is shorter than the length of the longest side.

e. 8 cm, 8 cm, 4 cm

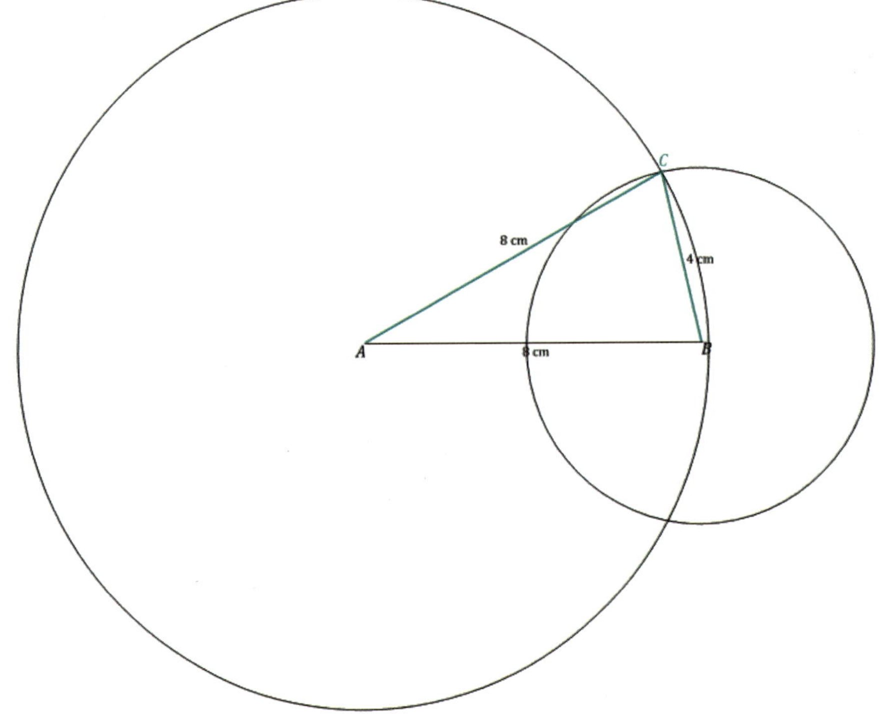

EUREKA MATH™

f. 4 cm, 4 cm, 4 cm

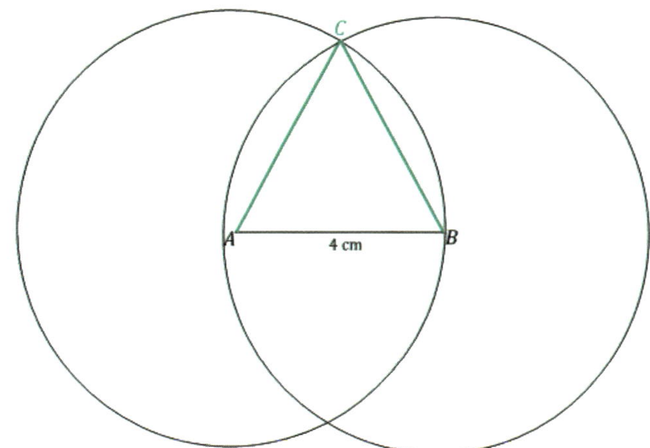

2. For each angle measurement below, provide one angle measurement that will determine a triangle and one that will not determine a triangle. Provide a brief justification for the angle measurements that will not form a triangle. Assume that the angles are being drawn to a horizontal segment AB; describe the position of the non-horizontal rays of angles $\angle A$ and $\angle B$.

$\angle A$	$\angle B$: A Measurement That Determines a Triangle	$\angle B$: A Measurement That *Does Not* Determine a Triangle	Justification for No Triangle
$40°$	*One possible answer:* $30°$	*One possible answer:* $150°$	*The non-horizontal rays do not intersect.*
$100°$	*One possible answer:* $30°$	*One possible answer:* $150°$	*The non-horizontal rays do not intersect.*
$90°$	*One possible answer:* $30°$	*One possible answer:* $90°$	*The non-horizontal rays do not intersect.*
$135°$	*One possible answer:* $30°$	*One possible answer:* $80°$	*The non-horizontal rays do not intersect.*

Note:

- Measurements that determine a triangle should be less than $180° -$ (the measurement of $\angle A$).
- Measurements that do not determine a triangle should be greater than $180° -$ (the measurement of $\angle A$).

3. For the given side lengths, provide the minimum and maximum whole number side lengths that determine a triangle.

Given Side Lengths	Minimum Whole Number Third Side Length	Maximum Whole Number Third Side Length
5 cm, 6 cm	**2 cm**	**10 cm**
3 cm, 7 cm	**5 cm**	**9 cm**
4 cm, 10 cm	**7 cm**	**13 cm**
1 cm, 12 cm	**12 cm**	**12 cm**

Lesson 12: Unique Triangles—Two Sides and a Non-Included Angle

Student Outcomes

- Students understand that two sides of a triangle and an acute angle not included between the two sides may not determine a unique triangle.
- Students understand that two sides of a triangle and a $90°$ angle (or obtuse angle) not included between the two sides determine a unique triangle.

Lesson Notes

A triangle drawn under the condition of two sides and a non-included angle is often thought of as a condition that does not determine a unique triangle. Lesson 12 breaks this idea down by sub-condition. Students see that the sub-condition, two sides and a non-included angle, provided the non-included angle is an *acute* angle, is the only sub-condition that does not determine a unique triangle. Furthermore, there is a maximum of two possible non-identical triangles that can be drawn under this sub-condition.

Classwork

Exploratory Challenge (30 minutes)

Ask students to predict, record, and justify whether they think the provided criteria will determine a unique triangle for each set of criteria.

Exploratory Challenge

1. **Use your tools to draw $\triangle ABC$ in the space below, provided $AB = 5$ cm, $BC = 3$ cm, and $\angle A = 30°$. Continue with the rest of the problem as you work on your drawing.**

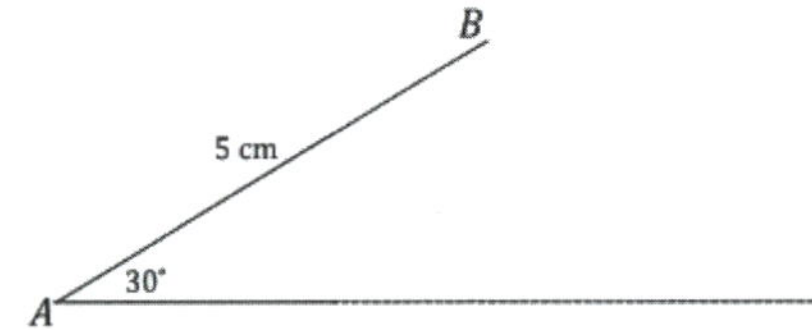

   a. **What is the relationship between the given parts of $\triangle ABC$?**

      Two sides and a non-included angle are provided.

   b. **Which parts of the triangle can be drawn without difficulty? What makes this drawing challenging?**

      The parts that are adjacent, $\overline{AB}$ and $\angle A$, are easiest to draw. It is difficult to position $\overline{BC}$.

c. A ruler and compass are instrumental in determining where C is located.

- ✓ Even though the length of segment AC is unknown, extend the ray AC in anticipation of the intersection with segment BC.
- ✓ Draw segment BC with length 3 cm away from the drawing of the triangle.
- ✓ Adjust your compass to the length of $\overline{BC}$.
- ✓ Draw a circle with center B and a radius equal to BC, or 3 cm.

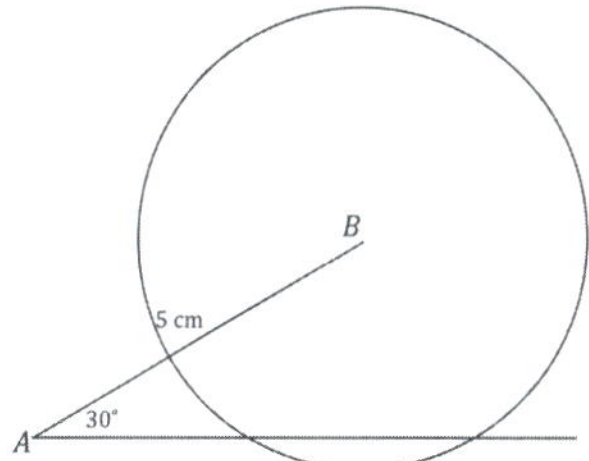

d. How many intersections does the circle make with segment AC? What does each intersection signify?

Two intersections; each intersection represents a possible location for vertex C.

As students arrive at part (e), recommend that they label the two points of intersection as C_1 and C_2.

e. Complete the drawing of $\triangle ABC$.

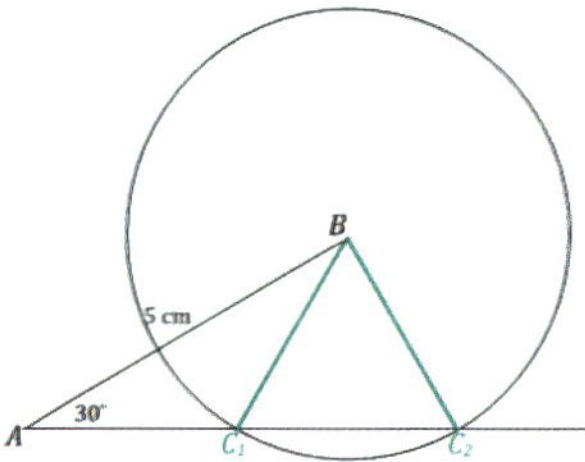

f. Did the results of your drawing differ from your prediction?

Answers will vary.

2. Now attempt to draw $\triangle DEF$ in the space below, provided $DE = 5$ cm, $EF = 3$ cm, and $\angle F = 90°$. Continue with the rest of the problem as you work on your drawing.

   a. How are these conditions different from those in Exercise 1, and do you think the criteria will determine a unique triangle?

   The provided angle was an acute angle in Exercise 1; now the provided angle is a right angle. Possible prediction: Since the same general criteria (two sides and a non-included angle) determined more than one triangle in Exercise 1, the same can happen in this situation.

Scaffolding:

For Exercise 2, part (a), remind students to draw the adjacent parts first (i.e., $\overline{EF}$ and $\angle F$).

b. **What is the relationship between the given parts of $\triangle DEF$?**

Two sides and a non-included angle are provided.

c. **Describe how you will determine the position of $\overline{DE}$.**

I will draw a segment equal in length to $\overline{DE}$, or 5 cm, and adjust my compass to this length. Then, I will draw a circle with center E and radius equal to DE. This circle should intersect with the ray FD.

d. **How many intersections does the circle make with $\overline{FD}$?**

Just one intersection.

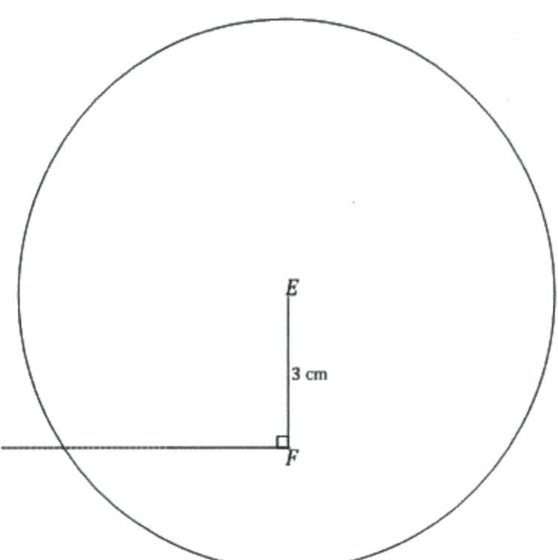

e. **Complete the drawing of $\triangle DEF$. How is the outcome of $\triangle DEF$ different from that of $\triangle ABC$?**

In drawing $\triangle ABC$, there are two possible locations for vertex C, but in drawing $\triangle DEF$, there is only one location for vertex D.

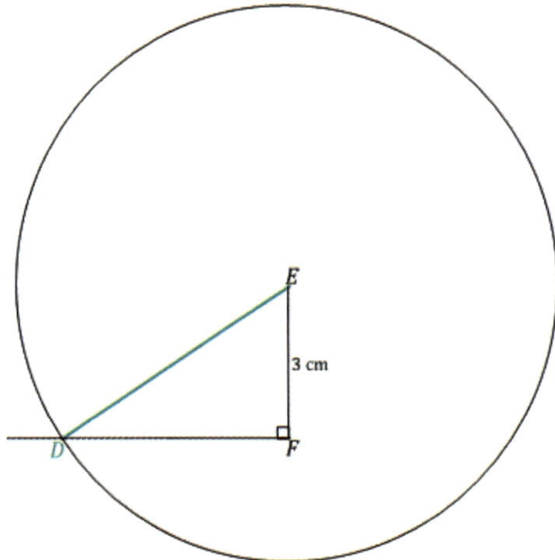

f. **Did your results differ from your prediction?**

Answers will vary.

3. **Now attempt to draw $\triangle JKL$, provided $KL = 8$ cm, $KJ = 4$ cm, and $\angle J = 120°$. Use what you drew in Exercises 1 and 2 to complete the full drawing.**

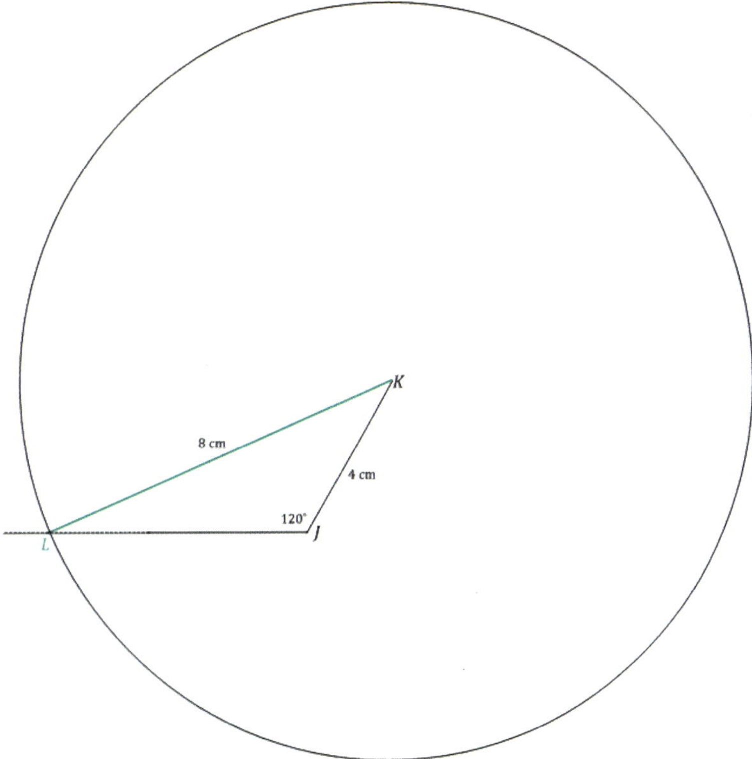

4. **Review the conditions provided for each of the three triangles in the Exploratory Challenge, and discuss the uniqueness of the resulting drawing in each case.**

   All three triangles are under the condition of two sides and a non-included angle. The non-included angle in $\triangle ABC$ is an acute angle, while the non-included angle in $\triangle DEF$ is $90°$, and the non-included angle in $\triangle JKL$ is obtuse. The triangles drawn in the latter two cases are unique because there is only one possible triangle that could be drawn for each. However, the triangle drawn in the first case is not unique because there are two possible triangles that could be drawn.

Discussion (8 minutes)

Review the results from each case of the two sides and non-included angle condition.

- Which of the three cases, or sub-conditions, of two sides and a non-included angle, determines a unique triangle?
  - *Unique triangles are determined when the non-included angle in this condition is $90°$ or greater.*
- How should we describe the case of two sides and a non-included angle that does not determine a unique triangle?
  - *The only case of the two sides and a non-included angle condition that does not determine a unique triangle is when the non-included angle is an acute angle.*
- Highlight how the radius in the figure in Exercise 1 part (e) can be pictured to be "swinging" between C_1 and C_2. Remind students that the location of C is initially unknown and that ray $\overrightarrow{AC}$ is extended to emphasize this.

Closing (2 minutes)

- A triangle drawn under the condition of two sides and a non-included angle, where the angle is acute, does not determine a unique triangle. This condition determines two non-identical triangles.

> **Lesson Summary**
>
> **Consider a triangle correspondence $\triangle ABC \leftrightarrow \triangle XYZ$ that corresponds to two pairs of equal sides and one pair of equal, non-included angles. If the triangles are not identical, then $\triangle ABC$ can be made to be identical to $\triangle XYZ$ by swinging the appropriate side along the path of a circle with a radius length of that side.**
>
> **A triangle drawn under the condition of two sides and a non-included angle, where the angle is $90°$ or greater, creates a unique triangle.**

Exit Ticket (5 minutes)

Name ______________________________ Date ______________

Lesson 12: Unique Triangles—Two Sides and a Non-Included Angle

Exit Ticket

So far, we have learned about four conditions that determine unique triangles: three sides, two sides and an included angle, two angles and an included side, and two angles and the side opposite a given angle.

a. In this lesson, we studied the criterion two sides and a non-included angle. Which case of this criterion determines a unique triangle?

b. Provided $\overline{AB}$ has length 5 cm, $\overline{BC}$ has length 3 cm, and the measurement of $\angle A$ is $30°$, draw $\triangle ABC$, and describe why these conditions do not determine a unique triangle.

Exit Ticket Sample Solutions

So far, we have learned about four conditions that determine unique triangles: three sides, two sides and an included angle, two angles and an included side, and two angles and the side opposite a given angle.

a. In this lesson, we studied the criterion two sides and a non-included angle. Which case of this criterion determines a unique triangle?

For the criterion two sides and a non-included angle, the case where the non-included angle is $90°$ or greater determines a unique triangle.

b. Provided $\overline{AB}$ has length 5 cm, $\overline{BC}$ has length 3 cm, and the measurement of $\angle A$ is $30°$, draw $\triangle ABC$, and describe why these conditions do not determine a unique triangle.

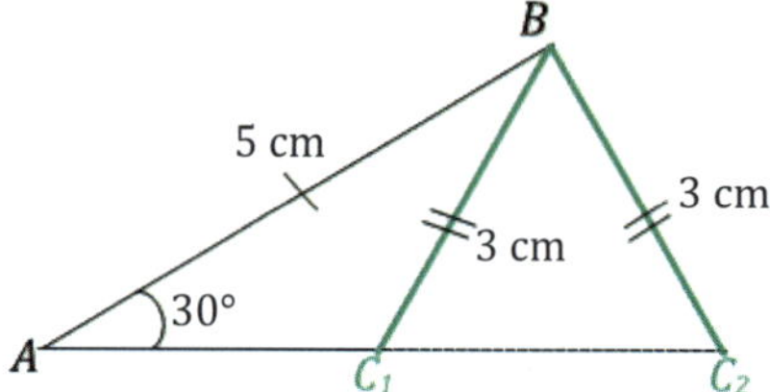

The non-included angle is an acute angle, and two different triangles can be determined in this case since $\overline{BC}$ can be in two different positions, forming a triangle with two different lengths of $\overline{AC}$.

Problem Set Sample Solutions

1. In each of the triangles below, two sides and a non-included angle are marked. Use a compass to draw a nonidentical triangle that has the same measurements as the marked angle and marked sides (look at Exercise 1, part (e) of the Exploratory Challenge as a reference). Draw the new triangle on top of the old triangle. What is true about the marked angles in each triangle that results in two non-identical triangles under this condition?

a.

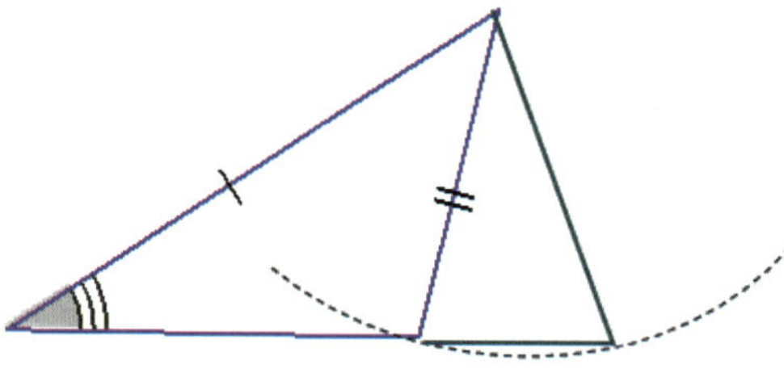

The non-included angle is acute.

b.

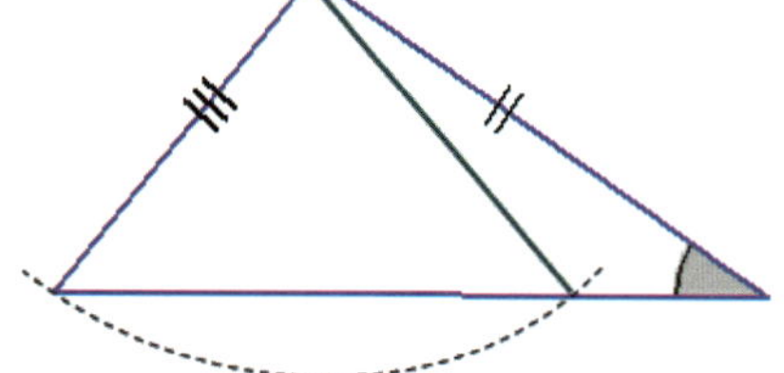

The non-included angle is acute.

c.

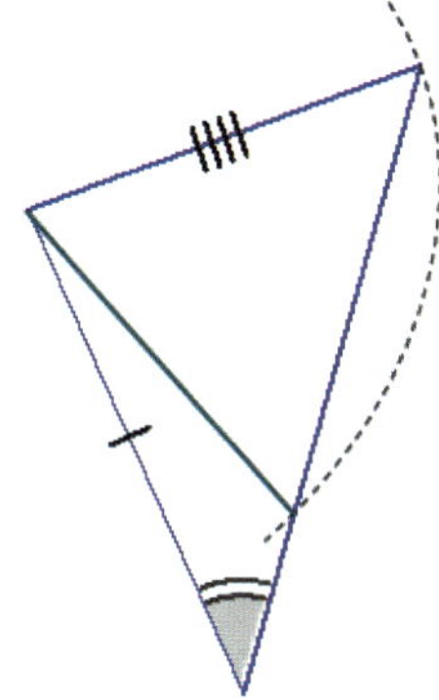

The non-included angle is acute.

2. Sometimes two sides and a non-included angle of a triangle determine a unique triangle, even if the angle is acute. In the following two triangles, copy the marked information (i.e., two sides and a non-included acute angle), and discover which determines a unique triangle. Measure and label the marked parts.

   In each triangle, how does the length of the marked side adjacent to the marked angle compare with the length of the side opposite the marked angle? Based on your drawings, specifically state when the two sides and acute non-included angle condition determines a unique triangle.

   While redrawing $\triangle ABC$, students will see that a unique triangle is not determined, but in redrawing $\triangle DEF$, a unique triangle is determined. In $\triangle ABC$, the length of the side opposite the angle is shorter than the side adjacent to the angle. However, in $\triangle DEF$, the side opposite the angle is longer than the side adjacent to the angle.

   The two sides and acute non-included angle condition determines a unique triangle if the side opposite the angle is longer than the side adjacent to the angle.

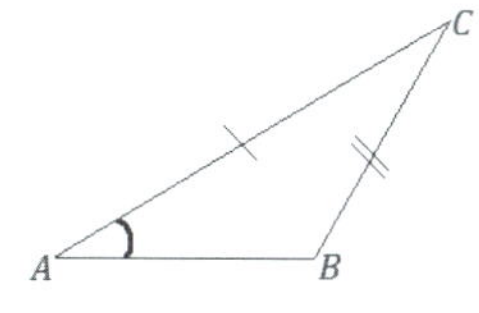

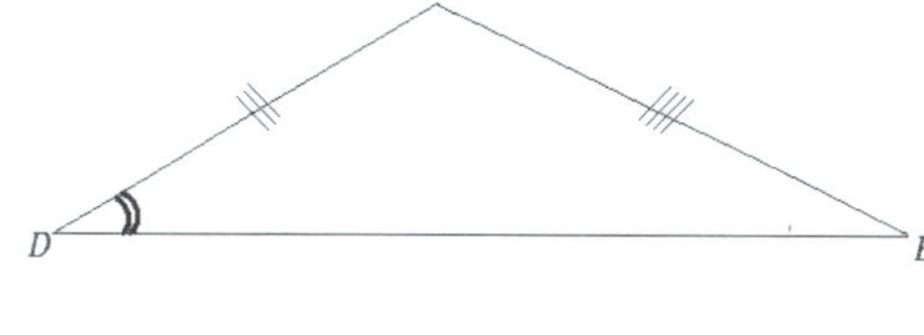

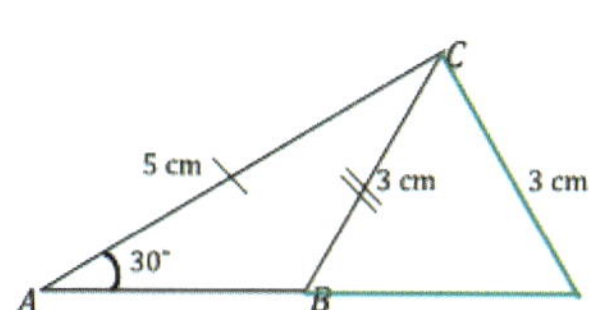

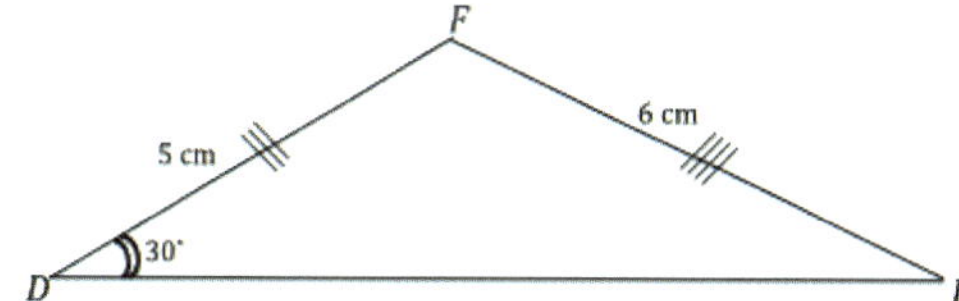

3. A sub-condition of the two sides and non-included angle is provided in each row of the following table. Decide whether the information determines a unique triangle. Answer with a *yes*, *no*, or *maybe* (for a case that may or may not determine a unique triangle).

	Condition	Determines a Unique Triangle?
1	Two sides and a non-included $90°$ angle.	*yes*
2	Two sides and an acute, non-included angle.	*maybe*
3	Two sides and a non-included $140°$ angle.	*yes*
4	Two sides and a non-included $20°$ angle, where the side adjacent to the angle is shorter than the side opposite the angle.	*yes*
5	Two sides and a non-included angle.	*maybe*
6	Two sides and a non-included $70°$ angle, where the side adjacent to the angle is longer than the side opposite the angle.	*no*

4. Choose one condition from the table in Problem 3 that does not determine a unique triangle, and explain why.

   Possible response: Condition 6 does not determine a unique triangle because the condition of two sides and an acute non-included angle determines two possible triangles when the side adjacent to the angle is longer than the side opposite the angle.

5. Choose one condition from the table in Problem 3 that does determine a unique triangle, and explain why.

   Possible response: Condition 1 determines a unique triangle because the condition of two sides and a non-included angle with a measurement of $90°$ or more has a ray that only intersects the circle once.

Lesson 13: Checking for Identical Triangles

Student Outcomes

- Students use conditions that determine a unique triangle to determine when two triangles are identical.
- Students construct viable arguments to explain why the given information can or cannot give a triangle correspondence between identical triangles.

Lesson Notes

Lessons 13 and 14 are application lessons for Topic B. Students must look at a pair of triangles and decide whether the triangles are identical based on what they know about conditions that determine unique triangles.

Classwork

Opening Exercise (5 minutes)

Opening Exercise

a. List all the conditions that determine unique triangles.

- *Three sides condition*
- *Two sides and included angle condition*
- *Two angles and included side condition*
- *Two angles and the side opposite a given angle condition*
- *Two sides and a non-included angle, provided the angle is* $90°$ *or greater*
- *Two sides and a non-included angle, provided the side adjacent to the angle is shorter than the side opposite the angle.*

b. How are the terms *identical* and *unique* related?

When drawing a triangle under a given condition, the triangle will either be identical or non-identical to the original triangle. If only one triangle can be drawn under the condition, we say the condition determines a unique triangle. A triangle drawn under a condition that is known to determine a unique triangle will be identical to the original triangle.

Discussion (2 minutes)

Students synthesize their knowledge of triangles and use what they have learned about correspondences and conditions that determine a unique triangle to explain whether each pair of triangles is identical or not. Hold students accountable for the same level of precision in their responses as the response provided in Example 1.

Follow the instructions below for Example 1 and Exercises 1–3.

Each of the following problems gives two triangles. State whether the triangles are *identical*, *not identical*, or *not necessarily identical*. If the triangles are identical, give the triangle conditions that explain why, and write a triangle correspondence that matches the sides and angles. If the triangles are not identical, explain why. If it is not possible to definitively determine whether the triangles are identical, write "the triangles are not necessarily identical," and explain your reasoning.

Example 1 (5 minutes)

Scaffolding:

For students struggling to visualize whether the triangles are identical or not, suggest that they trace one triangle, mark it with all tick and arc marks, and cut it out to try to map over the other triangle.

Example 1

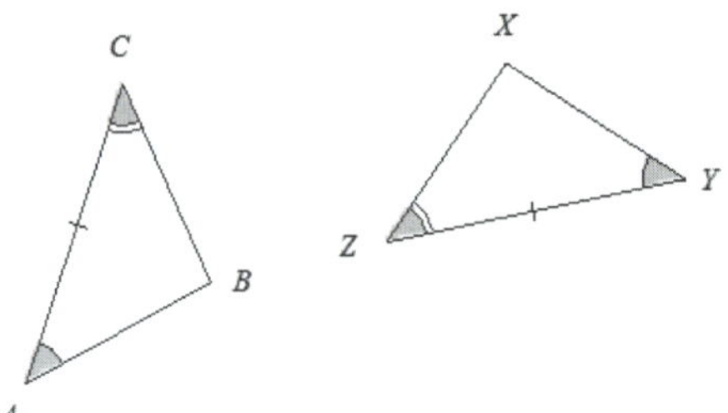

The triangles are identical by the two angles and the included side condition. The correspondence $\triangle ABC \leftrightarrow \triangle YXZ$ matches two pairs of angles and one pair of equal sides. Since both triangles have parts under the condition of the same measurement, the triangles must be identical.

Exercises 1–3 (10 minutes)

Exercises 1–3

1.

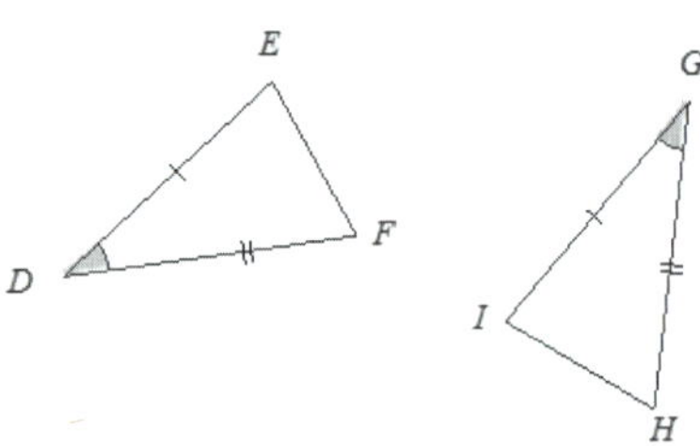

The triangles are identical by the two sides and the included angle condition. The correspondence $\triangle DEF \leftrightarrow \triangle GIH$ matches two equal pairs of sides and one equal pair of angles. Since both triangles have parts under the condition of the same measurement, the triangles must be identical.

2.

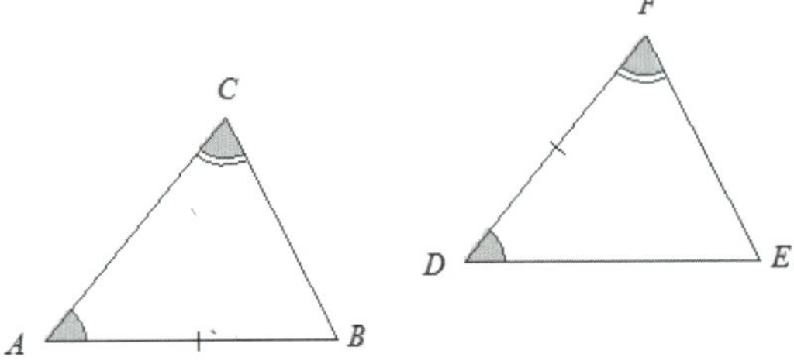

The triangles are not necessarily identical. Although the two angles in each triangle match, the marked sides do not correspond. In $\triangle ABC$, the marked side is not between the marked angles; whereas in $\triangle DEF$, the marked side is between the marked angles.

3.

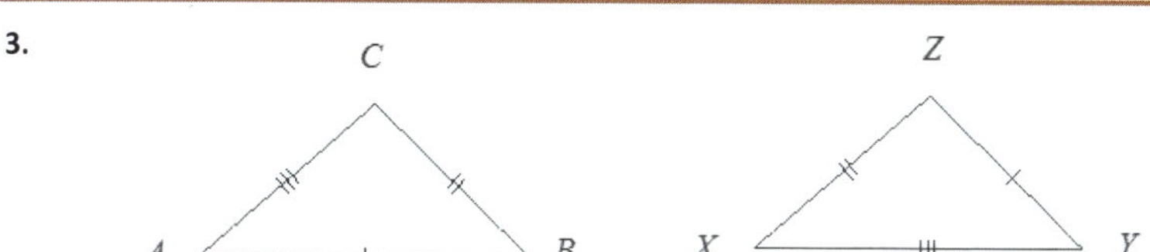

The triangles are identical by the three sides condition. The correspondence $\triangle ABC \leftrightarrow \triangle YZX$ matches three equal pairs of sides. Since both triangles have the same side lengths, the triangles must be identical.

Example 2 (5 minutes)

Follow the instructions below for Example 2 and Exercises 4–6.

In Example 2 and Exercises 4–6, three pieces of information are given for $\triangle ABC$ and $\triangle XYZ$. Draw, freehand, the two triangles (do not worry about scale), and mark the given information. If the triangles are identical, give a triangle correspondence that matches equal angles and equal sides. Explain your reasoning.

Example 2

$AB = XZ, AC = XY, \angle A = \angle X$

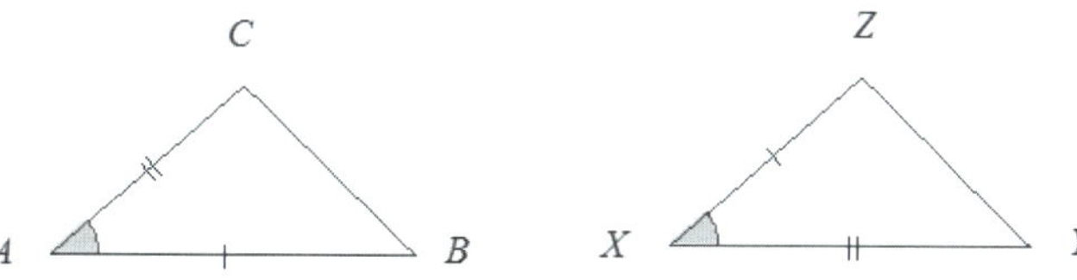

These triangles are identical by the two sides and included angle condition. The triangle correspondence $\triangle ABC \leftrightarrow \triangle XZY$ matches two pairs of equal sides and one pair of equal, included angles. Since both triangles have parts under the condition of the same measurement, the triangles must be identical.

Note: Students need not worry about exact drawings in these questions; the objective is to recognize that the triangles' matching parts fit the condition.

Exercises 4–6 (12 minutes)

Exercises 4–6

4. $\angle A = \angle Z, \angle B = \angle Y, AB = YZ$

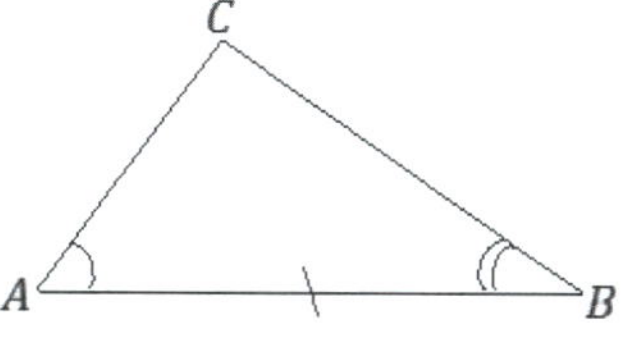

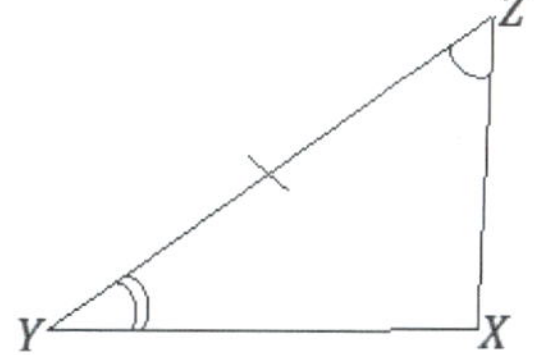

These triangles are identical by the two angles and included side condition. The triangle correspondence $\triangle ABC \leftrightarrow \triangle ZYX$ matches two pairs of equal angles and one pair of equal, included sides. Since both triangles have parts under the condition of the same measurement, the triangles must be identical.

5. $\angle A = \angle Z$, $\angle B = \angle Y$, $BC = XY$

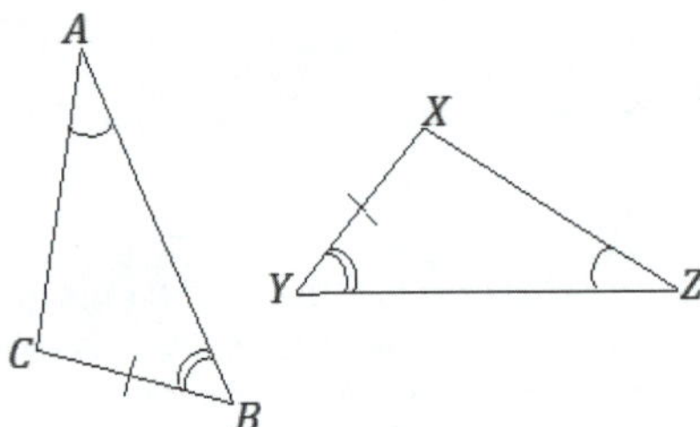

These triangles are identical by the two angles and side opposite a given angle condition. The triangle correspondence $\triangle ABC \leftrightarrow \triangle ZYX$ matches two pairs of equal angles and one pair of equal sides. Since both triangles have parts under the condition of the same measurement, the triangles must be identical.

6. $\angle A = \angle Z$, $\angle B = \angle Y$, $BC = XZ$

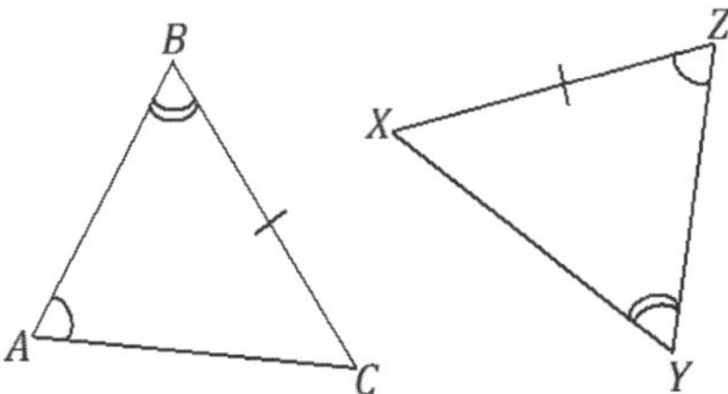

These triangles are not necessarily identical. In $\triangle ABC$, the marked side is opposite $\angle A$. In $\triangle XYZ$, the marked side is not opposite $\angle Z$, which is equal to $\angle A$. Rather, it is opposite $\angle Y$, which is equal to $\angle B$.

Closing (1 minutes)

Lesson Summary

The measurement and arrangement (and correspondence) of the parts in each triangle play a role in determining whether two triangles are identical.

Exit Ticket (5 minutes)

Name ______________________________ Date ______________

Lesson 13: Checking for Identical Triangles

Exit Ticket

$\angle A$ and $\angle D$ are equal in measure. Draw two triangles around each angle, and mark parts appropriately so that the triangles are identical; use $\angle A$ and $\angle D$ as part of the chosen condition. Write a correspondence for the triangles.

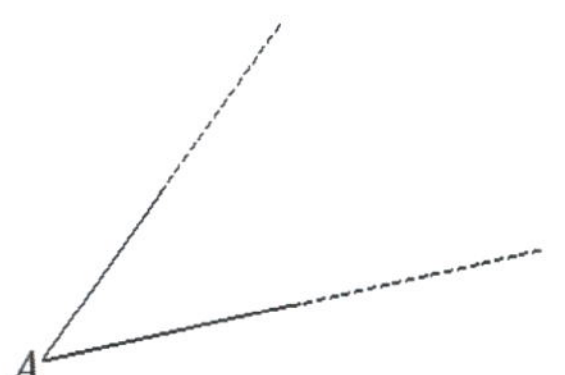

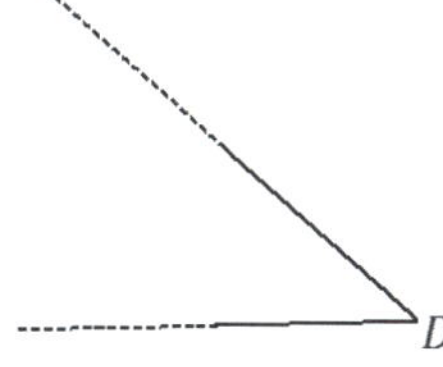

Exit Ticket Sample Solutions

$\angle A$ and $\angle D$ are equal in measure. Draw two triangles around each angle, and mark parts appropriately so that the triangles are identical; use $\angle A$ and $\angle D$ as part of the chosen condition. Write a correspondence for the triangles.

Answers will vary; students should select any condition except for the three sides condition and show the appropriate correspondence for their condition on the two triangles.

Problem Set Sample Solutions

In each of the following four problems, two triangles are given. State whether the triangles are *identical, not identical*, or *not necessarily identical*. If the triangles are identical, give the triangle conditions that explain why, and write a triangle correspondence that matches the sides and angles. If the triangles are not identical, explain why. If it is not possible to definitively determine whether the triangles are identical, write "the triangles are not necessarily identical" and explain your reasoning.

1.

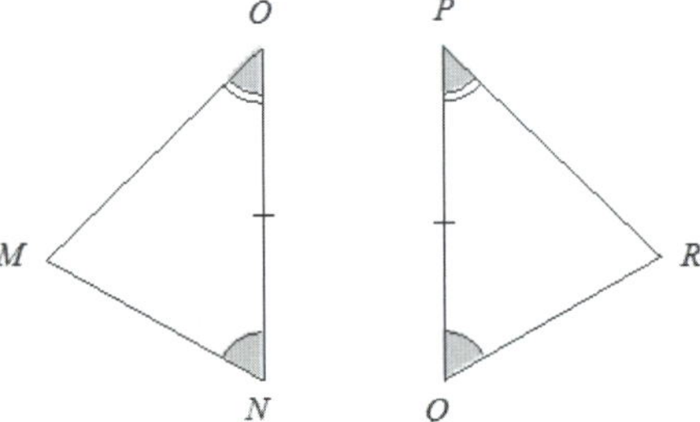

The triangles are identical by the two angles and included side condition. The correspondence $\triangle MNO \leftrightarrow \triangle RQP$ matches two equal pairs of angles and one equal pair of included sides. Since both triangles have parts under the condition of the same measurement, the triangles must be identical.

2.

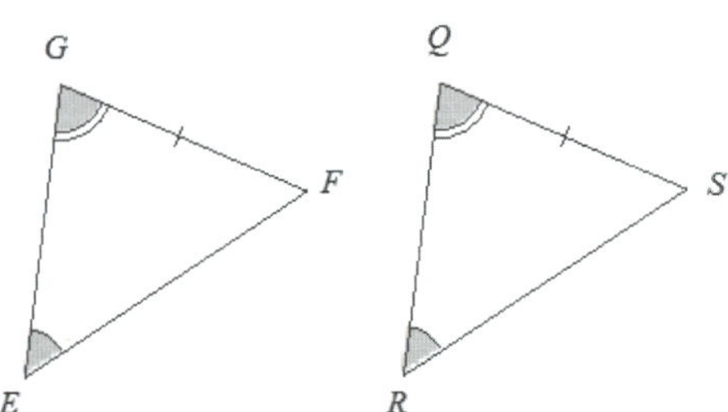

The triangles are identical by the two angles and side opposite a given angle condition. The correspondence $\triangle EGF \leftrightarrow \triangle RQS$ matches two equal pairs of angles and one equal pair of sides. Since both triangles have parts under the condition of the same measurement, the triangles must be identical.

3.

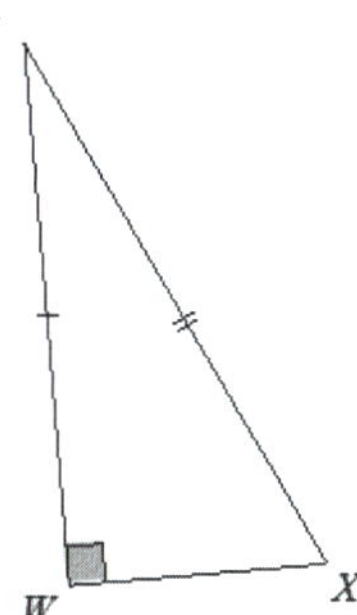

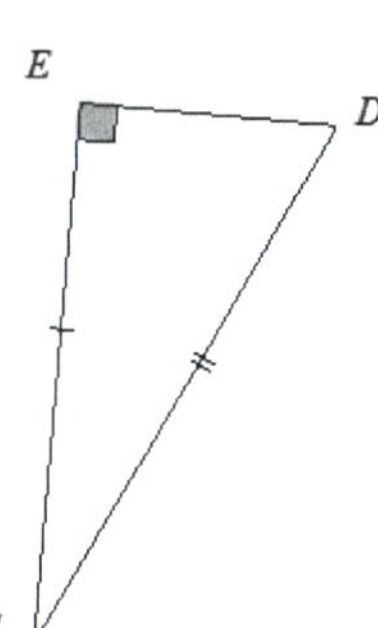

The triangles are identical by the two sides and non-included 90° (or greater) angle condition. The correspondence $\triangle WXY \leftrightarrow \triangle EDC$ matches two pairs of equal sides and one pair of equal angles. Since both triangles have parts under the condition of the same measurement, the triangles must be identical.

4.

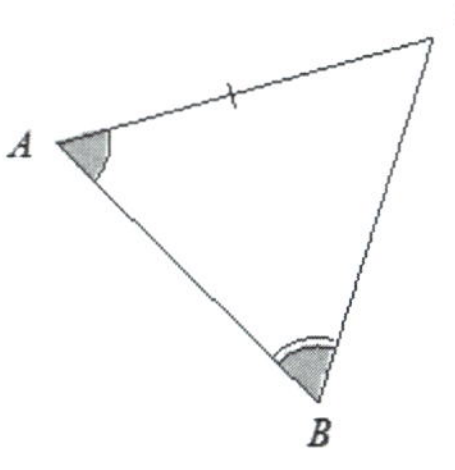

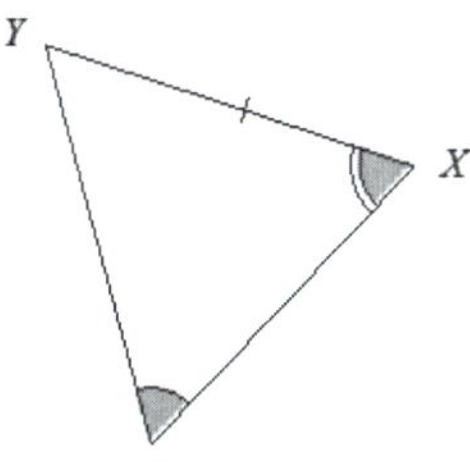

The triangles are not necessarily identical by the two angles and side opposite a given angle condition. In $\triangle ABC$, the marked side is adjacent to the angle marked with a single arc mark. In $\triangle WXY$, the marked side is not adjacent to the angle marked with a single arc mark.

For Problems 5–8, three pieces of information are given for $\triangle ABC$ and $\triangle YZX$. Draw, freehand, the two triangles (do not worry about scale), and mark the given information. If the triangles are identical, give a triangle correspondence that matches equal angles and equal sides. Explain your reasoning.

5. $AB = YZ, BC = ZX, AC = YX$

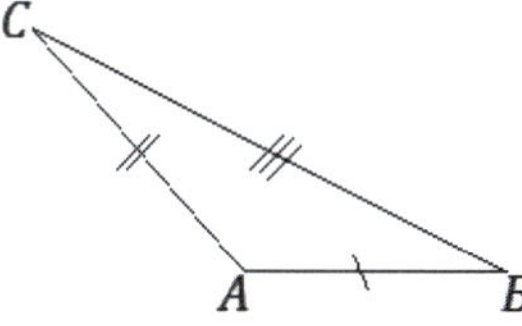

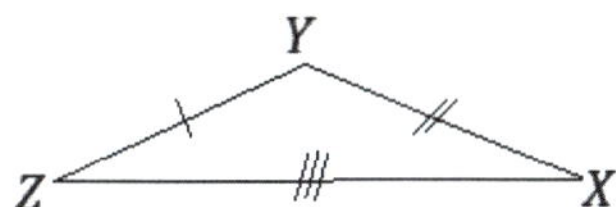

These triangles are identical by the three sides condition. The triangle correspondence $\triangle ABC \leftrightarrow \triangle YZX$ matches three pairs of equal sides. Since both triangles have parts under the condition of the same measurement, the triangles must be identical.

6. $AB = YZ, BC = ZX, \angle C = \angle Y$

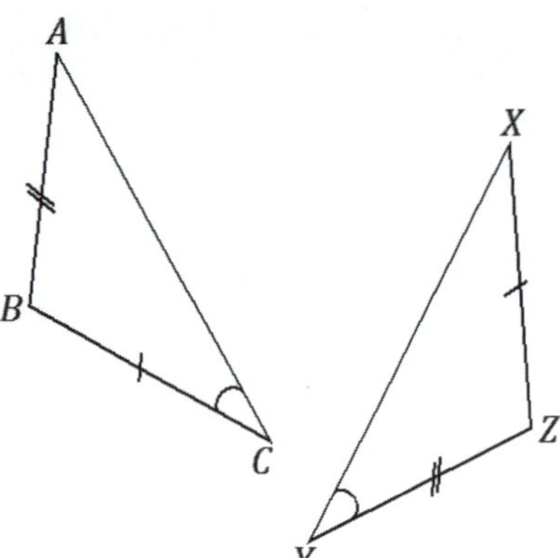

These triangles are not necessarily identical. In $\triangle ABC$, the marked angle is adjacent to $\overline{BC}$. In $\triangle YZX$, the marked angle is not adjacent to the side equal to ZX, which is equal to BC.

7. $AB = XZ, \angle A = \angle Z, \angle C = \angle Y$

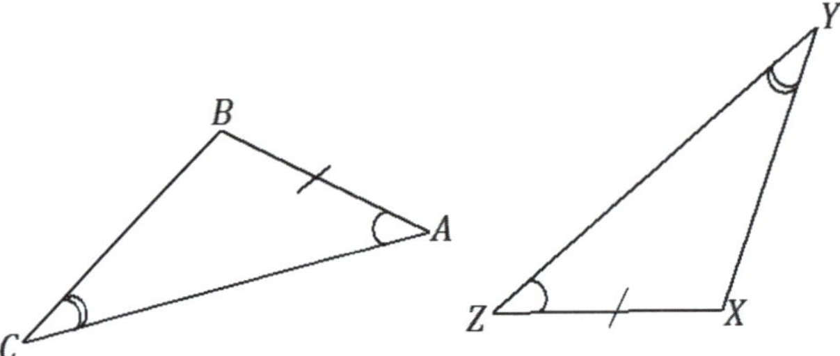

These triangles are identical by the two angles and a side opposite a given angle condition. The triangle correspondence $\triangle ABC \leftrightarrow \triangle ZXY$ matches two pairs of equal angles and one pair of equal sides. Since both triangles have parts under the condition of the same measurement, the triangles must be identical.

8. $AB = XY, AC = YZ, \angle C = \angle Z$ (Note that both angles are obtuse.)

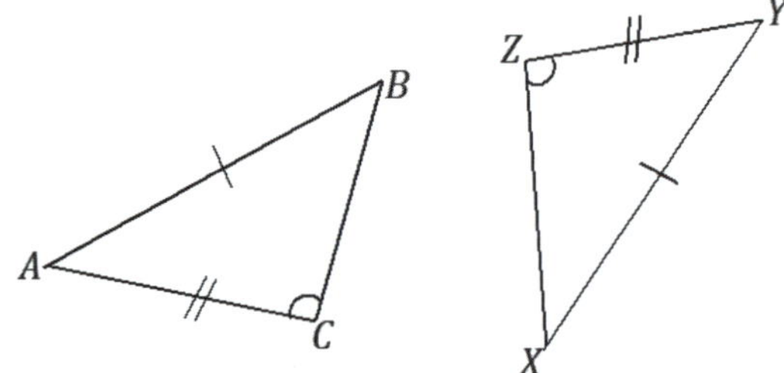

The triangles are identical by the two sides and non-included $90°$ (or greater) angle condition. The correspondence $\triangle ABC \leftrightarrow \triangle YXZ$ matches two pairs of equal sides and one pair of equal angles. Since both triangles have parts under the condition of the same measurement, the triangles must be identical.

Lesson 14: Checking for Identical Triangles

Student Outcomes

- Students use information, such as vertical angles and common sides in the structure of triangle diagrams, to establish whether conditions that determine a unique triangle exist.
- Students use conditions that determine a unique triangle to determine when two triangles are identical.
- Students construct viable arguments to explain why the given information can or cannot give a triangle correspondence between identical triangles.

Lesson Notes

In contrast to Lesson 13, where students had to examine pairs of distinct triangles, Lesson 14 presents the diagrams of triangles so that a relationship exists between the triangles due to the way they are positioned. For example, they may share a common side, may be arranged in a way so that two angles from the triangles are vertical angles, and so on. Students must use the structure of each diagram to establish whether a condition exists that renders the triangles identical.

Classwork

Opening (2 minutes)

MP.7

- Scan the figures in the next several problems. How are these diagrams different from the diagrams in Lesson 13?
  - *The triangles seem to be joined instead of being separated.*
- Does this change the way you figure out if a condition exists that determines whether the triangles are identical?
  - *You have to check the connection between the two triangles and determine if it shows whether two parts between the triangles are equal in measure.*

In each of the following problems, determine whether the triangles are *identical, not identical*, or *not necessarily identical*; justify your reasoning. If the relationship between the two triangles yields information that establishes a condition, describe the information. If the triangles are identical, write a triangle correspondence that matches the sides and angles.

Example 1 (5 minutes)

Example 1

What is the relationship between the two triangles below?

- What is the relationship between the two triangles below?
  - *The triangles share a common side.*

MP.7

- Imagine that $\triangle WXY$ and $\triangle WZY$ were pulled apart and separated. Sketch the triangles separately. Based on how they were joined, what kind of tick mark should be added to each triangle?
  - *$\overline{WY}$ is a common side. Since it belongs to each triangle, we should put a triple tick mark on $\overline{WY}$ to indicate that it is a part of equal measure in both triangles.*

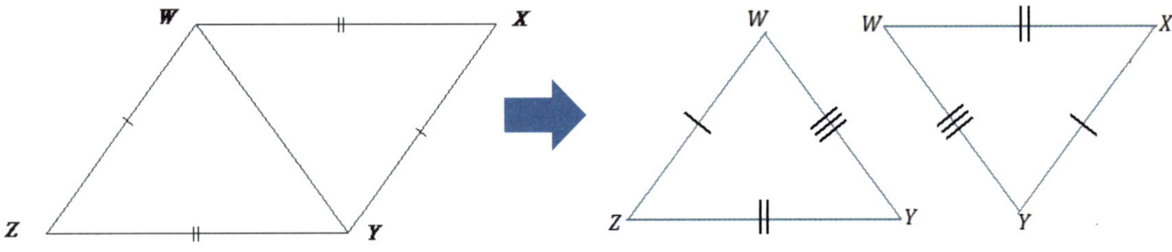

- Are the triangles identical? How do you know?
  - *The triangles are identical by the three sides condition. The correspondence that matches the three equal pairs of sides is $\triangle WXY \leftrightarrow \triangle YZW$.*

Exercises 1–2 (8 minutes)

Exercises 1–2

1. **Are the triangles identical? Justify your reasoning.**

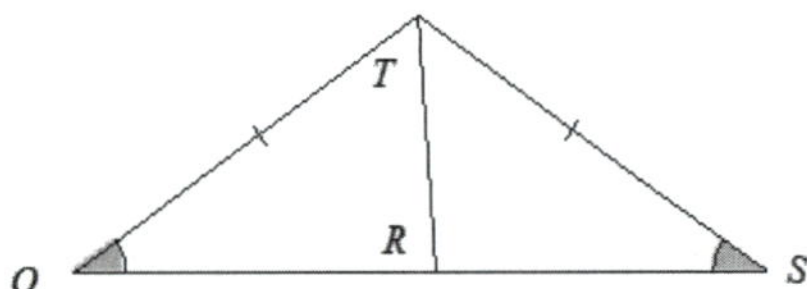

The triangles are not necessarily identical. The correspondence $\triangle QRT \leftrightarrow \triangle SRT$ matches a pair of equal angles and a pair of equal sides. The correspondence also matches a common side, $\overline{RT}$, to both triangles. Two sides and a non-included acute angle do not necessarily determine a unique triangle.

Scaffolding:

As shown above, consider modeling the process of sketching the triangles as *unattached* to one another.

2. **Are the triangles identical? Justify your reasoning.**

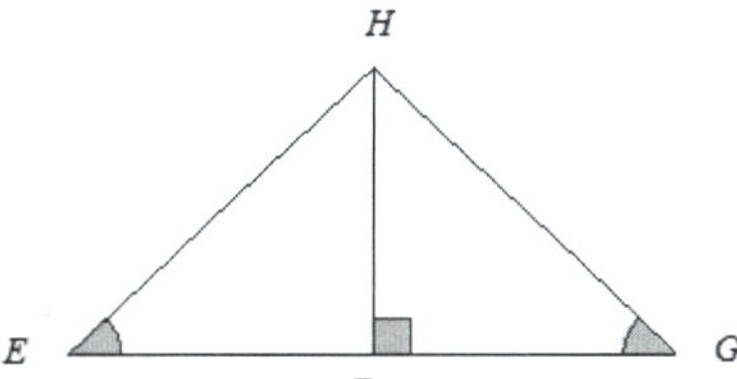

These triangles are identical by the two angles and side opposite a given angle condition. The correspondence $\triangle EFH \leftrightarrow \triangle GFH$ matches two pairs of equal angles and a pair of equal sides. $\angle HFE$ must be a right angle since $\angle HFG$ is a right angle, and they are both on a line. Also, $\overline{HF}$ is a common side to both triangles. Since both triangles have parts under the condition of the same measurement, the triangles must be identical.

Example 2 (5 minutes)

- What is the relationship between the two triangles below?
  - *The triangle is positioned so that there is a pair of vertical angles, $\angle AOC = \angle BOD$, in the diagram.*

Example 2

Are the triangles identical? Justify your reasoning.

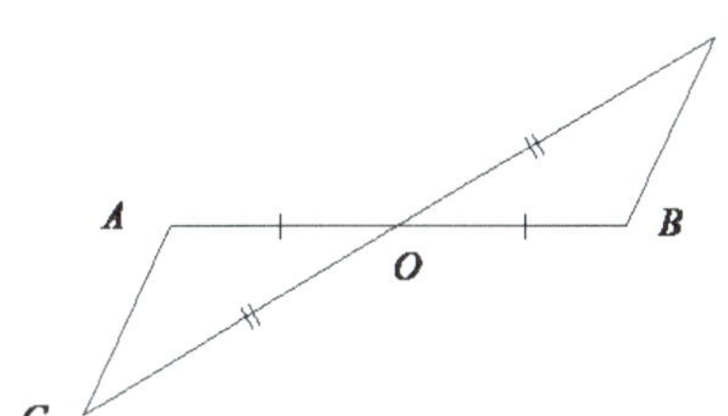

The triangles are identical by the two sides and the included angle condition. The correspondence $\triangle AOC \leftrightarrow \triangle BOD$ matches two equal pairs of sides and a pair of equal angles, $\angle AOC = \angle BOD$, which we know to be equal in measurement because they are vertical angles.

Exercises 3–4 (8 minutes)

Exercises 3–4

3. **Are the triangles identical? Justify your reasoning.**

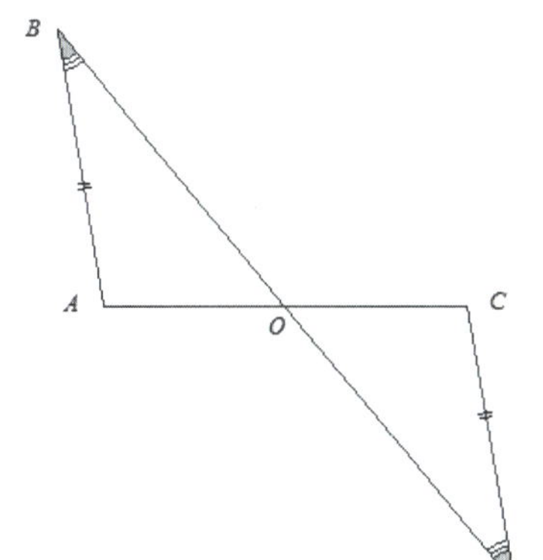

These triangles are identical by the two angles and side opposite a given angle condition. The correspondence $\triangle AOB \leftrightarrow \triangle COD$ matches the two pairs of equal angles and one pair of equal sides. There is a marked pair of equal sides and one pair of marked, equal angles. The second pair of equal angles, $\angle AOB = \angle COD$, are vertical angles.

4. Are the triangles identical? Justify your reasoning.

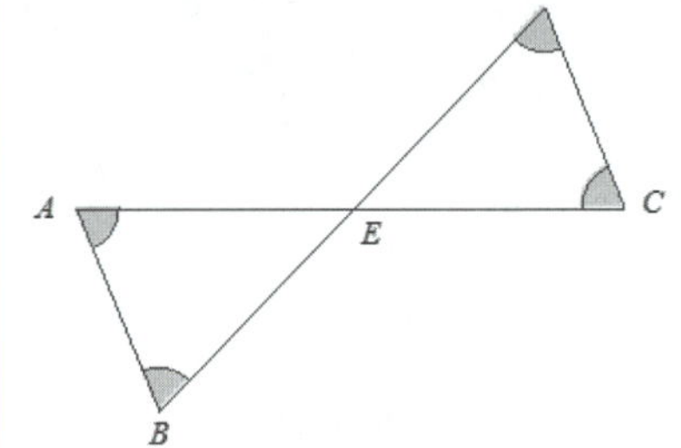

The triangles are not necessarily identical. The correspondence $\triangle AEB \leftrightarrow \triangle CED$ matches three pairs of equal angles, including the unmarked angles, $\angle AEB$ and $\angle CED$, which are equal in measurement because they are vertical angles. The triangles could have different side lengths; therefore, they are not necessarily identical.

Exercises 5–8 (10 minutes)

Exercises 5–8

5. Are the triangles identical? Justify your reasoning.

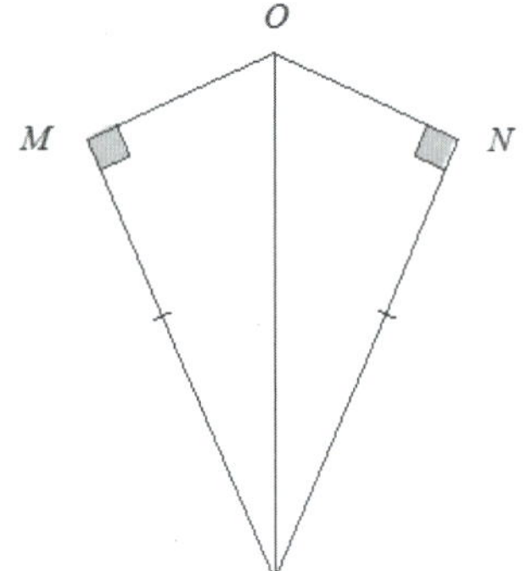

The triangles are identical by the two sides and non-included $90°$ (or greater) angle condition. The correspondence $\triangle MPO \leftrightarrow \triangle NPO$ matches two pairs of equal sides and one pair of equal angles. One of the two pairs of equal sides is side $\overline{OP}$, which is common to both triangles.

6. Are the triangles identical? Justify your reasoning.

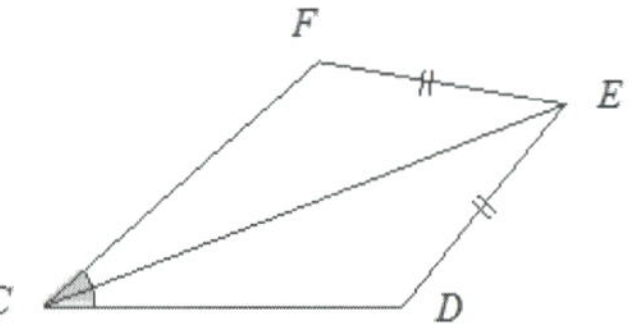

These triangles are not necessarily identical. The triangles have a pair of marked equal sides and equal angles; side $\overline{CE}$ is also common to both triangles. The triangles satisfy the two sides and non-included acute angle condition, which does not determine a unique triangle.

7. **Are the triangles identical? Justify your reasoning.**

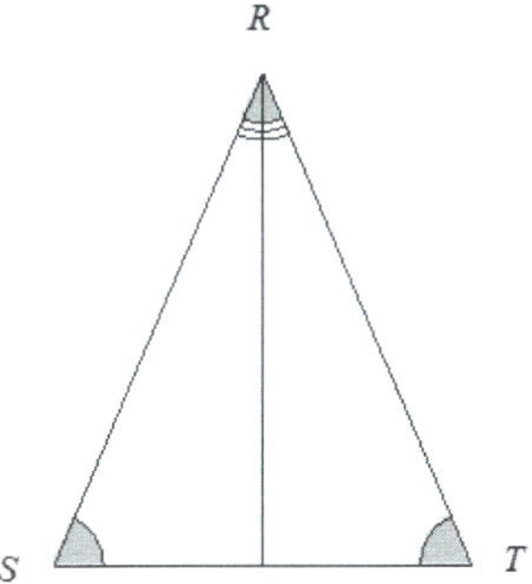

These triangles are identical by the two angles and a side opposite a given angle condition. The triangle correspondence $\triangle RWS \leftrightarrow \triangle RWT$ matches two pairs of equal angles and one pair of equal sides. The equal pair of sides, $\overline{RW}$, is common to both triangles.

8. **Create your own labeled diagram and set of criteria for a pair of triangles. Ask a neighbor to determine whether the triangles are identical based on the provided information.**

   Answers will vary.

Closing (2 minutes)

Lesson Summary

In deciding whether two triangles are identical, examine the structure of the diagram of the two triangles to look for a relationship that might reveal information about corresponding parts of the triangles. This information may determine whether the parts of the triangle satisfy a particular condition, which might determine whether the triangles are identical.

Exit Ticket (5 minutes)

Name ______________________________ Date ______________

Lesson 14: Checking for Identical Triangles

Exit Ticket

Are $\triangle DEF$ and $\triangle DGF$ identical, not identical, or not necessarily identical? Justify your reasoning. If the relationship between the two triangles yields information that establishes a condition, describe the information. If the triangles are identical, write a triangle correspondence that matches the sides and angles.

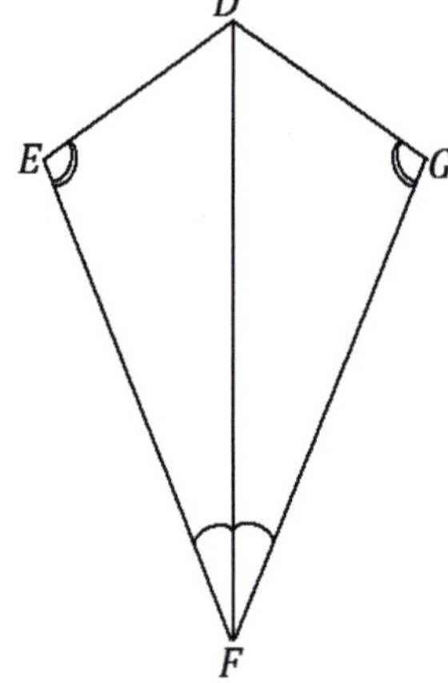

Exit Ticket Sample Solutions

Are $\triangle DEF$ and $\triangle DGF$ identical, not identical, or not necessarily identical? Justify your reasoning. If the relationship between the two triangles yields information that establishes a condition, describe the information. If the triangles are identical, write a triangle correspondence that matches the sides and angles.

These triangles are identical by the two angles and side opposite a given angle condition. The triangle correspondence $\triangle DEF \leftrightarrow \triangle DGF$ matches the two pairs of equal angles and one pair of equal sides condition. The pair of equal sides, $\overline{DF}$, is common to both triangles.

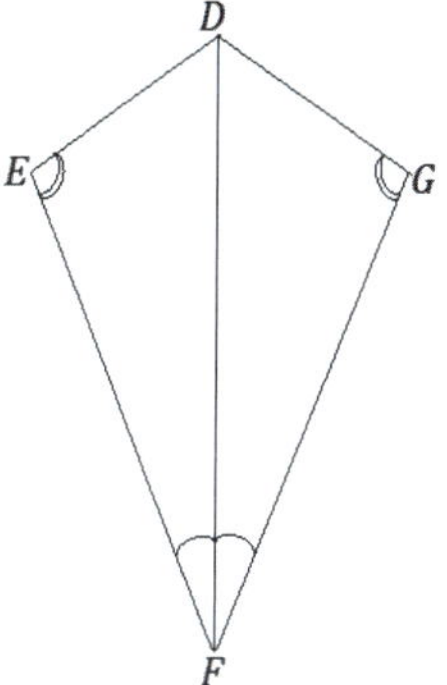

Problem Set Sample Solutions

In the following problems, determine whether the triangles are *identical*, *not identical*, or *not necessarily identical*; justify your reasoning. If the relationship between the two triangles yields information that establishes a condition, describe the information. If the triangles are identical, write a triangle correspondence that matches the sides and angles.

1. 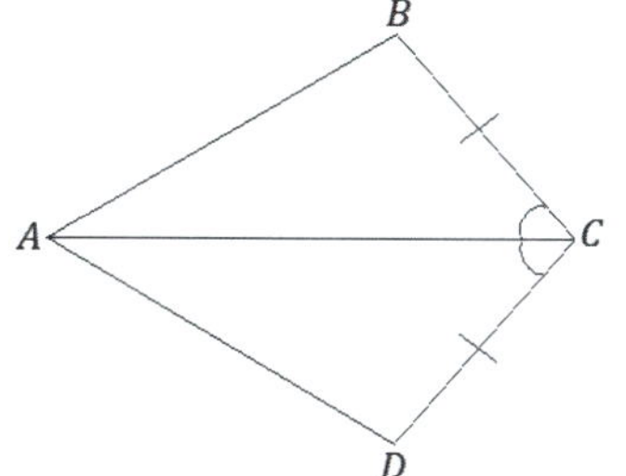

   These triangles are identical by the two sides and the included angle condition. The triangle correspondence $\triangle ABC \leftrightarrow \triangle ADC$ matches two pairs of equal sides and one pair of equal angles. One of the equal pairs of sides is shared side $\overline{AC}$.

2. 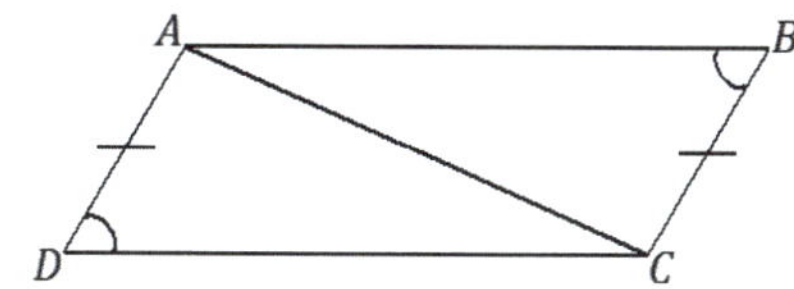

   These triangles are not necessarily identical. The triangles have a pair of marked equal sides, and a pair of marked, acute, equal angles; side $\overline{AC}$ is also common to both triangles. The triangles satisfy the two sides and non-included acute angle condition, which does not determine a unique triangle.

3.

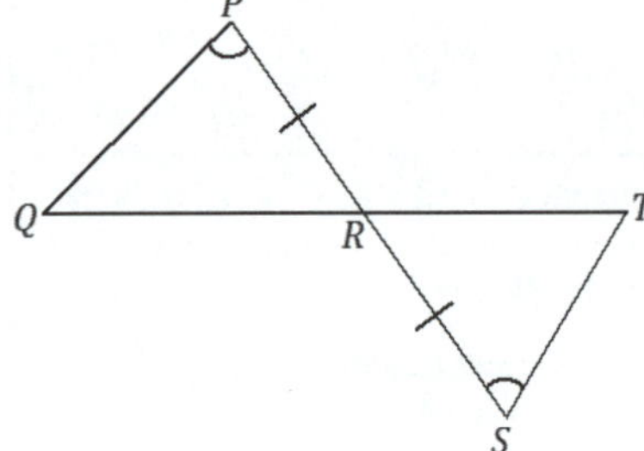

These triangles are identical by the two angles and included side condition. The triangle correspondence $\triangle QPR \leftrightarrow \triangle TSR$ matches the two pairs of equal angles and one pair of equal sides. One pair of equal angles is the vertical angle pair $\angle PRQ = \angle SRT$.

4.

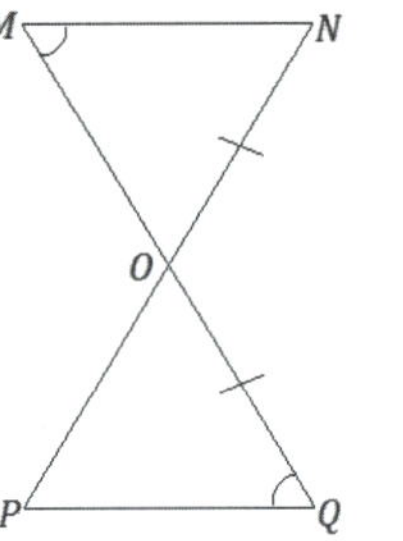

These triangles are not necessarily identical. A correspondence that matches up the equal pair of sides and the equal pair of vertical angles does not match the marked equal pair of angles.

5.

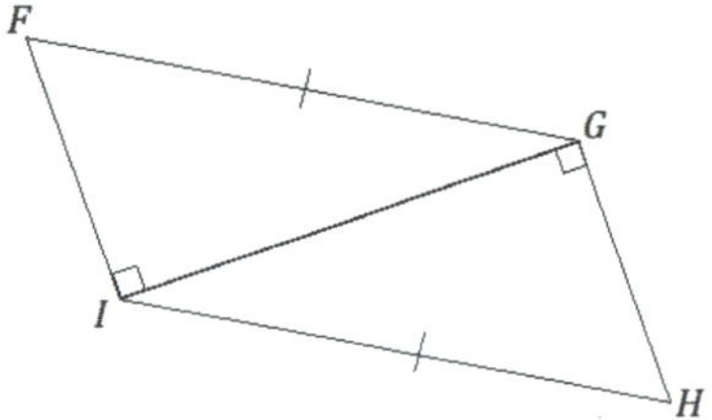

The triangles are identical by the two sides and non-included $90°$ (or greater) angle condition. The correspondence $\triangle IFG \leftrightarrow \triangle GHI$ matches two pairs of equal sides and one pair of equal angles. One of the two pairs of equal sides is side $\overline{IG}$, which is common to both triangles.

6.

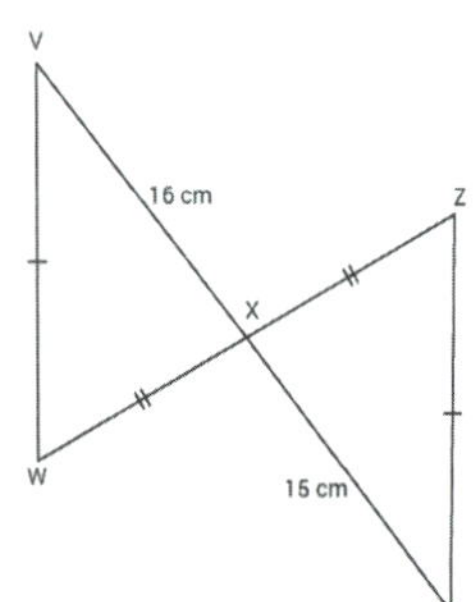

The triangles are not identical since a correspondence that matches the two marked equal pairs of sides also matches sides $\overline{VX}$ and $\overline{XY}$, which are not equal in length.

7.

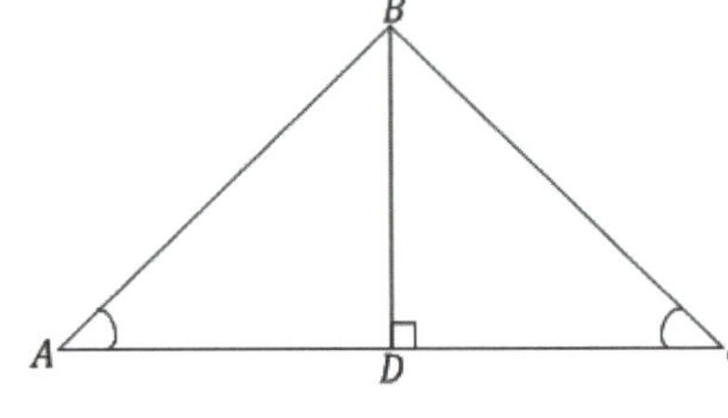

These triangles are identical by the two angles and side opposite a given angle condition. The correspondence $\triangle ABD \leftrightarrow \triangle CBD$ matches the two pairs of equal angles and one pair of equal sides. The pair of equal sides is the common side, $\overline{BD}$. We know $\angle ADB$ must be a right angle since $\angle CDB$ is a right angle; they are both on a line, and of course angles $\angle A$ and $\angle C$ are equal in measurement.

8. Are there any identical triangles in this diagram?

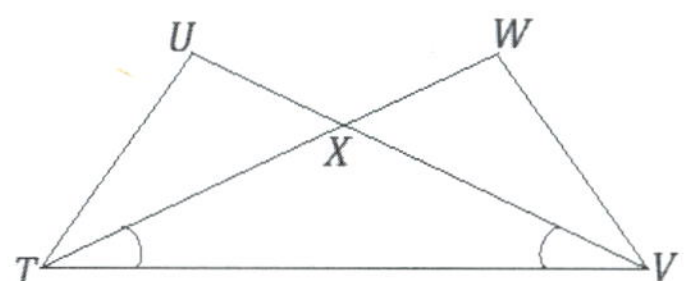

$\triangle TUX$ and $\triangle VWX$ are not necessarily identical; we only know about a single pair of equal angles in the triangles, which are vertical angles. $\triangle TUV$ and $\triangle VWT$ are also not necessarily identical. We only know about a single pair of equal angles and a side common to both triangles, which is not enough information to determine the triangles as identical or non-identical.

9.

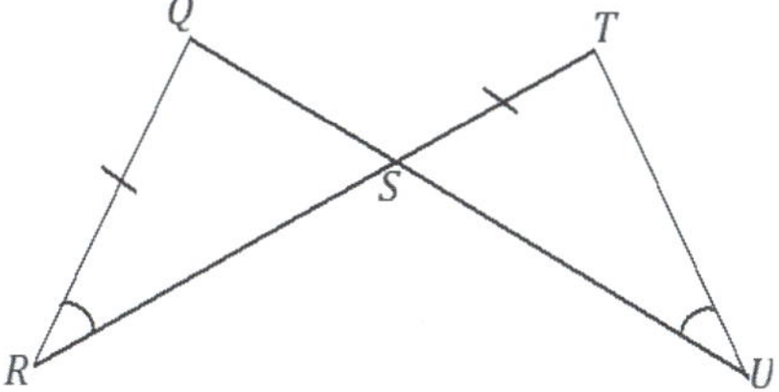

The triangles are not necessarily identical since there is no correspondence that matches the two marked equal pairs of sides as well as the two pairs of equal angles. One of the pairs of equal angles is the pair of vertical angles.

10.

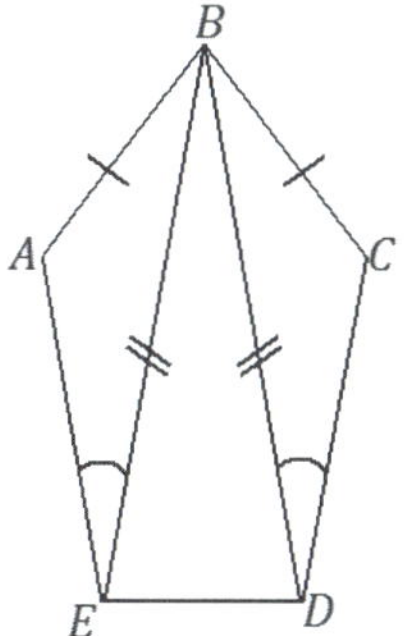

$\triangle ABE$ and $\triangle CBD$ are not necessarily identical. The triangles satisfy the two sides and non-included acute angle condition, which does not determine a unique triangle.

Lesson 15: Using Unique Triangles to Solve Real-World and Mathematical Problems

Student Outcomes

- Students use conditions that determine a unique triangle to construct viable arguments that angle measures and lengths are equal between triangles.

Lesson Notes

In Lesson 15, students continue to apply their understanding of the conditions that determine a unique triangle. Lesson 14 introduced students to diagrams of triangles with pre-existing relationships, in contrast to the diagrams in Lesson 13 that showed distinct triangles with three matching marked parts. This added a new challenge to the task of determining whether triangles were identical because some information had to be assessed from the diagram to establish a condition that would determine triangles as identical. Lesson 15 exposes students to yet another challenge where they are asked to determine whether triangles are identical and then to show how this information can lead to further conclusions about the diagram (i.e., showing why a given point must be the midpoint of a segment). Problems in this lesson are both real-world and mathematical. All problems require an explanation that logically links given knowledge, a correspondence, and a condition that determines triangles to be identical; some problems require these links to yield one more conclusion. This lesson is an opportunity to highlight Mathematical Practice 1, giving students an opportunity to build perseverance in solving problems.

Classwork

Example 1 (5 minutes)

MP.1

Example 1

A triangular fence with two equal angles, $\angle S = \angle T$, is used to enclose some sheep. A fence is constructed inside the triangle that exactly cuts the other angle into two equal angles: $\angle SRW = \angle TRW$. Show that the gates, represented by $\overline{SW}$ and $\overline{WT}$, are the same width.

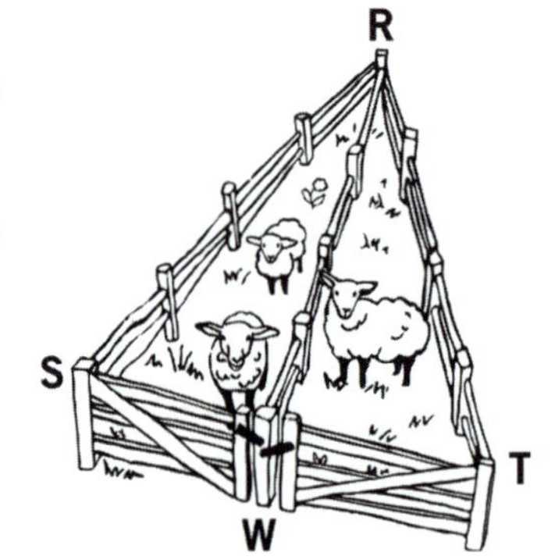

There is a correspondence $\triangle SRW \leftrightarrow \triangle TRW$ that matches two pairs of angles of equal measurement, $\angle S = \angle T$ and $\angle SRW = \angle TRW$, and one pair of sides of equal length shared, side $\overline{RW}$. The triangles satisfy the two angles and side opposite a given angle condition. From the correspondence, we can conclude that $SW = WT$, or that the gates are of equal width.

EUREKA MATH™

Example 2 (5 minutes)

As students work through Example 2, remind them that this question is addressed in an easier format in Grade 4, when students folded the triangle so that $\overline{AC}$ folded onto $\overline{BC}$.

> *Scaffolding:*
> - If needed, provide a hint: Draw the segments AC and BC.
> - Also, provide students with compasses so that they can mimic the construction marks in the diagram.

Example 2

In $\triangle ABC$, $AC = BC$, and $\triangle ABC \leftrightarrow \triangle B'A'C'$. John says that the triangle correspondence matches two sides and the included angle and shows that $\angle A = \angle B'$. Is John correct?

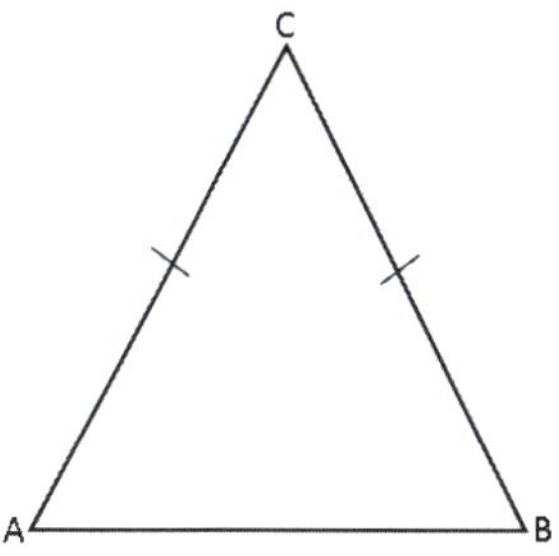

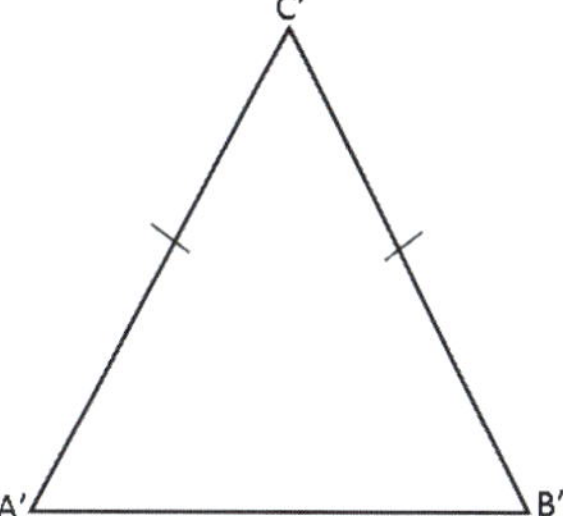

We are told that $AC = BC$. The correspondence $\triangle ABC \leftrightarrow \triangle B'A'C'$ tells us that $BC \leftrightarrow A'C'$, $CA \leftrightarrow C'B'$, and $\angle C \leftrightarrow \angle C'$, which means $\triangle ABC$ is identical to $\triangle B'A'C'$ by the two sides and included angle condition. From the correspondence, we can conclude that $\angle A = \angle B'$; therefore, John is correct.

Exercises 1–4 (20 minutes)

Exercises 1–4

1. **Mary puts the center of her compass at the vertex O of the angle and locates points A and B on the sides of the angle. Next, she centers her compass at each of A and B to locate point C. Finally, she constructs the ray $\overrightarrow{OC}$. Explain why $\angle BOC = \angle AOC$.**

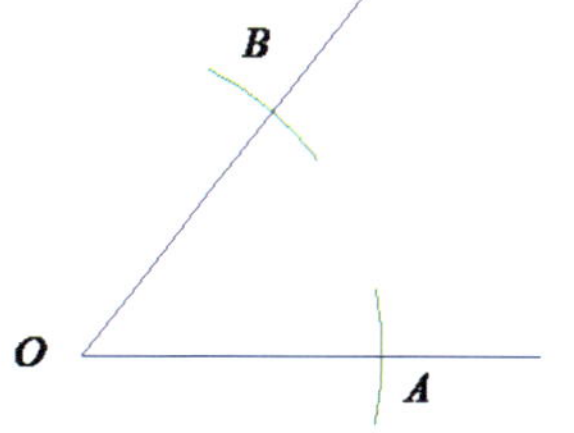

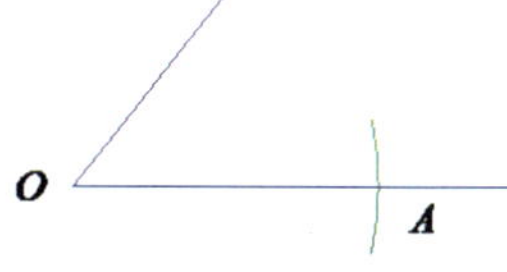

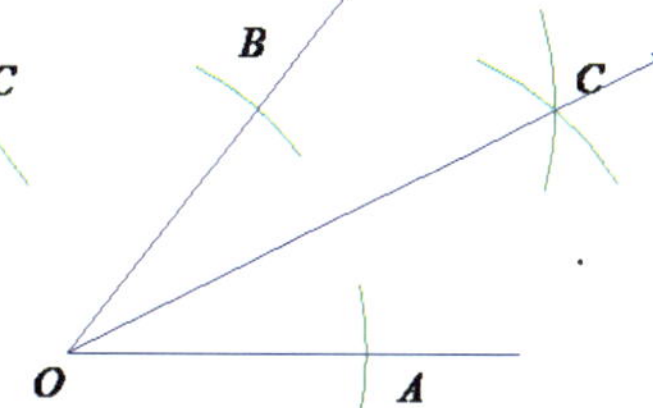

Since Mary uses one compass adjustment to determine points A and B, $OA = OB$. Mary also uses the same compass adjustment from B and A to find point C; this means $BC = AC$. Side $\overline{OC}$ is common to both the triangles, $\triangle OBC$ and $\triangle OAC$. Therefore, there is a correspondence $\triangle OBC \leftrightarrow \triangle OAC$ that matches three pairs of equal sides, and the triangles are identical by the three sides condition. From the correspondence, we conclude that $\angle BOC = \angle AOC$.

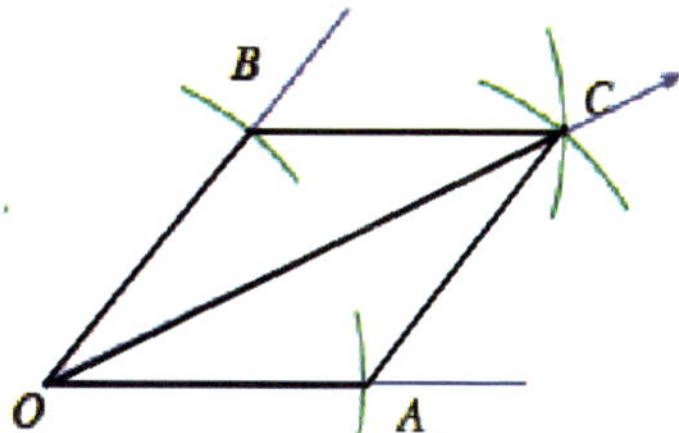

2. Quadrilateral $ACBD$ is a model of a kite. The diagonals $\overline{AB}$ and $\overline{CD}$ represent the sticks that help keep the kite rigid.

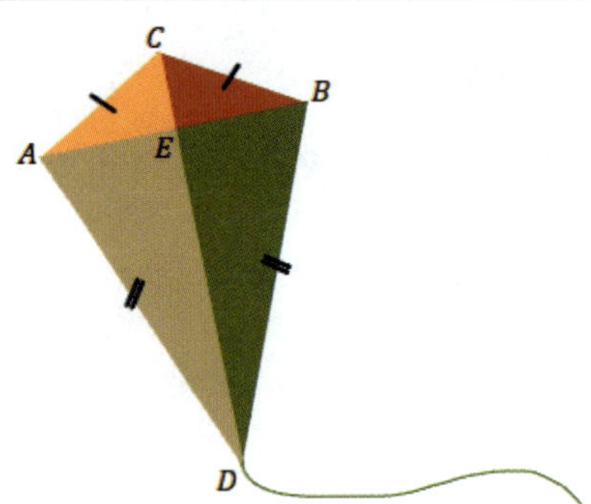

a. John says that $\angle ACD = \angle BCD$. Can you use identical triangles to show that John is correct?

From the diagram, we see that $AC = BC$, and $AD = BD$. $\overline{CD}$ is a common side to both triangles, $\triangle ACD$ and $\triangle BCD$. There is a correspondence $\triangle ACD \leftrightarrow \triangle BCD$ that matches three pairs of equal sides; the two triangles are identical by the three sides condition. From the correspondence, we conclude that $\angle ACD = \angle BCD$. John is correct.

b. Jill says that the two sticks are perpendicular to each other. Use the fact that $\angle ACD = \angle BCD$ and what you know about identical triangles to show $\angle AEC = 90°$.

Since we know that $AC = BC$ and $\angle ACD = \angle BCD$, and that $\triangle ACE$ and $\triangle BCE$ share a common side, $\overline{CE}$, we can find a correspondence that matches two pairs of equal sides and a pair of equal, included angles. The triangles are identical by the two sides and included angle condition. We can then conclude that $\angle AEC = \angle BEC$. Since both angles are adjacent to each other on a straight line, we also know their measures must sum to $180°$. We can then conclude that each angle measures $90°$.

c. John says that Jill's triangle correspondence that shows the sticks are perpendicular to each other also shows that the sticks cross at the midpoint of the horizontal stick. Is John correct? Explain.

Since we have established that $\triangle ACE$ and $\triangle BCE$ are adjacent to each other, we know that $AE = BE$. This means that E is the midpoint of $\overline{AB}$, by definition.

3. In $\triangle ABC$, $\angle A = \angle B$, and $\triangle ABC \leftrightarrow \triangle B'A'C'$. Jill says that the triangle correspondence matches two angles and the included side and shows that $AC = B'C'$. Is Jill correct?

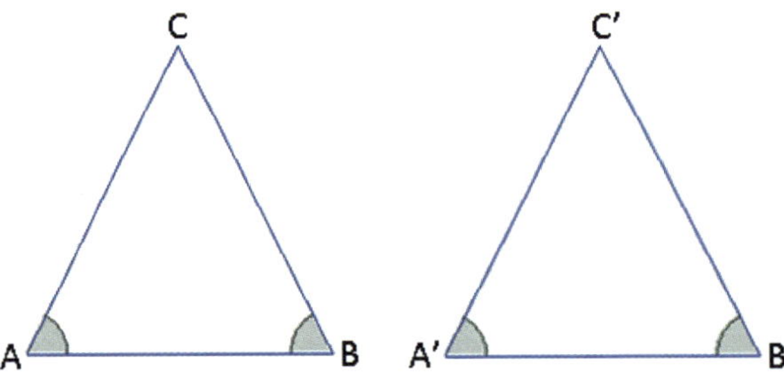

We are told that $\angle A = \angle B$. The correspondence $\triangle ABC \leftrightarrow \triangle B'A'C'$ tells us that $\angle A = \angle B'$, $\angle B = \angle A'$, and $AB = B'A'$, which means $\triangle ABC$ is identical to $\triangle B'A'C'$ by the two angles and included side condition. From the correspondence, we can conclude that $AC = B'C'$; therefore, Jill is correct.

4. Right triangular corner flags are used to mark a soccer field. The vinyl flags have a base of 40 cm and a height of 14 cm.

   a. Mary says that the two flags can be obtained by cutting a rectangle that is $40 \text{ cm} \times 14 \text{ cm}$ on the diagonal. Will that create two identical flags? Explain.

   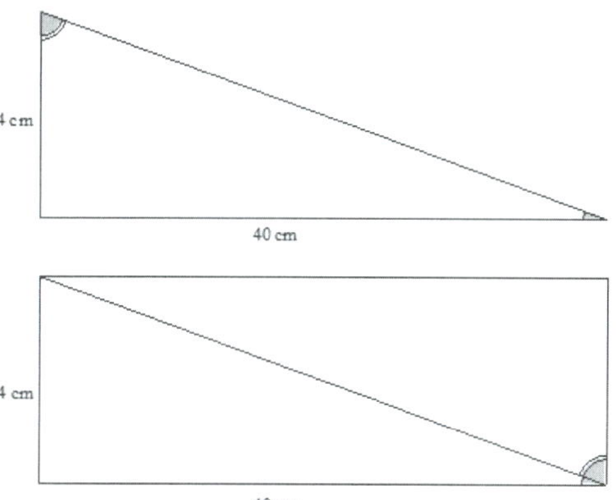

   If the flag is to be cut from a rectangle, both triangles will have a side of length 40 cm, a length of 14 cm, and a right angle. There is a correspondence that matches two pairs of equal sides and an included pair of equal angles to the corner flag; the two triangles are identical to the corner flag as well as to each other.

   b. Will measures the two non-right angles on a flag and adds the measurements together. Can you explain, without measuring the angles, why his answer is $90°$?

   The two non-right angles of the flags are adjacent angles that together form one angle of the four angles of the rectangle. We know that a rectangle has four right angles, so it must be that the two non-right angles of the flag together sum to $90°$.

Discussion (8 minutes)

Consider a gallery walk to review responses to each exercise.

- Hold students accountable for providing evidence in their responses that logically progresses to a conclusion.
- Offer opportunities for students to share and compare their solution methods.

Closing (2 minutes)

Lesson Summary

- **In deciding whether two triangles are identical, examine the structure of the diagram of the two triangles to look for a relationship that might reveal information about corresponding parts of the triangles. This information may determine whether the parts of the triangle satisfy a particular condition, which might determine whether the triangles are identical.**
- **Be sure to identify and label all known measurements, and then determine if any other measurements can be established based on knowledge of geometric relationships.**

Exit Ticket (5 minutes)

Name ______________________________ Date ______________

Lesson 15: Using Unique Triangles to Solve Real-World and Mathematical Problems

Exit Ticket

Alice is cutting wrapping paper to size to fit a package. How should she cut the rectangular paper into two triangles to ensure that each piece of wrapping paper is the same? Use your knowledge of conditions that determine unique triangles to justify that the pieces resulting from the cut are the same.

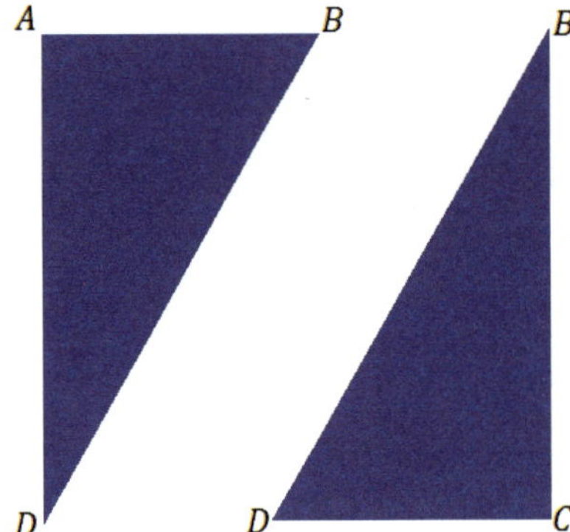

Exit Ticket Sample Solutions

Alice is cutting wrapping paper to size to fit a package. How should she cut the rectangular paper into two triangles to ensure that each piece of wrapping paper is the same? Use your knowledge of conditions that determine unique triangles to prove that the pieces resulting from the cut are the same.

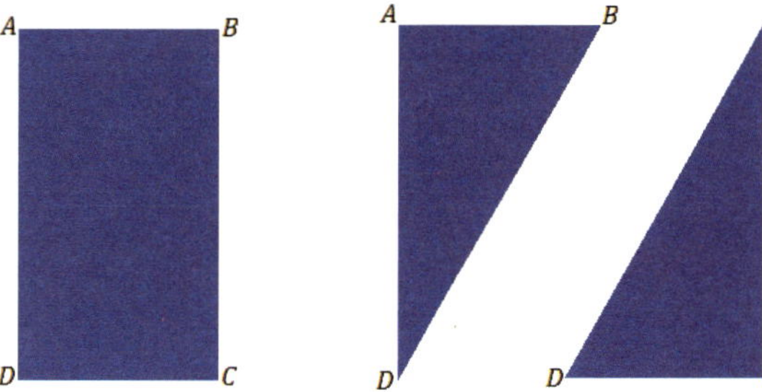

Alice should cut along the diagonal of rectangle $ABCD$. Since $ABCD$ is a rectangle, the opposite sides will be equal in length, or $AB = DC$ and $AD = BC$. A rectangle also has four right angles, which means a cut along the diagonal will result in each triangle with one $90°$ angle. The correspondence $\triangle ABD \leftrightarrow \triangle CDB$ matches two equal pairs of sides and an equal, included pair of angles; the triangles are identical by the two sides and included angle condition.

Problem Set Sample Solutions

1. **Jack is asked to cut a cake into 8 equal pieces. He first cuts it into equal fourths in the shape of rectangles, and then he cuts each rectangle along a diagonal.**

   Did he cut the cake into 8 equal pieces? Explain.

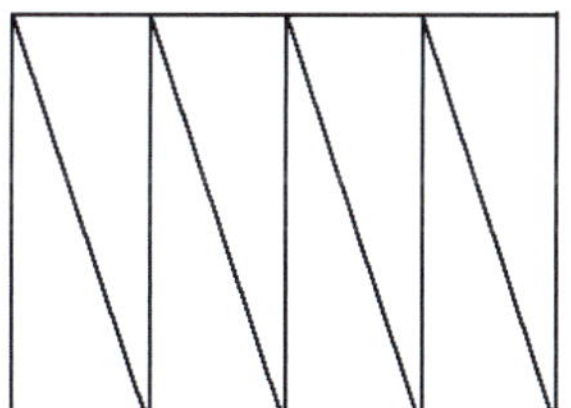
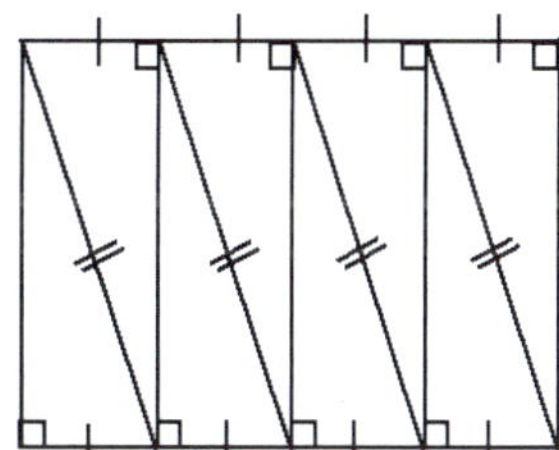

Yes, Jack cut the cake into 8 equal pieces. Since the first series of cuts divided the cake into equal fourths in the shape of rectangles, we know that the opposite sides of the rectangles are equal in length; that means all 8 triangles have two sides that are equal in length to each other. Each of the triangular pieces also has one right angle because we know that rectangles have four right angles. Therefore, there is a correspondence between all 8 triangles that matches two pairs of equal sides and an equal, $90°$ non-included angle, determining 8 identical pieces of cake.

2. The bridge below, which crosses a river, is built out of two triangular supports. The point M lies on $\overline{BC}$. The beams represented by $\overline{AM}$ and $\overline{DM}$ are equal in length, and the beams represented by $\overline{AB}$ and $\overline{DC}$ are equal in length. If the supports were constructed so that $\angle A$ and $\angle D$ are equal in measurement, is point M the midpoint of $\overline{BC}$? Explain.

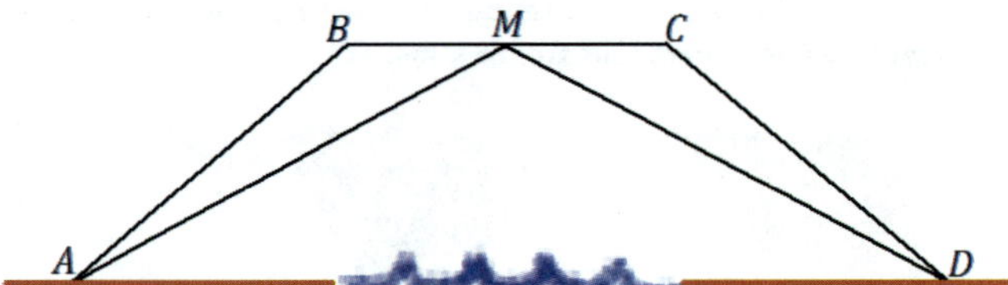

Yes, M is the midpoint of $\overline{BC}$. The triangles are identical by the two sides and included angle condition. The correspondence $\triangle ABM \leftrightarrow \triangle DCM$ matches two pairs of equal sides and one pair of included equal angles. Since the triangles are identical, we can use the correspondence to conclude that $BM = CM$, which makes M the midpoint, by definition.

EUREKA MATH™

Name ______________________________ Date ______________

1. In each problem, set up and solve an equation for the unknown angles.

   a. Four lines meet at a point. Find the measures m° and n°.

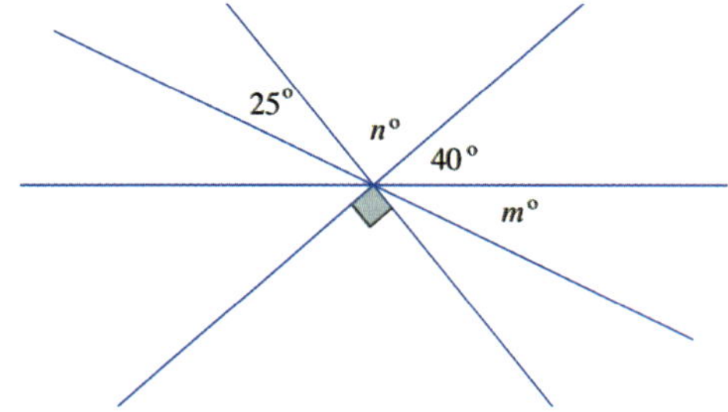

   b. Two lines meet at the vertex of two rays. Find the measures m° and n°.

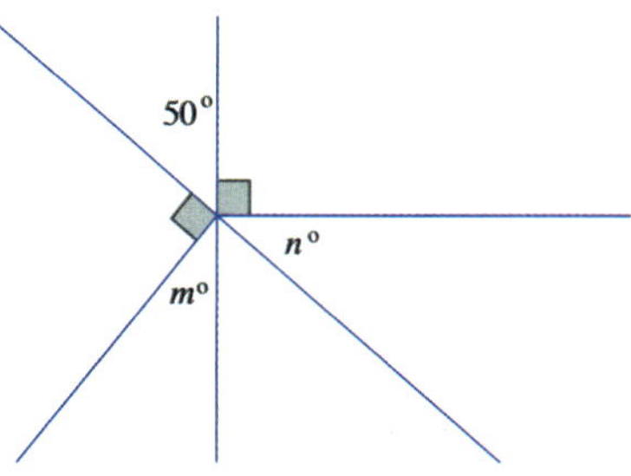

   c. Two lines meet at a point that is the vertex of two rays. Find the measures m° and n°.

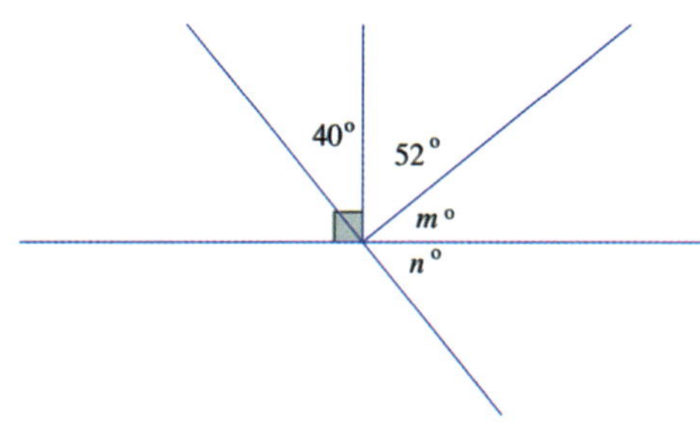

d. Three rays have a common vertex on a line. Find the measures $m°$ and $n°$.

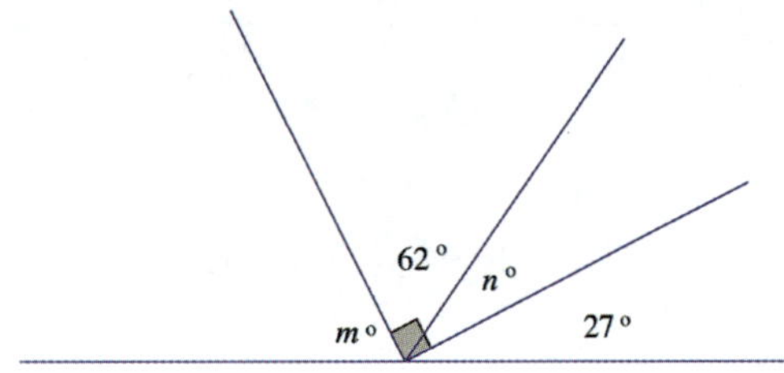

2. Use tools to construct a triangle based on the following given conditions.

   a. If possible, use your tools to construct a triangle with angle measurements $20°$, $55°$, and $105°$, and leave evidence of your construction. If it is not possible, explain why.

   b. Is it possible to construct two different triangles that have the same angle measurements? If it is, construct examples that demonstrate this condition, and label all angle and length measurements. If it is not possible, explain why.

3. In each of the following problems, two triangles are given. For each: (1) state if there are sufficient or insufficient conditions to show the triangles are identical, and (2) explain your reasoning.

a.

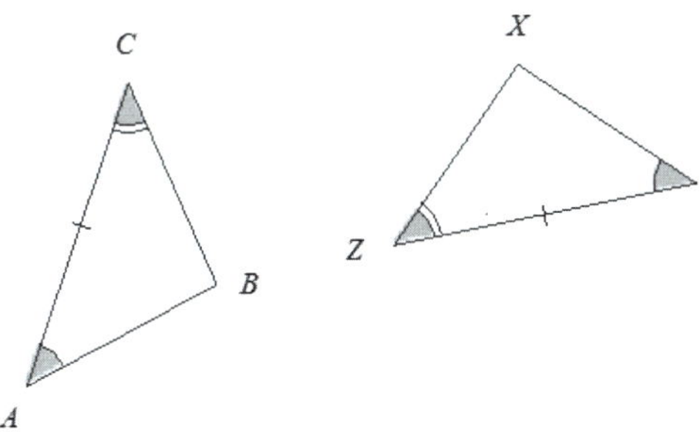

b.

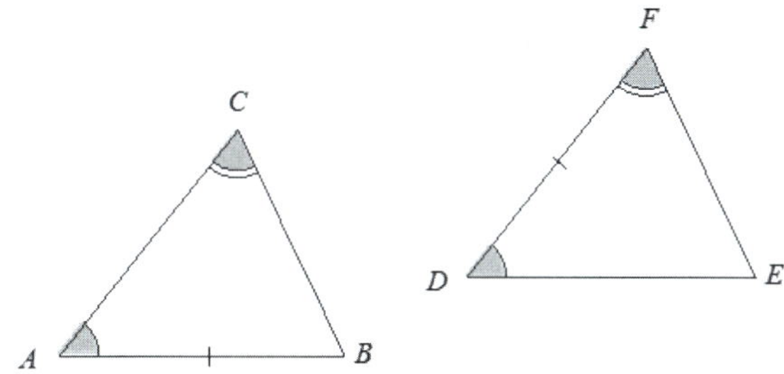

c.

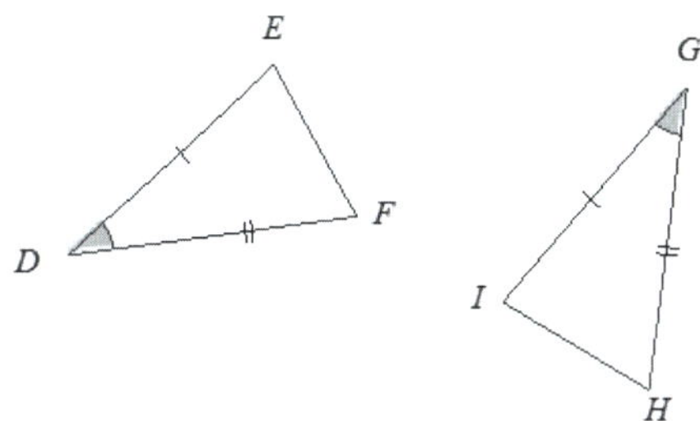

d.

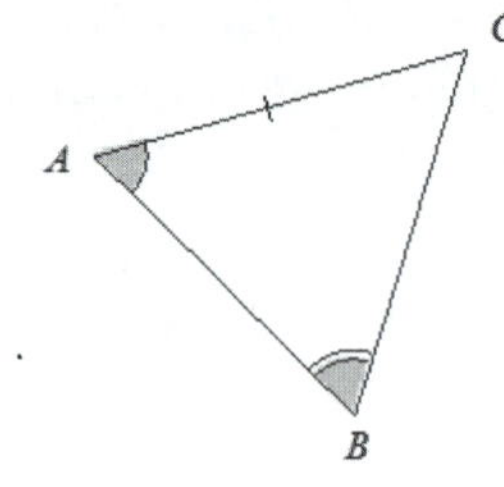

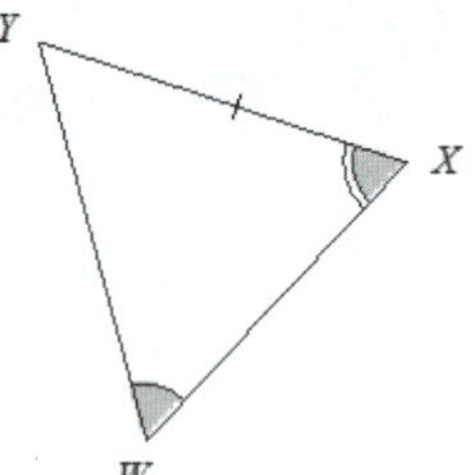

4. Use tools to draw rectangle $ABCD$ with $AB = 2$ cm and $BC = 6$ cm. Label all vertices and measurements.

5. The measures of two complementary angles have a ratio of $3:7$. Set up and solve an equation to determine the measurements of the two angles.

6. The measure of the supplement of an angle is $12°$ less than the measure of the angle. Set up and solve an equation to determine the measurements of the angle and its supplement.

7. Three angles are at a point. The ratio of two of the angles is $2:3$, and the remaining angle is $32°$ more than the larger of the first two angles. Set up and solve an equation to determine the measures of all three angles.

8. Draw a right triangle according to the following conditions, and label the provided information. If it is not possible to draw the triangle according to the conditions, explain why. Include a description of the kind of figure the current measurements allow. Provide a change to the conditions that makes the drawing feasible.

   a. Construct a right triangle ABC so that $AB = 3\text{ cm}$, $BC = 4\text{ cm}$, and $CA = 5\text{ cm}$; the measure of angle B is $90°$.

   b. Construct triangle DEF so that $DE = 4\text{ cm}$, $EF = 5\text{ cm}$, and $FD = 11\text{ cm}$; the measure of angle D is $50°$.

A Progression Toward Mastery

Assessment Task Item		STEP 1 Missing or incorrect answer and little evidence of reasoning or application of mathematics to solve the problem.	STEP 2 Missing or incorrect answer but evidence of some reasoning or application of mathematics to solve the problem.	STEP 3 A correct answer with some evidence of reasoning or application of mathematics to solve the problem OR an incorrect answer with substantial evidence of solid reasoning or application of mathematics to solve the problem.	STEP 4 A correct answer supported by substantial evidence of solid reasoning or application of mathematics to solve the problem.
1	a 7.G.B.5	Student sets up correct equations to solve for $m°$ and $n°$, but no further evidence is shown.	Student finds incorrect values for $m°$ and $n°$, but complete supporting work is shown; conceptual errors, such as an equation that does not reflect the angle relationship, lead to incorrect answers.	Student finds one correct value for either $m°$ or $n°$. Complete supporting work is shown, but a calculation error, such as an arithmetic error, leads to one incorrect answer.	Student finds $m° = 25°$ and $n° = 90°$, shows complete supporting work, including an equation that appropriately models the angle relationship(s), and gives a correct algebraic solution.
	b 7.G.B.5	Student sets up the correct equations to solve for $m°$ and $n°$, but no further evidence is shown.	Student finds incorrect values for $m°$ and $n°$, but complete supporting work is shown; conceptual errors, such as an equation that does not reflect the angle relationship, lead to incorrect answers.	Student finds one correct value for either $m°$ or $n°$. Complete supporting work is shown, but a calculation error, such as an arithmetic error, leads to one incorrect answer.	Student finds $m° = 40°$ and $n° = 40°$, shows complete supporting work, including an equation that appropriately models the angle relationship(s), and gives a correct algebraic solution.
	c 7.G.B.5	Student sets up the correct equations to solve for $m°$ and $n°$, but no further evidence is shown.	Student finds incorrect values for $m°$ and $n°$, but complete supporting work is shown; conceptual errors, such as an equation that does not reflect the angle relationship, lead to incorrect answers.	Student finds one correct value for either $m°$ or $n°$. Complete supporting work is shown, but a calculation error, such as an arithmetic error, leads to one incorrect answer.	Student finds $m° = 38°$ and $n° = 50°$, shows complete supporting work, including an equation that appropriately models the angle relationship(s), and gives a correct algebraic solution.

	d 7.G.B.5	Student sets up the correct equations to solve for $m°$ and $n°$, but no further evidence is shown.	Student finds incorrect values for $m°$ and $n°$, but complete supporting work is shown; conceptual errors, such as an equation that does not reflect the angle relationship, lead to incorrect answers.	Student finds one correct value for either $m°$ or $n°$. Complete supporting work is shown, but a calculation error, such as an arithmetic error, leads to one incorrect answer.	Student finds $m° = 63°$ and $n° = 28°$, shows complete supporting work, including an equation that appropriately models the angle relationship(s), and gives a correct algebraic solution.
2	a 7.G.A.2	Student constructs a triangle with angle measurements that are off by more than $3°$ of the given measurements, with the intersection of two extended sides of two angles shown as the location of the last vertex.	Student constructs a triangle with the given angle measurements, but no evidence of the construction is provided.	Student constructs a triangle with angle measurements that are not exact but are within $3°$ of the given measurements, with the intersection of two extended sides of two angles shown as the location of the last vertex.	Student constructs a triangle with the given angle measurements, with the intersection of two extended sides of two angles shown as the location of the last vertex.
	b 7.G.A.2	Student constructs two triangles that have corresponding angle measurements that are off by more than $3°$ of each other and different corresponding side lengths.	Student provides no examples, but the answer does contain a verbal description stating that triangles that are scale drawings of each other have the same angle measurements and corresponding side lengths that are proportional.	Student constructs two triangles; however, the corresponding angle measurements are not exactly equal but are within $3°$ of each other and have different corresponding side lengths.	Student constructs two triangles that both have the same set of angle measurements but different corresponding side lengths.
3	a 7.G.A.2	Student does not provide a response. OR Student fails to provide evidence of comprehension.	Student correctly identifies triangles as identical or not identical, but no further evidence is provided.	Student correctly identifies triangles as identical or not identical but with the incorrect supporting evidence, such as the incorrect condition by which they are identical.	Student correctly identifies triangles as identical or not identical and supports with appropriate evidence, such as the condition by which they are identical or the information that prevents them from being identical.

	b 7.G.A.2	Student does not provide a response. OR Student fails to provide evidence of comprehension.	Student correctly identifies triangles as identical or not identical, but no further evidence is provided.	Student correctly identifies triangles as identical or not identical but with the incorrect supporting evidence, such as the incorrect condition by which they are identical.	Student correctly identifies triangles as identical or not identical and supports with appropriate evidence, such as the condition by which they are identical or the information that prevents them from being identical.
	c 7.G.A.2	Student does not provide a response. OR Student fails to provide evidence of comprehension.	Student correctly identifies triangles as identical or not identical, but no further evidence is provided.	Student correctly identifies triangles as identical or not identical but with the incorrect supporting evidence, such as the incorrect condition by which they are identical.	Student correctly identifies triangles as identical or not identical and supports with appropriate evidence, such as the condition by which they are identical or the information that prevents them from being identical.
	d 7.G.A.2	Student does not provide a response. OR Student fails to provide evidence of comprehension.	Student correctly identifies triangles as identical or not identical, but no further evidence is provided.	Student correctly identifies triangles as identical or not identical but with the incorrect supporting evidence, such as the incorrect condition by which they are identical.	Student correctly identifies triangles as identical or not identical and supports with appropriate evidence, such as the condition by which they are identical or the information that prevents them from being identical.
4	7.G.A.2	Student provides a response that has inaccurate measurements and is missing labeling.	Student provides a drawing with errors in the measurements of the provided sides and angle, but the figure has all the provided information labeled.	Student provides a drawing that is accurate in measurements, but the figure is missing labeling.	Student provides an accurately drawn rectangle with dimensions 2 cm and 6 cm and all provided information labeled.
5	7.G.B.5	Student finds incorrect angle measurements because the equations are incorrectly set up, and the supporting work is incorrect.	Student finds one or both angle measurements incorrectly due to conceptual errors such as an equation that does not reflect the angle relationship. All other supporting work is correctly shown.	Student finds one or both angle measurements incorrectly due to errors in calculation such as an arithmetic error, but all other supporting work is correctly shown.	Student finds the two angle measurements to be $27°$ and $63°$ and shows complete supporting work, including an equation that appropriately models the angle relationship(s) and a correct algebraic solution.

6	**7.G.B.5**	Student finds incorrect angle measurements because the equations are incorrectly set up, and the supporting work is incorrect.	Student finds one or both angle measurements incorrectly due to conceptual errors such as an equation that does not reflect the angle relationship. All other supporting work is correctly shown.	Student finds one or both angle measurements incorrectly due to errors in calculation such as an arithmetic error, but all other supporting work is correctly shown.	Student finds the two angle measurements to be $96°$ and $84°$ and shows complete supporting work, including an equation that appropriately models the angle relationship(s) and a correct algebraic solution.
7	**7.G.B.5**	Student finds incorrect angle measurements because the equations are incorrectly set up, and the supporting work is incorrect.	Student finds one, two, or all three angle measures incorrectly due to conceptual errors such as an equation that does not reflect the angle relationship. All other supporting work is correctly shown.	Student finds one, two, or all three angle measurements incorrectly due to errors in calculation such as an arithmetic error, but all other supporting work is correctly shown.	Student finds the three angle measurements to be $82°$, $123°$, and $155°$ and shows complete supporting work, including an equation that appropriately models the angle relationship(s) and a correct algebraic solution.
8	**a** **7.G.A.2**	Student provides a response that has inaccurate measurements and is missing labeling.	Student provides a drawing that has errors in the measurements of the provided sides and angle, but the figure has all the provided information labeled.	Student provides a drawing that is accurate in measurements, but the figure is missing labeling.	Student provides an accurately drawn right triangle with lengths 3 cm, 4 cm, and 5 cm, and $\angle B = 90°$; all provided information is labeled.

	b **7.G.B.5**	Student indicates the triangle cannot be drawn but provides no further evidence of understanding. AND Student fails to include a description of the figure formed by the current measurements and provides no alteration to one of the measurements so that the triangle can be drawn.	Student clearly indicates that the triangle cannot be drawn because the sum of the lengths of the two smaller sides, $\overline{DE}$ and $\overline{EF}$, is less than the length of the third side, $\overline{FD}$; however, student fails to include a description of the figure formed by the current measurements and provides no alteration to one of the measurements so that the triangle can be drawn.	Student clearly indicates that the triangle cannot be drawn because the sum of the lengths of the two smaller sides, $\overline{DE}$ and $\overline{EF}$, is less than the length of the third side, $\overline{FD}$. AND Student includes a brief description of the figure formed by the current measurements. OR Student provides an alteration to one of the measurements so that the triangle can be drawn.	Student clearly indicates that the triangle cannot be drawn because the sum of the lengths of the two smaller sides, $\overline{DE}$ and $\overline{EF}$, is less than the length of the third side, $\overline{FD}$. The response should include some description of the two smaller sides as being unable to meet because their lengths are less than that of the third side. OR The response includes an alteration to one of the measurements so that the triangle can be drawn (i.e., one of the two smaller side lengths is increased so that the sum of the two smaller lengths is greater than 11 cm, or the side $\overline{FD}$ is decreased to less than 9 cm).

Name ______________________________ Date ______________

1. In each problem, set up and solve an equation for the unknown angles.

   a. Four lines meet at a point. Find the measures $m°$ and $n°$.

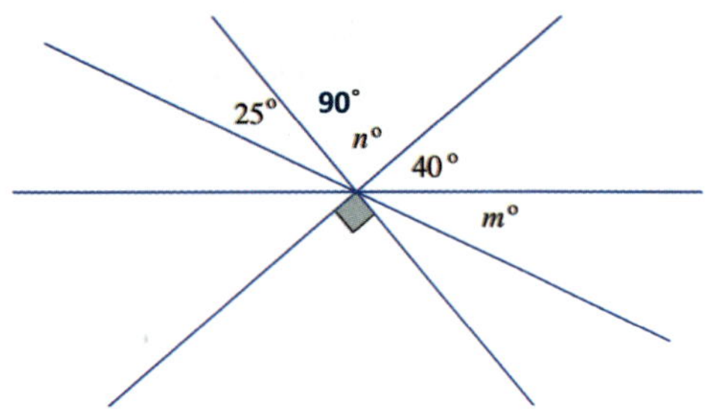

$n° = 90°$, vertical angles

$$25° + (90°) + 40° + m° = 180°$$
$$155° + m° = 180°$$
$$155° - 155° + m° = 180° - 155°$$
$$m° = 25°$$

   b. Two lines meet at the vertex of two rays. Find the measures $m°$ and $n°$.

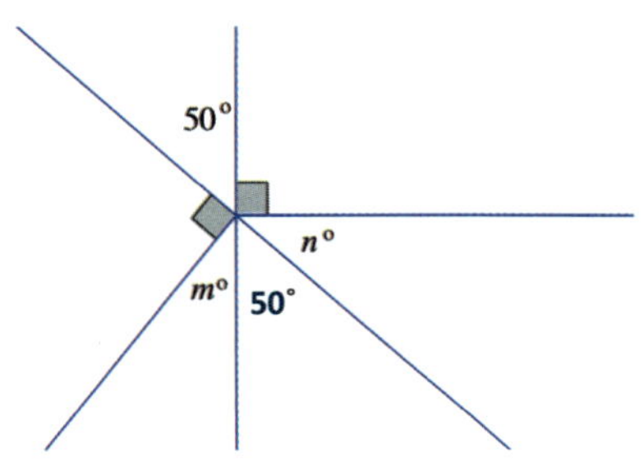

$$50° + 90° + n° = 180°$$
$$140° + n° = 180°$$
$$140° - 140° + n° = 180° - 140°$$
$$n° = 40°$$

$$m° + 50° = 90°$$
$$m° + 50° - 50° = 90° - 50°$$
$$m° = 40°$$

   c. Two lines meet at a point that is the vertex of two rays. Find the measures $m°$ and $n°$.

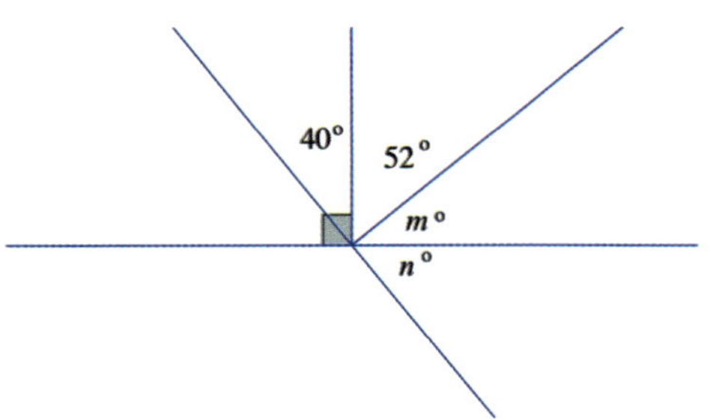

$$m° + 52° = 90°$$
$$m° + 52° - 52° = 90° - 52°$$
$$m° = 38°$$

$$40 + 52 + (38) + n° = 180$$
$$130 + n° = 180$$
$$130 - 130 + n° = 180 - 130$$
$$n° = 50°$$

d. Three rays have a common vertex on a line. Find the measures $m°$ and $n°$.

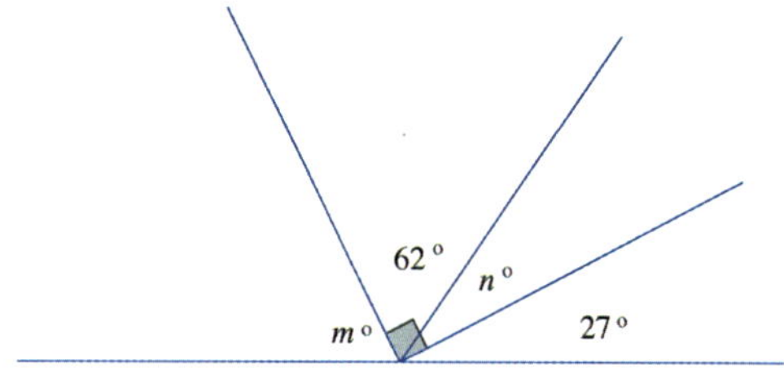

$$n° + 62° = 90°$$
$$n° + 62° - 62° = 90° - 62°$$
$$n° = 28°$$

$$m° + 62° + (28°) + 27° = 180°$$
$$m° + 117° = 180$$
$$m° + 117° - 117° = 180° - 117°$$
$$m° = 63°$$

2. Use tools to construct a triangle based on the following given conditions.

   a. If possible, use your tools to construct a triangle with angle measurements $20°$, $55°$, and $105°$, and leave evidence of your construction. If it is not possible, explain why.

   Solutions will vary. An example of a correctly constructed triangle is shown here.

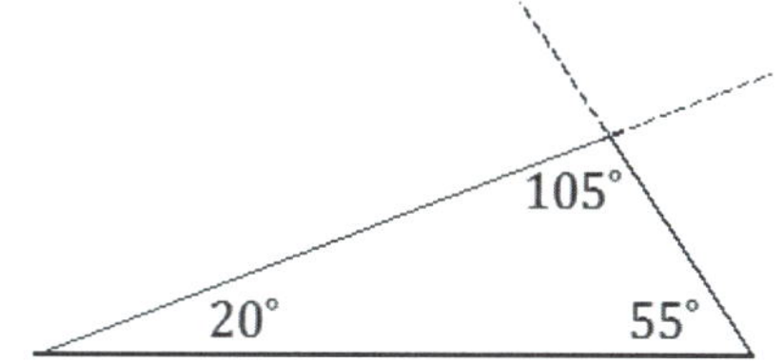

   b. Is it possible to construct two different triangles that have the same angle measurements? If it is, construct examples that demonstrate this condition, and label all angle and length measurements. If it is not possible, explain why.

   Solutions will vary; refer to the rubric.

3. In each of the following problems, two triangles are given. For each: (1) state if there are sufficient or insufficient conditions to show the triangles are identical, and (2) explain your reasoning.

a. *The triangles are identical by the two angles and included side condition. The marked side is between the given angles.*
$\triangle ABC \leftrightarrow \triangle YXZ$

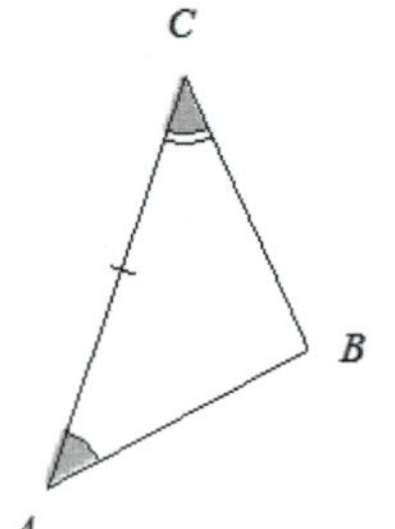

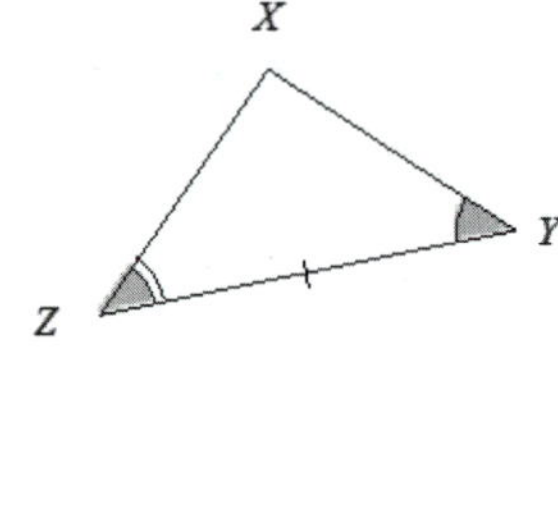

b. *There is insufficient evidence to determine that the triangles are identical. In $\triangle DEF$, the marked side is between the marked angles, but in $\triangle ABC$, the marked side is not between the marked angles.*

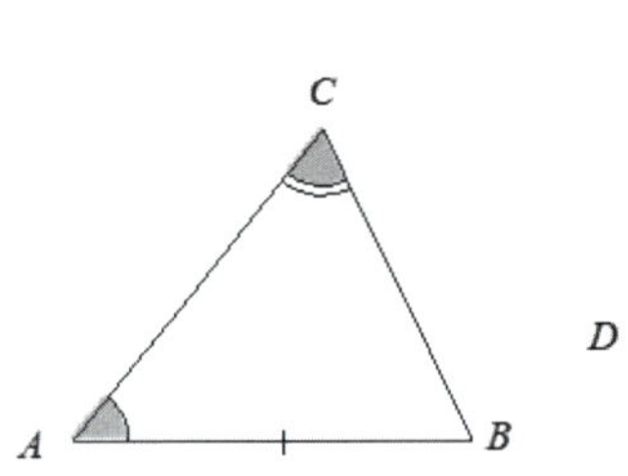

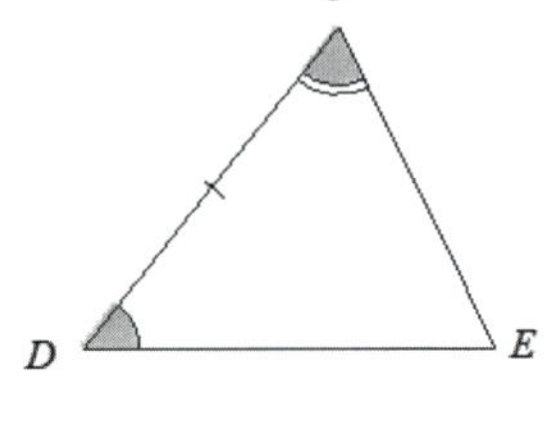

c. *The triangles are identical by the two sides and included angle condition.*
$\triangle DEF \leftrightarrow \triangle GIH$

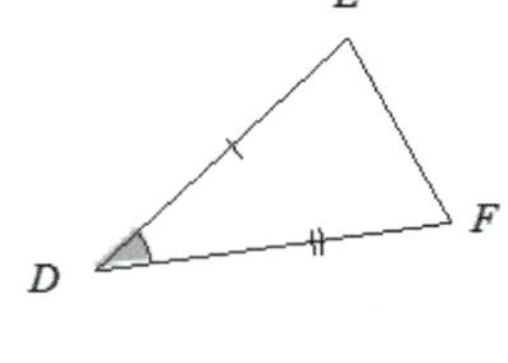

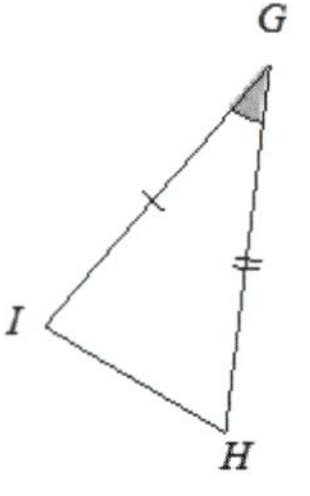

d. *The triangles are not identical. In $\triangle ABC$, the marked side is opposite $\angle B$. In $\triangle WXY$, the marked side is opposite $\angle W$. $\angle B$ and $\angle W$ are not necessarily equal in measure.*

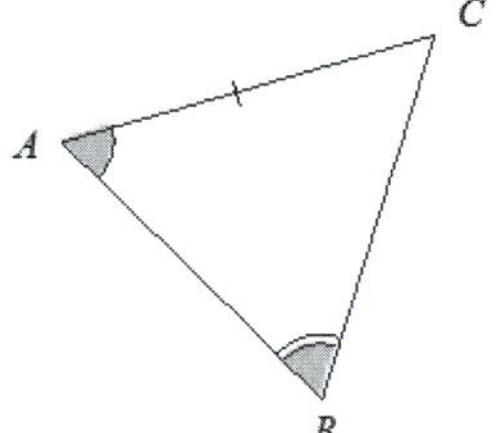

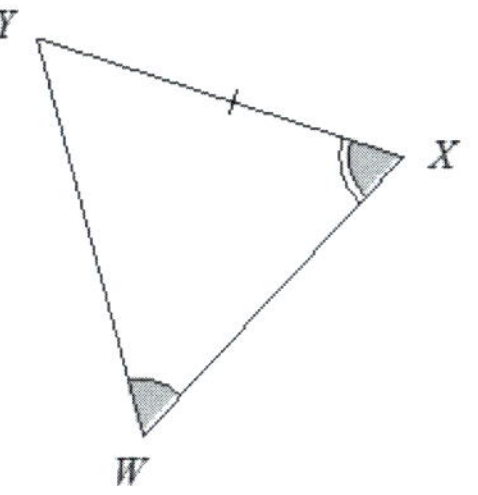

4. Use tools to draw rectangle $ABCD$ with $AB = 2$ cm and $BC = 6$ cm. Label all vertices and measurements.

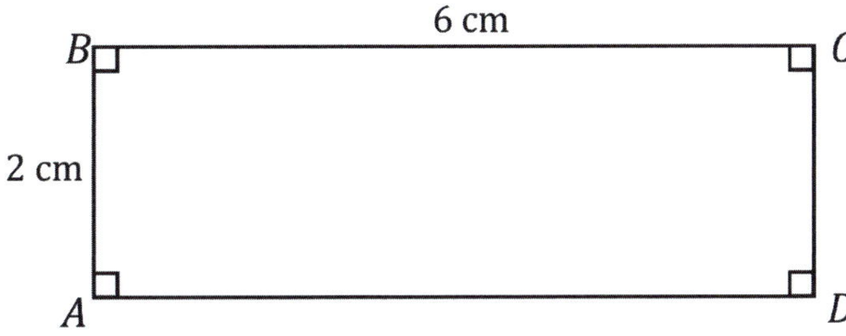

5. The measures of two complementary angles have a ratio of $3:7$. Set up and solve an equation to determine the measurements of the two angles.

$$3x + 7x = 90$$
$$10x = 90$$
$$\left(\frac{1}{10}\right)10x = \left(\frac{1}{10}\right)90$$
$$x = 9$$

Measure of Angle 1: $3(9) = 27$. The measure of the first angle is $27°$.

Measure of Angle 2: $7(9) = 63$. The measure of the second angle is $63°$.

6. The measure of the supplement of an angle is $12°$ less than the measure of the angle. Set up and solve an equation to determine the measurements of the angle and its supplement.

Let $y°$ be the number of degrees in the angle.

$$y + (y - 12) = 180$$
$$2y - 12 = 180$$
$$2y - 12 + 12 = 180 + 12$$
$$2y = 192$$
$$\left(\frac{1}{2}\right)2y = \left(\frac{1}{2}\right)192$$
$$y = 96$$

Measure of the angle: $96°$

Measure of its supplement: $(96)° - 12° = 84°$

7. Three angles are at a point. The ratio of two of the angles is $2:3$, and the remaining angle is $32°$ more than the larger of the first two angles. Set up and solve an equation to determine the measures of all three angles.

$$2x + 3x + (3x + 32) = 360$$
$$8x + 32 = 360$$
$$8x + 32 - 32 = 360 - 32$$
$$8x = 328$$
$$\left(\frac{1}{8}\right)8x = \left(\frac{1}{8}\right)328$$
$$x = 41$$

Measure of Angle 1: $2(41)° = 82°$

Measure of Angle 2: $3(41)° = 123°$

Measure of Angle 3: $3(41)° + 32° = 155°$

EUREKA MATH™

8. Draw a right triangle according to the following conditions, and label the provided information. If it is not possible to draw the triangle according to the conditions, explain why. Include a description of the kind of figure the current measurements allow. Provide a change to the condition that makes the drawing feasible.

   a. Construct a right triangle ABC so that $AB = 3$ cm, $BC = 4$ cm, and $CA = 5$ cm; the measure of angle B is $90°$.

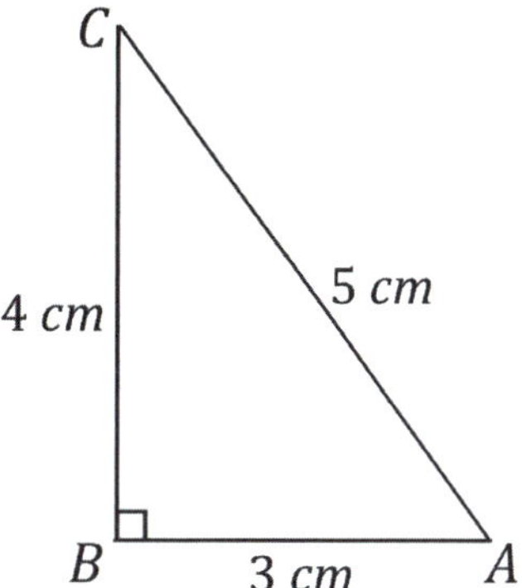

   b. Construct triangle DEF so that $DE = 4$ cm, $EF = 5$ cm, and $FD = 11$ cm; the measure of angle D is $50°$.

   It is not possible to draw this triangle because the lengths of the two shorter sides do not sum to be greater than the longest side. In this situation, the total lengths of $\overline{DE}$ and $\overline{EF}$ are less than the length of $\overline{FD}$; there is no way to arrange $\overline{DE}$ and $\overline{EF}$ so that they meet. If they do not meet, there is no arrangement of three non-collinear vertices of a triangle; therefore, a triangle cannot be formed. I would change $\overline{EF}$ to 9 cm instead of 5 cm so that the three sides would form a triangle.

Topic C

Slicing Solids

7.G.A.3

Focus Standard:	7.G.A.3	Describe the two-dimensional figures that result from slicing three-dimensional figures, as in plane sections of right rectangular prisms and right rectangular pyramids.
Instructional Days:	4	
Lesson 16:	Slicing a Right Rectangular Prism with a Plane (P)[1]	
Lesson 17:	Slicing a Right Rectangular Pyramid with a Plane (S)	
Lesson 18:	Slicing on an Angle (P)	
Lesson 19:	Understanding Three-Dimensional Figures (P)	

In Topic C, students begin exploring cross sections, or slices, of three-dimensional shapes and examining the two-dimensional results of different kinds of slices. In Lesson 16, students learn what it means to slice a three-dimensional figure with a plane and examine slices made parallel to the base of right rectangular prisms and pyramids. In Lesson 17, students slice the prisms and pyramids with vertical slices so that the plane meets the base in a line segment and contrast these cross sections to ones from the previous lessons. In Lesson 18, students experiment with skewed slices and try to predict how to slice figures to yield particular shapes; for example, they experiment with how to slice a cube in order to get a cross section shaped like a triangle or a pentagon. Finally, in Lesson 19, students study three-dimensional figures created from unit cubes but from the perspective of horizontal whole-unit slices. Students learn that the slices can be used to determine the number of cubes in the figure.

[1]Lesson Structure Key: **P**-Problem Set Lesson, **M**-Modeling Cycle Lesson, **E**-Exploration Lesson, **S**-Socratic Lesson

Lesson 16: Slicing a Right Rectangular Prism with a Plane

Student Outcomes

- Students describe rectangular regions that result from slicing a right rectangular prism by a plane perpendicular to one of the faces.

Lesson Notes

Students examine the cross sections of solid figures in the next four lessons. In Lessons 16 and 17, students examine slices made parallel or perpendicular to a face of a solid before moving to angled slices in Lesson 18. To help students visualize slices, provide them with the right rectangular prism nets included after Lesson 27 (and later, the right rectangular pyramid nets) to build and refer to as they complete the lesson.

Classwork

Discussion (8 minutes)

Provide context to the concept of taking slices of a solid by discussing what comes to mind when we think of *taking a slice* of something.

- In our next topic, we examine the slices of solid figures. What do you think of when you hear the word *slice*?
  - *Answers will vary. Some examples include a slice of cake, a slice of pizza, a slice of bread, and "carrot coins."*
- We want to make sure everyone thinks of *slice* in the same way. Let's begin to narrow our idea of *slice* by imagining that the actual cut of a slice can be done in a single motion (unlike the cuts that a wedge-shaped slice from a cylindrical cake would need or a cut that is jagged in any way).
- Perhaps you have been in a deli or a grocery store where cured meat and cheese are often sold in slices. These are examples of a slice made by a single-motion cut. A slice of bread from a loaf of bread is another example of such a slice.
- Can you think of a non-food related example that models the concept of a slice?
  - *Answers will vary. Some examples include a card from a deck of playing cards, a quarter from a roll of quarters, or coasters from a stack of coasters.*
- We must further distinguish whether a slice is the physical piece that has been cut (e.g., a single "carrot coin") or if it is the resulting surface from the cut (i.e., the region left on the carrot by the cut). Consider demonstrating the difference with a real carrot.
- We will answer this question after a discussion about the plane.

Remind students what a plane is and how it relates to a slice.

- Recall that a plane is a building block of geometry that does not have a definition (as it is an undefinable term); rather, we know what a representation of it looks like. How would you describe a representation of a plane?
  - *A representation of a plane is a flat surface, one that extends without edges; it can be thought of as a large sheet of paper.*

- Two planes are parallel if they do not meet (Figure 1).
- Two planes are perpendicular if one plane contains a line that is perpendicular to the other plane (Figure 2).
- Consider a right rectangular prism. Any two opposite faces of the right rectangular prism are parallel; any two adjacent faces are perpendicular. Model parallel and perpendicular faces of rectangular prisms with the walls of a classroom or the surfaces of a tissue box.

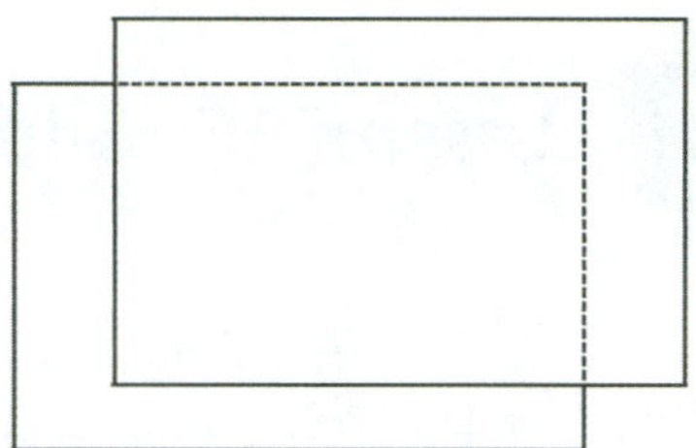

Figure 1. Parallel Planes

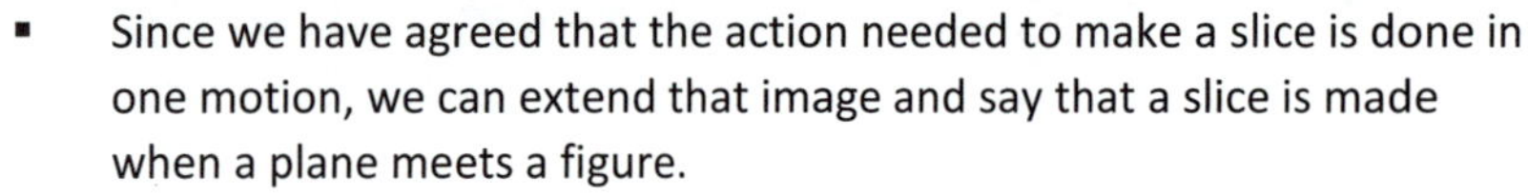

- Since we have agreed that the action needed to make a slice is done in one motion, we can extend that image and say that a slice is made when a plane meets a figure.
- The *plane section* of the figure, with respect to the plane, consists of all points where the plane meets the figure. We also call the plane section the *slice*. We call the plane the *slicing plane*.

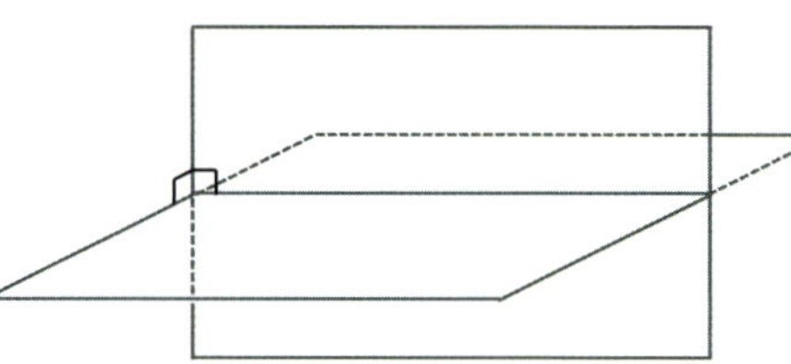

Figure 2. Perpendicular Planes

Example 1 (5 minutes)

Scaffolding:

All exercises in this lesson are made more accessible when physical models are used (e.g., a ball for the sphere, a wood block, or a prism constructed from the net provided after Lesson 27 for the others). Consider showing these or providing them for students to handle as they determine what the slice would look like.

MP.7

Example 1

Consider a ball B. Figure 3 shows one possible slice of B.

a. **What figure does the slicing plane form? Students may choose their method of representation of the slice (e.g., drawing a 2D sketch, a 3D sketch, or describing the slice in words).**

A circle (or disc)

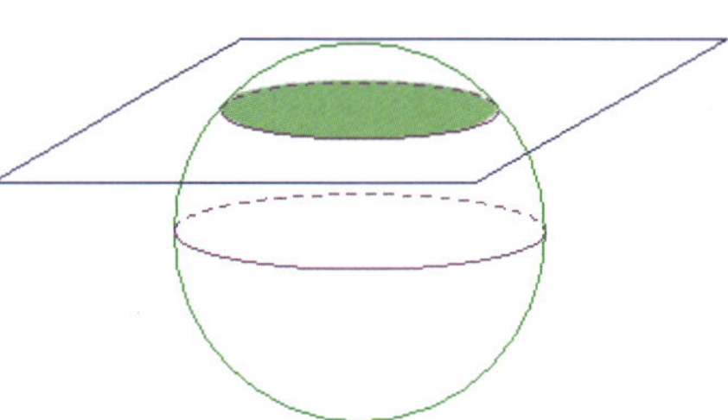

Figure 3. A Slice of Ball B

b. **Will all slices that pass through B be the same size? Explain your reasoning.**

No, different slices can result in circles of different sizes; it will depend on where the slicing plane meets the ball.

c. **How will the plane have to meet the ball so that the plane section consists of just one point?**

If you picture the ball and the plane as distinct but being brought toward each other, the plane section of just one point occurs when the plane just makes contact with the ball.

Examples 2 and 3 highlight slices made to a rectangular prism that make the plane section parallel to a face and perpendicular to a face, respectively. Angled slices are explored in another lesson. Point out that planar regions, such as the rectangular regions in the figures below, are parallel if the planes containing them are parallel.

Consider taking time here to build the rectangular prisms from the nets located at the end of this module. Ask students to imagine different slices that could be made perpendicular or parallel to a face and to sketch what these slices might look like.

Example 2 (5 minutes)

Example 2

The right rectangular prism in Figure 4 has been sliced with a plane parallel to face $ABCD$. The resulting slice is a rectangular region that is identical to the parallel face.

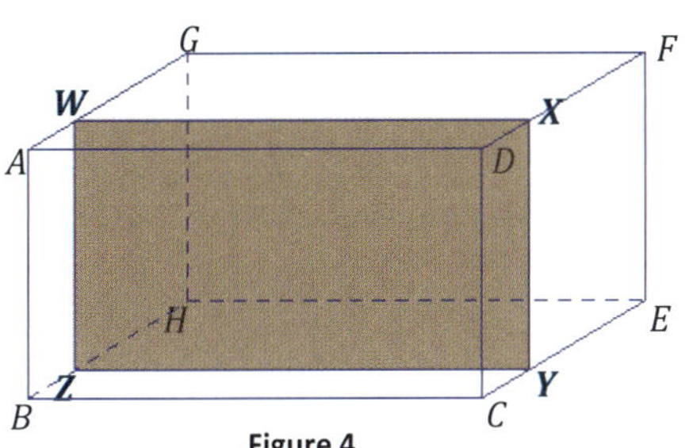

Figure 4

a. Label the vertices of the rectangular region defined by the slice as $WXYZ$.

b. To which other face is the slice parallel and identical?

The slice is parallel and identical to the face $EFGH$.

c. Based on what you know about right rectangular prisms, which faces must the slice be perpendicular to?

Since the slice is parallel to two faces, it will be perpendicular to whichever sides those faces are perpendicular to. Therefore, the slice is perpendicular to faces $ABHG$, $CDFE$, $BCEH$, and $ADFG$.

Exercise 1 (5 minutes)

Exercise 1

Discuss the following questions with your group.

MP.3

1. The right rectangular prism in Figure 5 has been sliced with a plane parallel to face $LMON$.

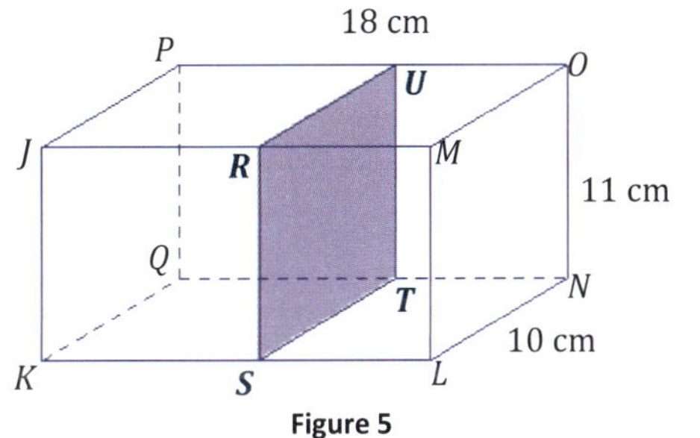

Figure 5

a. Label the vertices of the rectangle defined by the slice as $RSTU$.

b. What are the dimensions of the slice?

$10 \text{ cm} \times 11 \text{ cm}$

c. Based on what you know about right rectangular prisms, which faces must the slice be perpendicular to?

Since the slice is parallel to two faces, it will be perpendicular to whichever sides those faces are perpendicular to. Therefore, the slice is perpendicular to faces $JKLM$, $JMOP$, $NOPQ$, and $KLNQ$.

Example 3 (5 minutes)

Example 3

The right rectangular prism in Figure 6 has been sliced with a plane perpendicular to $BCEH$. The resulting slice is a rectangular region with a height equal to the height of the prism.

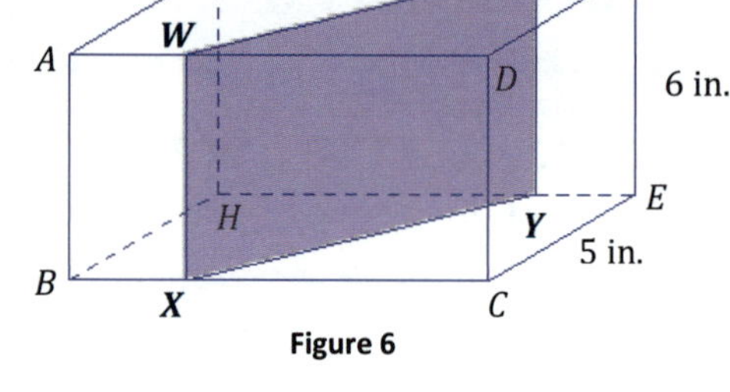

Figure 6

a. **Label the vertices of the rectangle defined by the slice as $WXYZ$.**

b. **To which other face is the slice perpendicular?**

   The slice is perpendicular to the face $ADFG$.

c. **What is the height of rectangle $WXYZ$?**

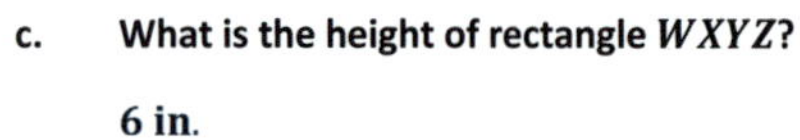

   6 in.

d. **Joey looks at $WXYZ$ and thinks that the slice may be a parallelogram that is not a rectangle. Based on what is known about how the slice is made, can he be right? Justify your reasoning.**

   The slice was made perpendicular to face $BCEH$. Then we know that the angles in the slice, $\angle X$ and $\angle Y$, formed by the slicing plane and face $BCEH$, are right angles. If we focus on $\angle X$ of the slice, since it is a right angle, we know that $\overline{WX}$ must be perpendicular to face $BCEH$. $\overline{WX}$ lies in face $ABCD$, which is perpendicular to both $BCEH$ and to $ADFG$, so $\overline{WX}$ is perpendicular to $ABCD$. This means that $\overline{WX}$ must also be perpendicular to $\overline{WZ}$. A similar argument can be made for $\angle Y$ of the slice, making all four angles of $WXYZ$ right angles and making $WXYZ$ a rectangle.

Exercises 2–6 (10 minutes)

Exercises 2–6

In the following exercises, the points at which a slicing plane meets the edges of the right rectangular prism have been marked. Each slice is either parallel or perpendicular to a face of the prism. Use a straightedge to join the points to outline the rectangular region defined by the slice, and shade in the rectangular slice.

2. **A slice parallel to a face**

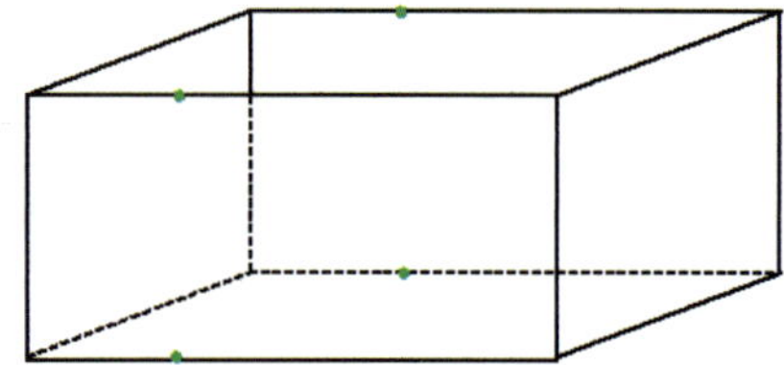

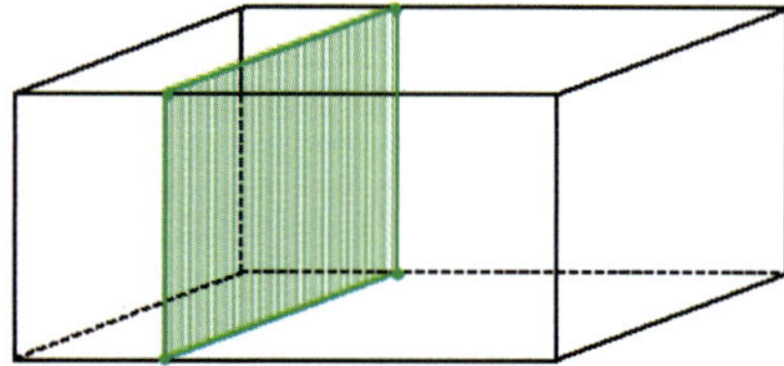

3. **A slice perpendicular to a face**

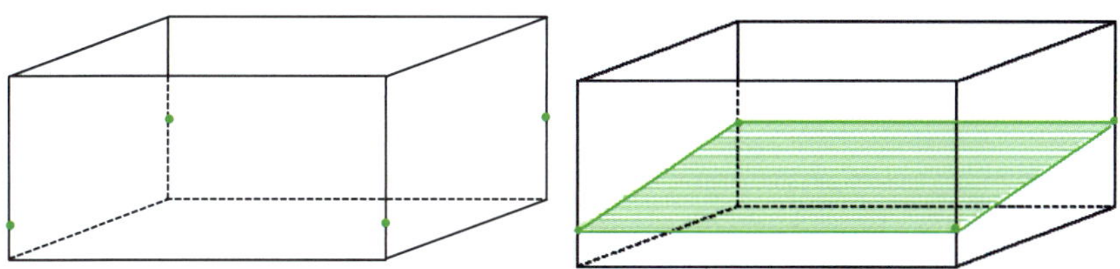

4. **A slice perpendicular to a face**

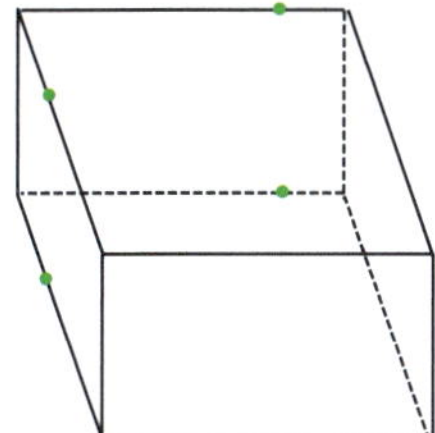

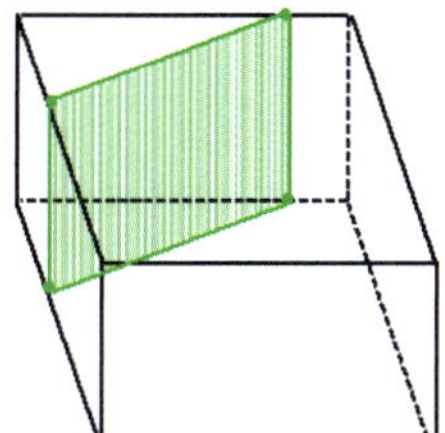

In Exercises 5–6, the dimensions of the prisms have been provided. Use the dimensions to sketch the slice from each prism, and provide the dimensions of each slice.

5. **A slice parallel to a face**

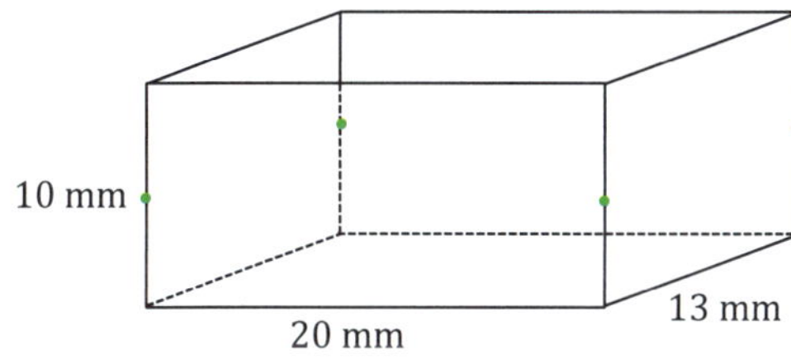

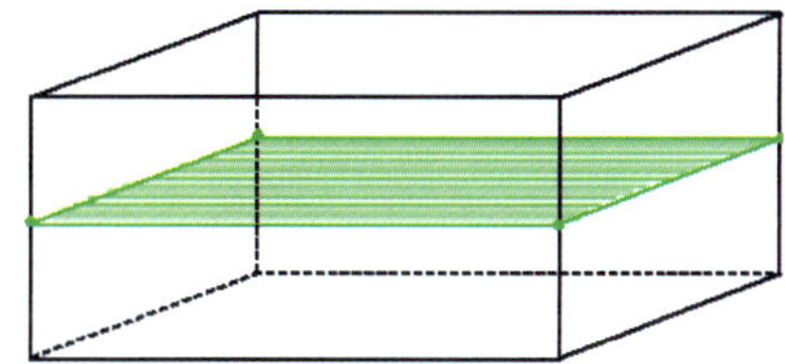

6. **A slice perpendicular to a face**

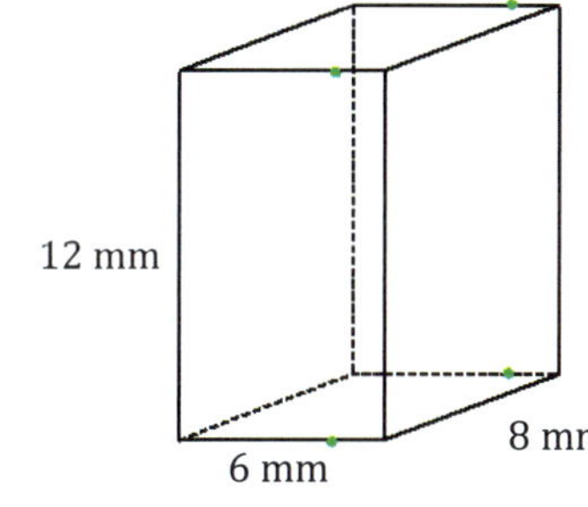

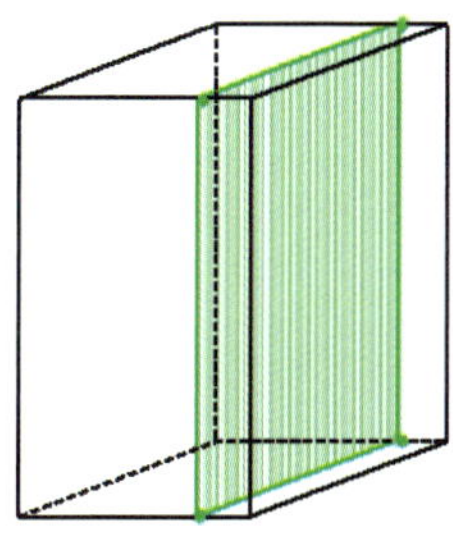

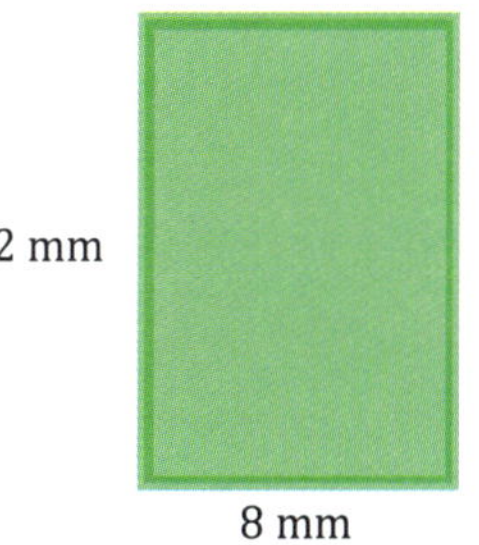

Closing (2 minutes)

> **Lesson Summary**
>
> - **A slice, also known as a plane section, consists of all the points where the plane meets the figure.**
> - **A slice made parallel to a face in a right rectangular prism will be parallel and identical to the face.**
> - **A slice made perpendicular to a face in a right rectangular prism will be a rectangular region with a height equal to the height of the prism.**

Exit Ticket (5 minutes)

Name ______________________________ Date ______________

Lesson 16: Slicing a Right Rectangular Prism with a Plane

Exit Ticket

In the following figures, use a straightedge to join the points where a slicing plane meets with a right rectangular prism to outline the slice.

i. Label the vertices of the rectangular slice $WXYZ$.

ii. State any known dimensions of the slice.

iii. Describe two relationships slice $WXYZ$ has in relation to faces of the right rectangular prism.

1.

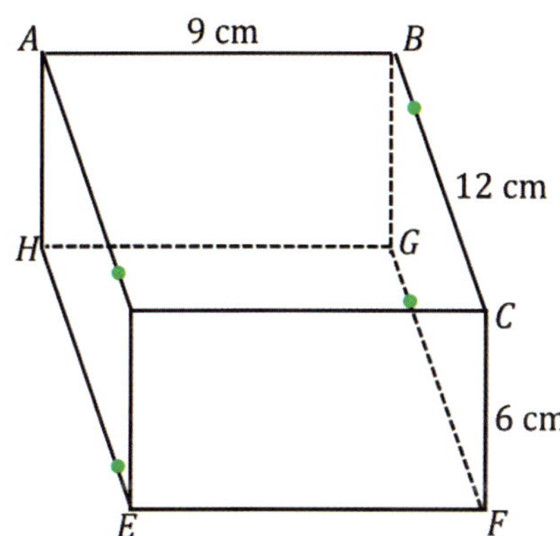

2.

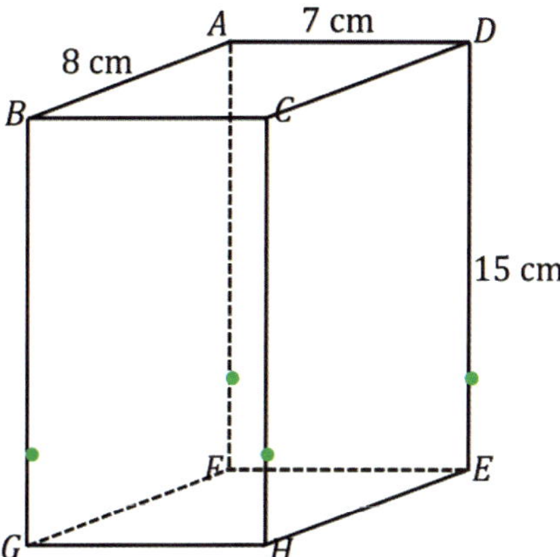

Exit Ticket Sample Solutions

In the following figures, use a straightedge to join the points where a slicing plane meets with a right rectangular prism to outline the slice.

i. **Label the vertices of the rectangular slice $WXYZ$.**

ii. **State any known dimensions of the slice.**

iii. **Describe two relationships slice $WXYZ$ has in relation to faces of the right rectangular prism.**

1.

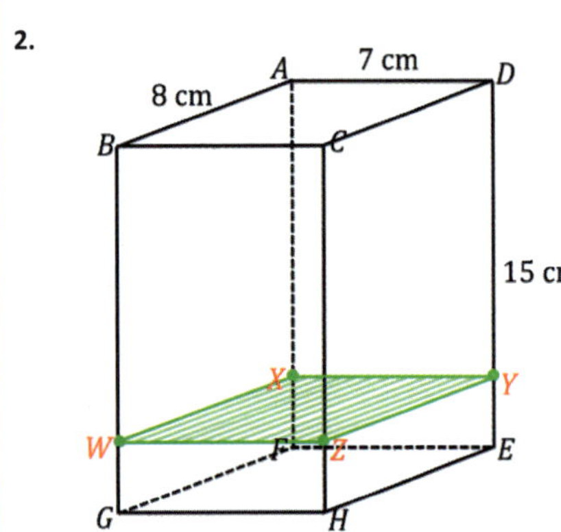

Sides $\overline{WZ}$ and $\overline{XY}$ are 6 cm in length. Slice $WXYZ$ is perpendicular to faces $ABCD$ and $EFGH$.

2.

A 7 cm D
8 cm
B C
15 cm
X Y
W F Z E
G H

Sides $\overline{WZ}$ and $\overline{XY}$ are 7 cm in length. Sides $\overline{WX}$ and $\overline{ZY}$ are 8 cm in length. Slice $WXYZ$ is parallel to faces $ABCD$ and $EFGH$ and perpendicular to faces $CDEH$, $ADEF$, $ABGF$, and $BCHG$.

Note: Students are only required to state two of the relationships the slice has with the faces of the prism.

EUREKA MATH

G7-M6-TE-B6-1.3.0-10.2015

Problem Set Sample Solutions

Note: Students have not yet studied the Pythagorean theorem; thus, the answers provided for the missing length of each line segment are only possible answers based on rough approximations.

A right rectangular prism is shown along with line segments that lie in a face. For each line segment, draw and give the approximate dimensions of the slice that results when the slicing plane contains the given line segment and is perpendicular to the face that contains the line segment.

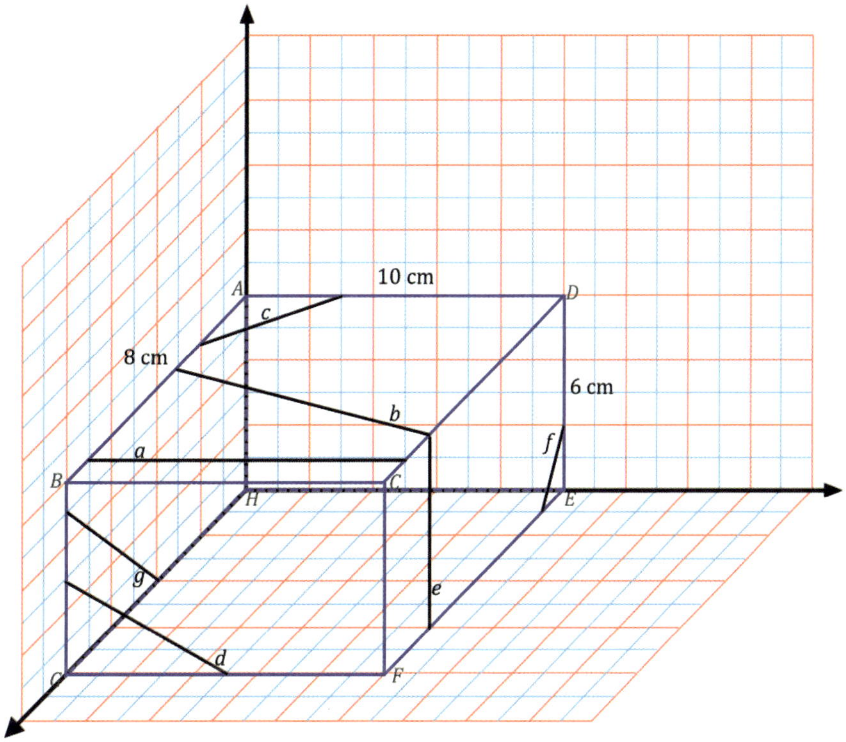

10 cm
8 cm
6 cm
A B C D E F G H
a b c d e f g

a.

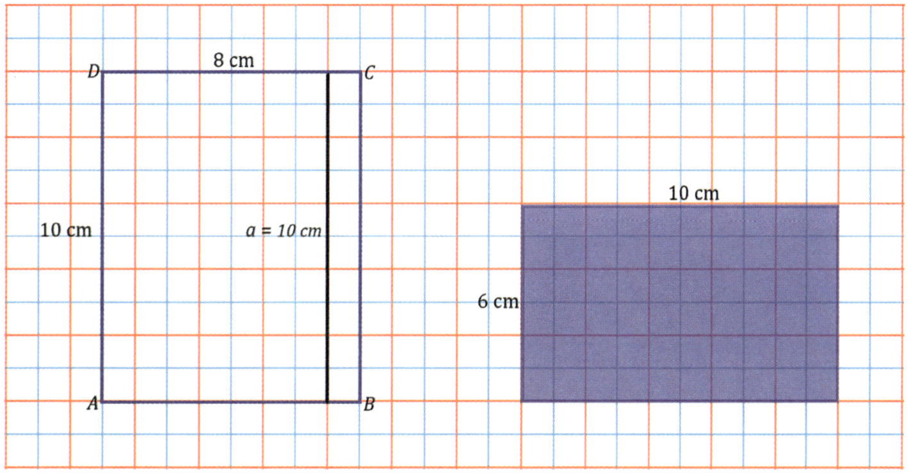

8 cm
D C
10 cm
$a = 10$ cm
A B

10 cm
6 cm

b.

8 cm
D
C
10 cm
$b \approx 11$ cm
11 cm
6 cm
A
B

c.

8 cm
D
C
10 cm
3.5 cm
6 cm
$c \approx 3.5$ cm
A
B

d.

B
10 cm
C
6 cm
6 cm
$d \approx 6$ cm
8 cm
G
F

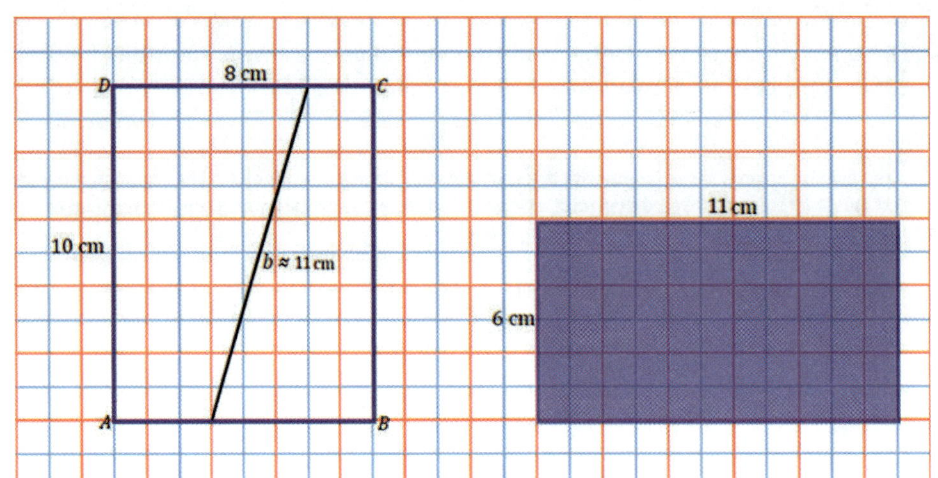

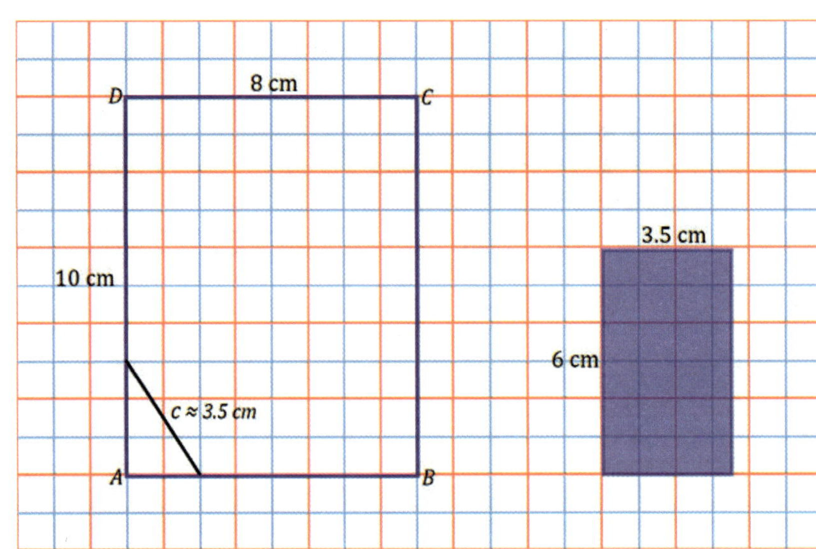

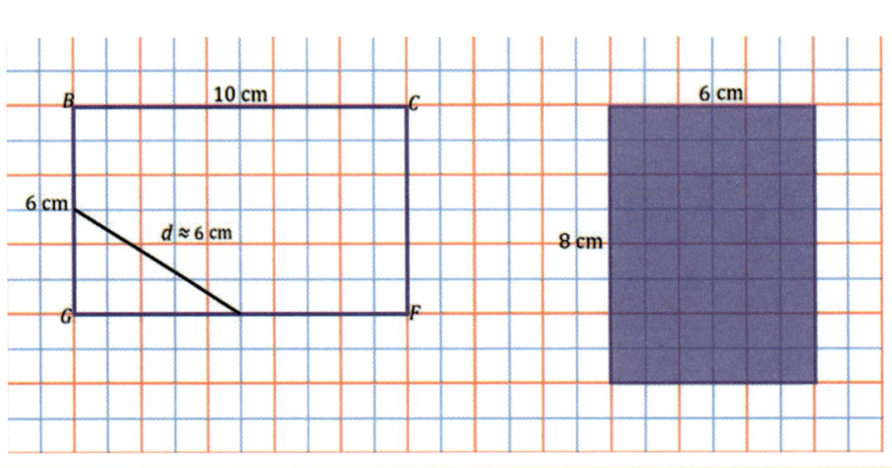

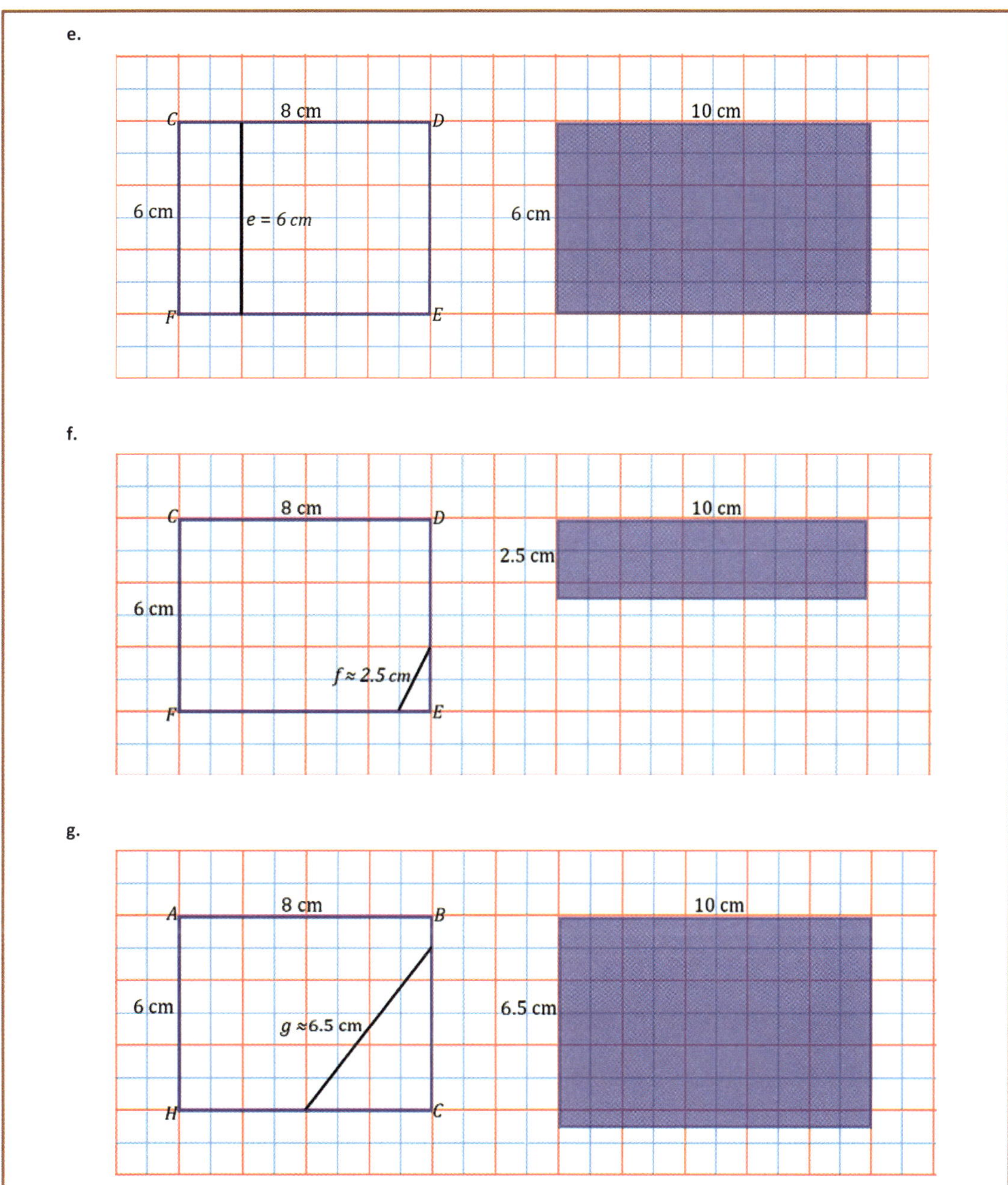
e.
8 cm
C
D
6 cm
e = 6 cm
F
E
10 cm
6 cm
f.
8 cm
C
D
6 cm
f ≈ 2.5 cm
F
E
10 cm
2.5 cm
g.
8 cm
A
B
6 cm
g ≈6.5 cm
H
C
10 cm
6.5 cm

Lesson 17: Slicing a Right Rectangular Pyramid with a Plane

Student Outcomes

- Students describe polygonal regions that result from slicing a right rectangular pyramid by a plane perpendicular to the base and by another plane parallel to the base.

Lesson Notes

In contrast to Lesson 16, Lesson 17 studies slices made to a right rectangular pyramid rather than a right rectangular prism. However, the slices are still made perpendicular and parallel to the base. Students have had some experience with pyramids in Module 3 (Lesson 22), but it was in the context of surface area. This lesson gives students the opportunity to build pyramids from nets as they study the formal definition of pyramid. (Nets for the pyramids are provided at the end of the module.)

Classwork

Opening (10 minutes)

Have students build right rectangular pyramids from the provided nets in their groups. Once the pyramids are built, lead a discussion that elicits a student description of what a right rectangular pyramid is.

- How would you describe a pyramid?
  - *Responses will vary. Students may remark on the existence of a rectangular base, that the sides are isosceles triangles, and that the edges of the isosceles triangles all meet at a vertex.*

Then, introduce the formal definition of a *rectangular pyramid,* and use the series of images that follow to make sense of the definition.

Opening

RECTANGULAR PYRAMID: Given a rectangular region B in a plane E, and a point V not in E, the *rectangular pyramid with base B and vertex V* is the collection of all segments VP for any point P in B. It can be shown that the planar region defined by a side of the base B and the vertex V is a triangular region called a *lateral face*.

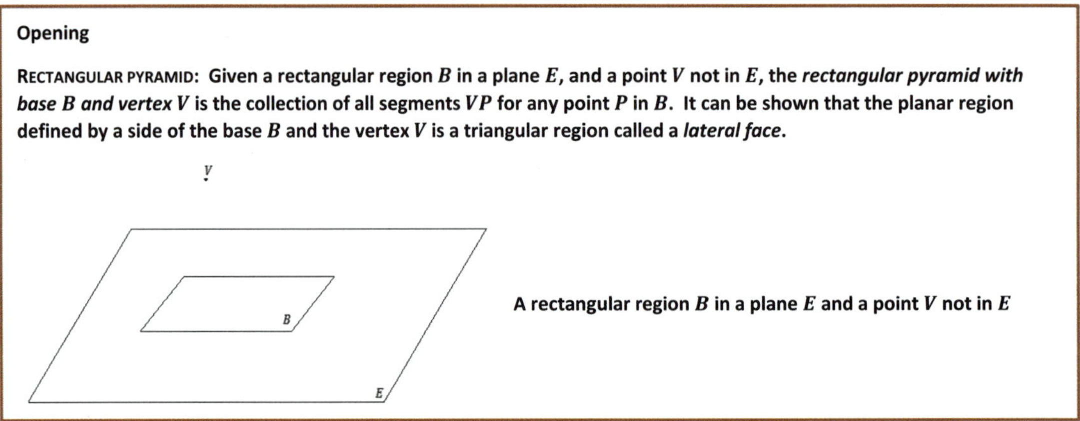

A rectangular region B in a plane E and a point V not in E

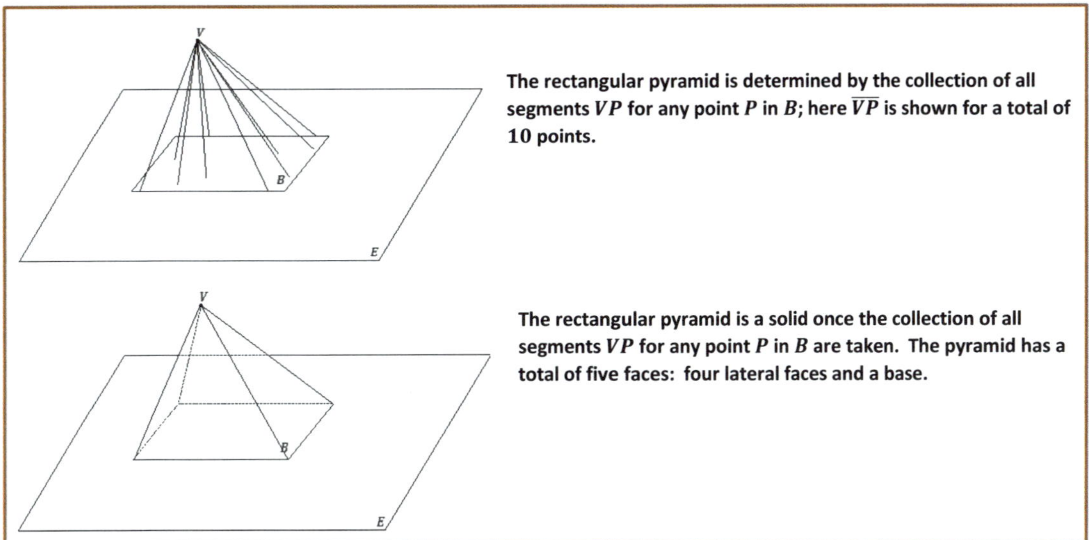

Students should understand that a rectangular pyramid is a solid figure and not a hollow shell like the pyramids they built from the nets, so the nets are not a perfect model in this sense. The collection of all segments renders the pyramid to be solid.

If the vertex lies on the line perpendicular to the base at its center (i.e., the intersection of the rectangle's diagonals), the pyramid is called a *right rectangular pyramid*. The example of the rectangular pyramid above is not a right rectangular pyramid, as evidenced in this image. The perpendicular from V does not meet at the intersection of the diagonals of the rectangular base B.

The following image is an example of a right rectangular pyramid. The opposite lateral faces are identical isosceles triangles.

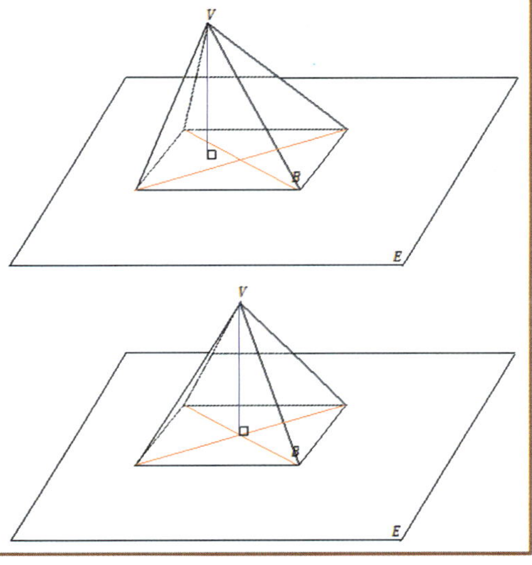

Visualizing slices made to a pyramid can be challenging. To build up to the task of taking slices of a pyramid, have students take time to sketch a pyramid from different perspectives. In Example 1, students sketch one of the models they built from any vantage point. In Example 2, students sketch a pyramid from particular vantage points.

Example 1 (5 minutes)

Example 1

Use the models you built to assist in a sketch of a pyramid. Though you are sketching from a model that is opaque, use dotted lines to represent the edges that cannot be seen from your perspective.

Sketches will vary; emphasize the distinction between the pyramids by asking how each must be drawn.

Students may struggle with this example; remind them that attempting these sketches is not a test of artistic ability but rather an exercise in becoming more familiar with the structure of a pyramid. They are working toward visualizing how a plane slices a rectangular pyramid perpendicular and parallel to its base.

Example 2 (5 minutes)

Example 2

Sketch a right rectangular pyramid from three vantage points: (1) from directly over the vertex, (2) from facing straight on to a lateral face, and (3) from the bottom of the pyramid. Explain how each drawing shows each view of the pyramid.

Possible sketches:

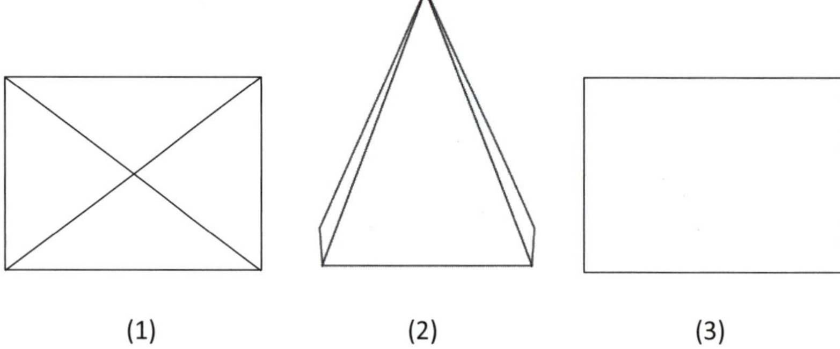

1. *From directly overhead, all four lateral faces are visible.*
2. *From facing a lateral face, the one entire lateral face is visible, as well as a bit of the adjacent lateral faces. If the pyramid were transparent, I would be able to see the entire base.*
3. *From the bottom, only the rectangular base is visible.*

Example 3 (6 minutes)

Example 3

Assume the following figure is a top-down view of a rectangular pyramid. Make a reasonable sketch of any two adjacent lateral faces. What measurements must be the same between the two lateral faces? Mark the equal measurements. Justify your reasoning for your choice of equal measurements.

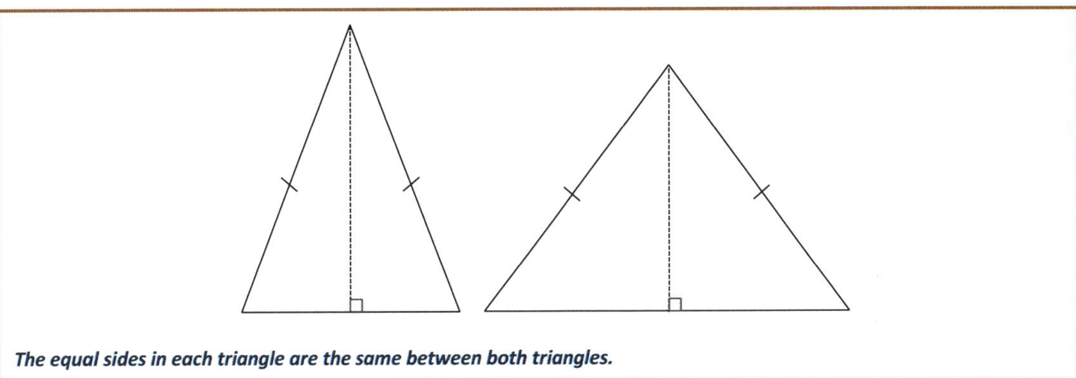

The equal sides in each triangle are the same between both triangles.

Students may think that the heights of the triangles are equal in length, when in fact they are not, unless the base is a square. The triangle with the shorter base has a height greater than that of the triangle with the longer base. An easy way of making this point is by looking at a right rectangular pyramid with a rectangular base of exaggerated dimensions: a very long length in contrast to a very short width. Though students do not yet have the Pythagorean theorem at their disposal to help them quantify the difference in heights of the lateral faces, an image should be sufficiently persuasive.

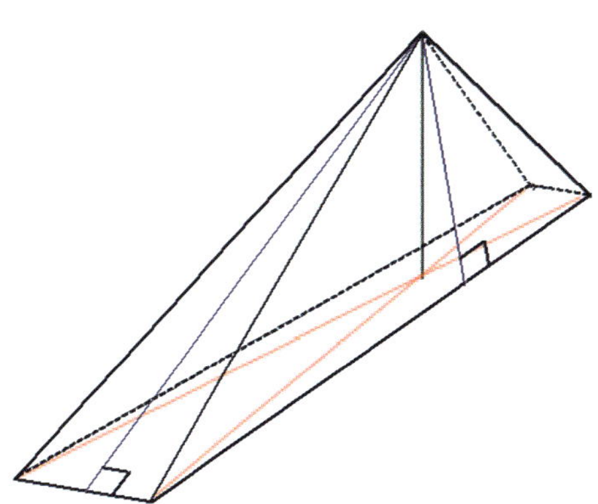

Example 4 (6 minutes)

Remind students of the types of slices taken in Lesson 15: slices parallel to a face and slices perpendicular to a face. This lesson examines slices made parallel and perpendicular to the rectangular base of the pyramid.

> *Scaffolding:*
>
> As with the last lesson, understanding the slices is made easier when students are able to view and handle physical models. Consider using the figures constructed from the nets in the Opening throughout these exercises.

Example 4

a. **A slicing plane passes through segment a parallel to base B of the right rectangular pyramid below. Sketch what the slice will look like into the figure. Then sketch the resulting slice as a two-dimensional figure. Students may choose how to represent the slice (e.g., drawing a 2D or 3D sketch or describing the slice in words).**

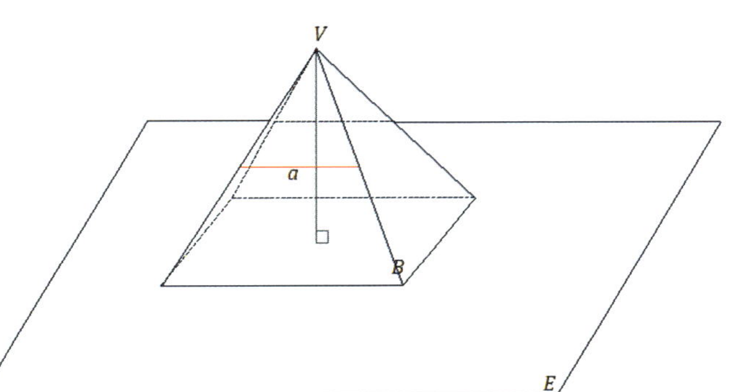

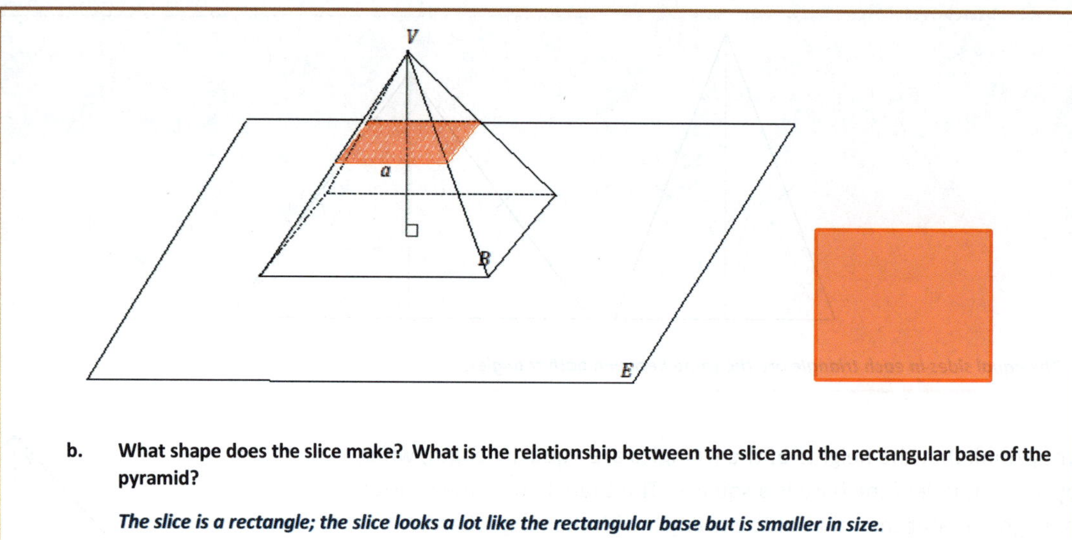

b. **What shape does the slice make? What is the relationship between the slice and the rectangular base of the pyramid?**

The slice is a rectangle; the slice looks a lot like the rectangular base but is smaller in size.

Students study similar figures in Grade 8, so they do not have the means to determine that a slice made parallel to the base is in fact a rectangle similar to the rectangular base. Students have, however, studied scale drawings in Module 4.

Tell students that a slice made parallel to the base of a right rectangular pyramid is a scale drawing, a reduction, of the base.

- How can the scale factor be determined?
  - *The scale factor can be calculated by dividing the side length of the slice by the corresponding side length of the base.*

Example 5 (7 minutes)

Example 5

A slice is to be made along segment a perpendicular to base B of the right rectangular pyramid below.

a. **Which of the following figures shows the correct slice? Justify why each of the following figures is or is not a correct diagram of the slice.**

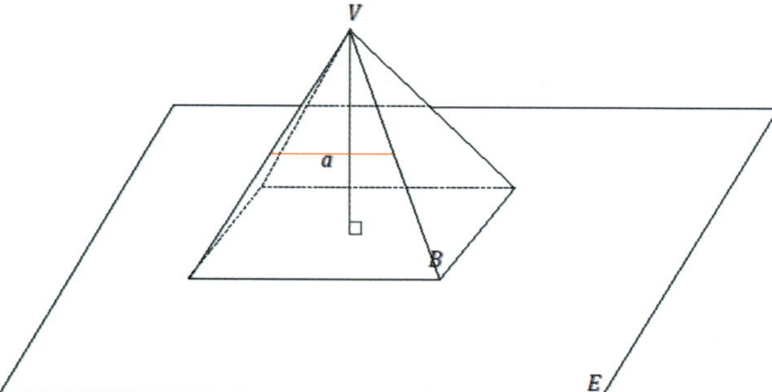

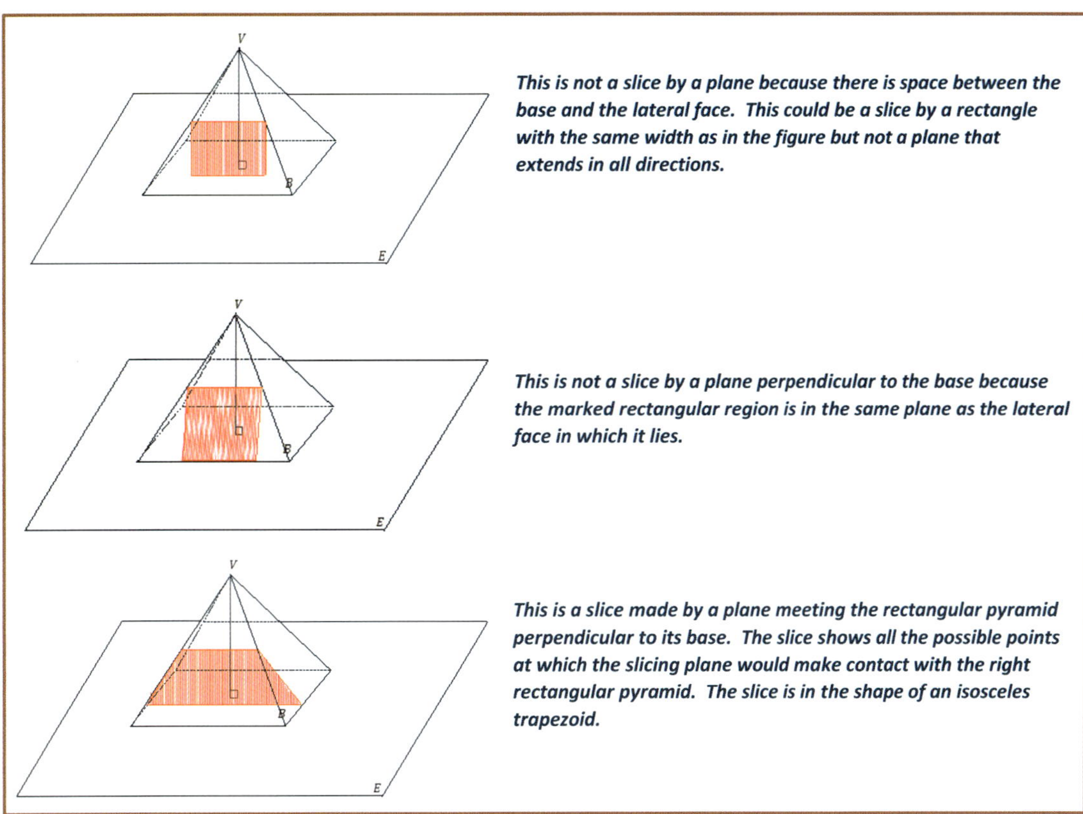

This is not a slice by a plane because there is space between the base and the lateral face. This could be a slice by a rectangle with the same width as in the figure but not a plane that extends in all directions.

This is not a slice by a plane perpendicular to the base because the marked rectangular region is in the same plane as the lateral face in which it lies.

This is a slice made by a plane meeting the rectangular pyramid perpendicular to its base. The slice shows all the possible points at which the slicing plane would make contact with the right rectangular pyramid. The slice is in the shape of an isosceles trapezoid.

It may help students to visualize the third figure by taking one of the model pyramids and tracing the outline of the slice. Ask students to visualize cutting along the outline and looking at what remains of the cut-open lateral face. For any slice made perpendicular to the base, ask students to visualize a plane (or, say, a piece of paper) moving perpendicularly toward the base through a marked segment on a lateral face. Ask them to think about where all the points of the paper would meet the pyramid.

b. A slice is taken through the vertex of the pyramid perpendicular to the base. Sketch what the slice will look like into the figure. Then, sketch the resulting slice itself as a two-dimensional figure.

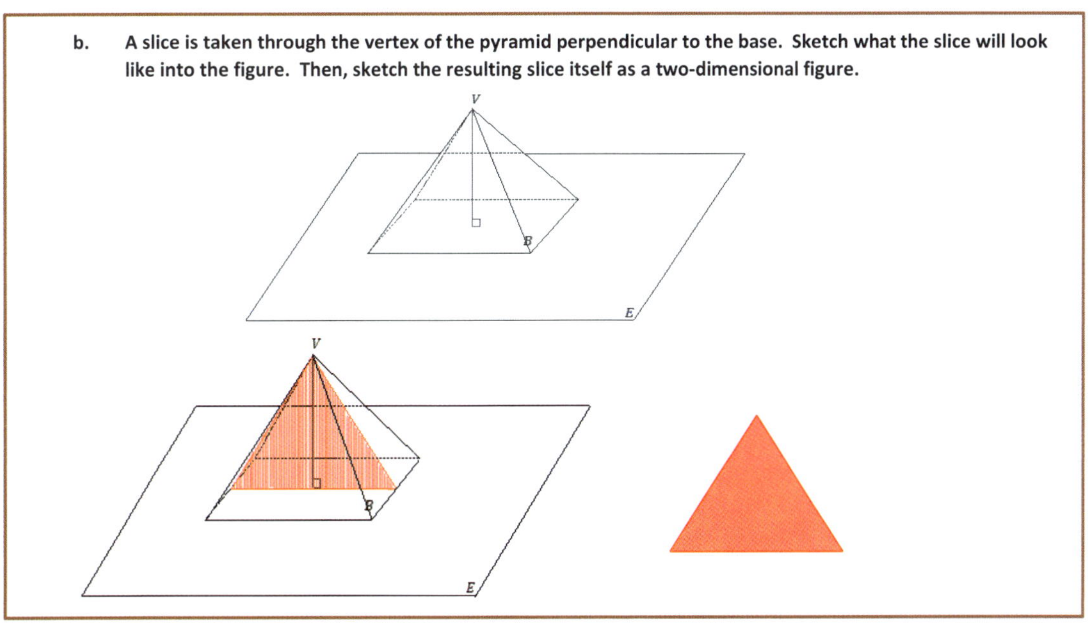

Closing (1 minute)

- How is a rectangular pyramid different from a right rectangular pyramid?
  - *The vertex of a right rectangular pyramid lies on the line perpendicular to the base at its center (the intersection of the rectangle base's diagonals); a pyramid that is not a right rectangular pyramid will have a vertex that is not on the line perpendicular to the base at its center.*

Students should visualize slices made perpendicular to the base of a pyramid by imagining a piece of paper passing through a given segment on a lateral face perpendicularly toward the base. Consider the outline the slice would make on the faces of the pyramid.

Slices made parallel to the base of a right rectangular pyramid are scale drawings (i.e., reductions) of the rectangular base of the pyramid.

> **Lesson Summary**
>
> - **A rectangular pyramid differs from a right rectangular pyramid because the vertex of a right rectangular pyramid lies on the line perpendicular to the base at its center whereas a pyramid that is not a right rectangular pyramid will have a vertex that is not on the line perpendicular to the base at its center.**
> - **Slices made parallel to the base of a right rectangular pyramid are scale drawings of the rectangular base of the pyramid.**

Exit Ticket (5 minutes)

Name ________________________________ Date ________________

Lesson 17: Slicing a Right Rectangular Pyramid with a Plane

Exit Ticket

Two copies of the same right rectangular pyramid are shown below. Draw in the slice along segment c perpendicular to the base and the slice along segment c parallel to the base. Then, sketch the resulting slices as two-dimensional figures.

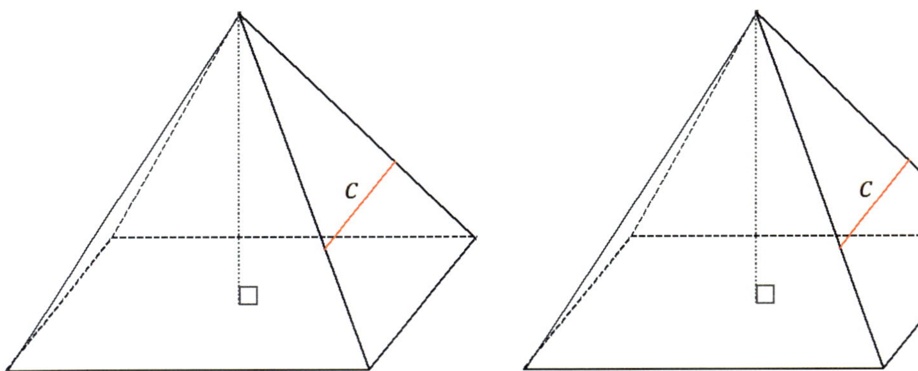

Slice Perpendicular to Base

Slice Parallel to Base

Exit Ticket Sample Solutions

Two copies of the same right rectangular pyramid are shown below. Draw in the slice along segment c perpendicular to the base and the slice along segment c parallel to the base. Then, sketch the resulting slices as two-dimensional figures.

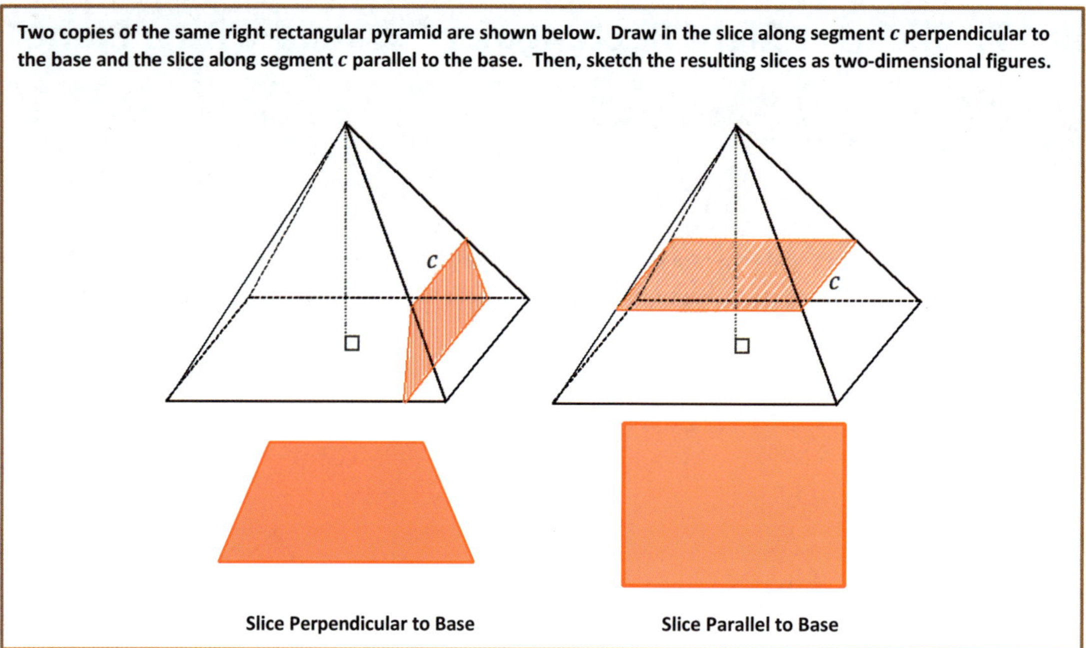

Problem Set Sample Solutions

A side view of a right rectangular pyramid is given. The line segments lie in the lateral faces.

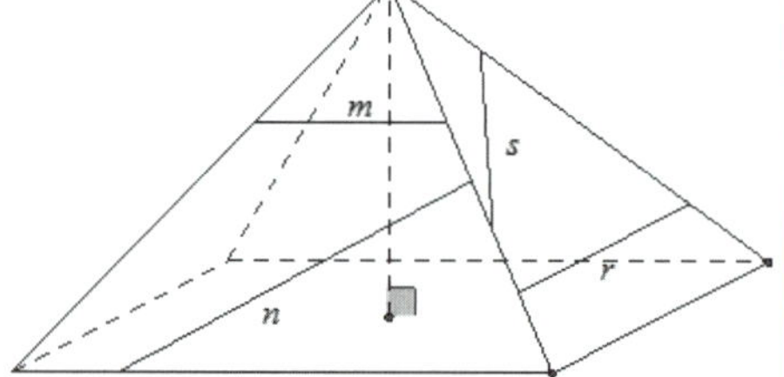

a. For segments n, s, and r, sketch the resulting slice from slicing the right rectangular pyramid with a slicing plane that contains the line segment and is perpendicular to the base.

b. For segment m, sketch the resulting slice from slicing the right rectangular pyramid with a slicing plane that contains the segment and is parallel to the base.

Note: To challenge yourself, you can try drawing the slice into the pyramid.

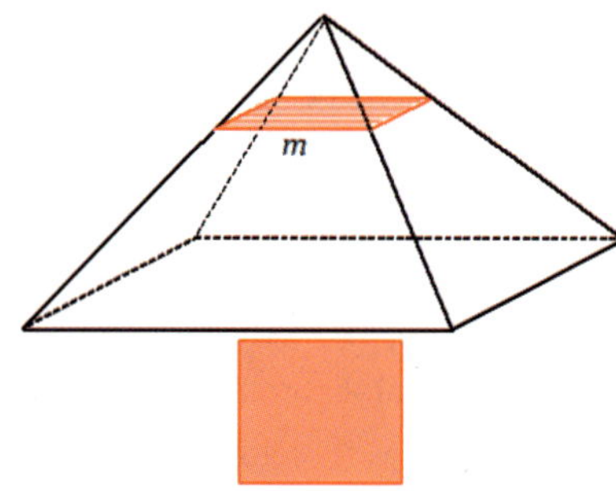

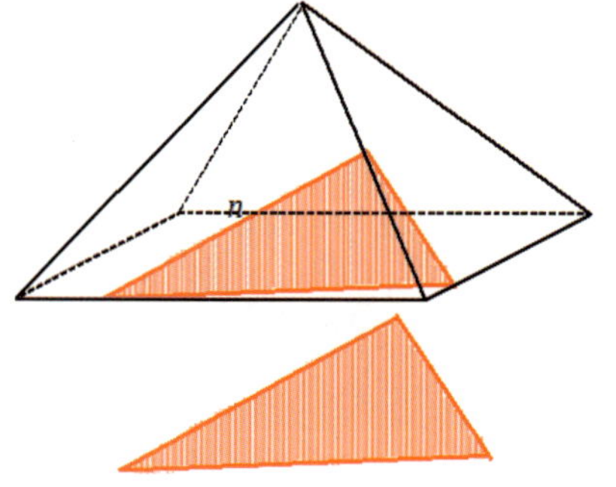

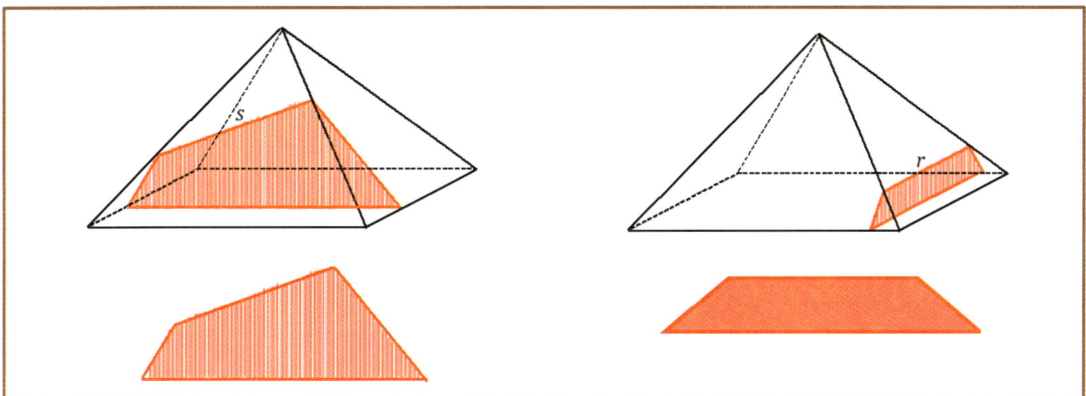

Note that the diagram for the slice made through s is from a perspective different from the one in the original pyramid. From the original perspective, the slice itself would not be visible and would appear as follows:

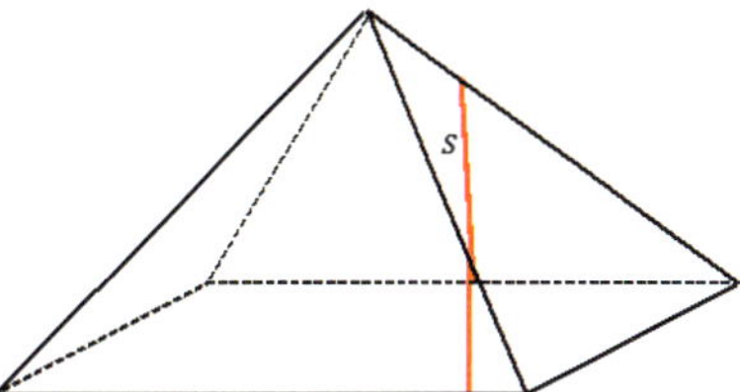

c. **A top view of a right rectangular pyramid is given. The line segments lie in the base face. For each line segment, sketch the slice that results from slicing the right rectangular pyramid with a plane that contains the line segment and is perpendicular to the base.**

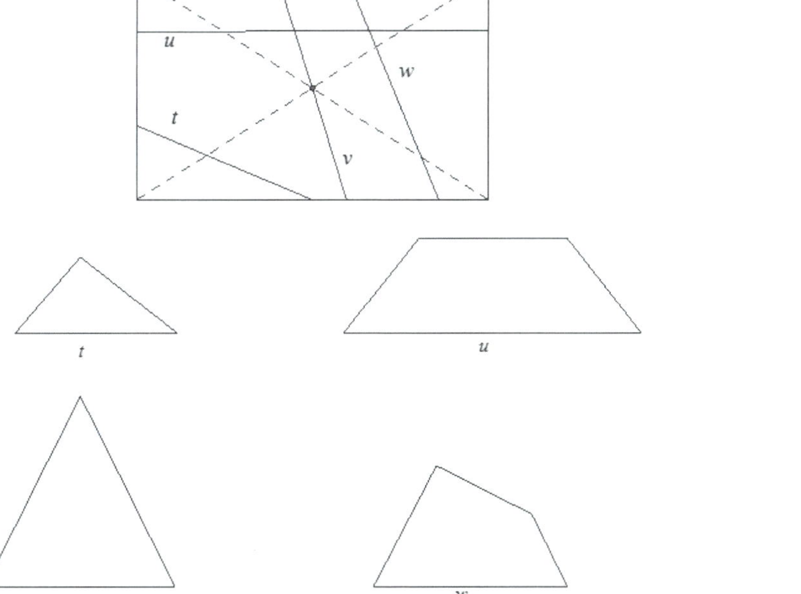

Lesson 18: Slicing on an Angle

Student Outcomes

- Students describe polygonal regions that result from slicing a right rectangular prism or pyramid by a plane that is not necessarily parallel or perpendicular to a base.

Lesson Notes

In Lessons 16 and 17, slices are made parallel or perpendicular to the base and/or faces of a right rectangular prism and to the base of a right rectangular pyramid. In this lesson, students examine the slices resulting from cuts made that are not parallel or perpendicular to a base or face.

Classwork

Discussion (3 minutes)

> *Scaffolding:*
>
> As with the last lesson, understanding the slices is made easier when students are able to view and handle physical models. Consider using the figures constructed from the nets at the end of the module after Lesson 27.

Lead students through a discussion regarding the change in the kinds of slices made in Lessons 16 and 17 versus those being made in Lesson 18. It may be useful to have models of right rectangular prisms, right rectangular pyramids, and cubes available throughout the classroom.

- What did the slices made in Lessons 16 and 17 have in common?
  - *The slices in both lessons were made either parallel or perpendicular to a base (or, in the case of a right rectangular prism, a face) of a right rectangular prism and right rectangular pyramid.*

MP.7

- In this lesson, we examine slices made at an angle, or, in other words, slices that are neither parallel nor perpendicular to any face of the prism or pyramid. Which of the following could be a slice of a right rectangular prism? Of a cube? Of a right rectangular pyramid? Students may choose their method of how to represent the slice (e.g., drawing a 2D or 3D sketch or describing the slice in words).

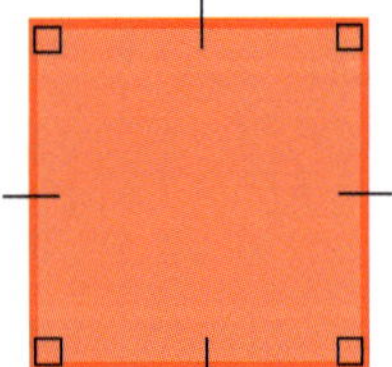   

  - *The square-shaped slice can be made from a slice parallel to the base of a right rectangular prism, cube, and right rectangular pyramid with a square base. The triangle-shaped slice and isosceles trapezoid-shaped slice can be made with a slice made perpendicular to the base of a right rectangular pyramid. A pentagon-shaped slice cannot be made from a slice made parallel or perpendicular to the base of either a right rectangular prism, right rectangular pyramid, or cube.*

Example 1 (7 minutes)

Pose Example 1 as a question to be discussed in small groups. In this problem, students must visualize the different types of triangles that can be sliced from a right rectangular prism.

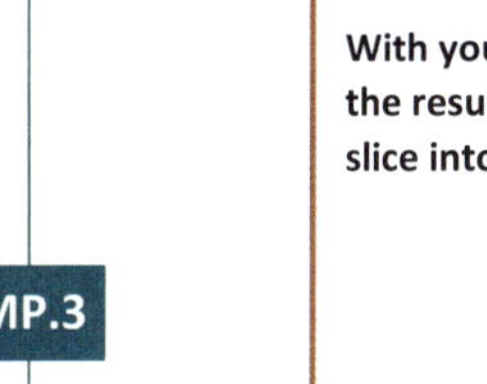

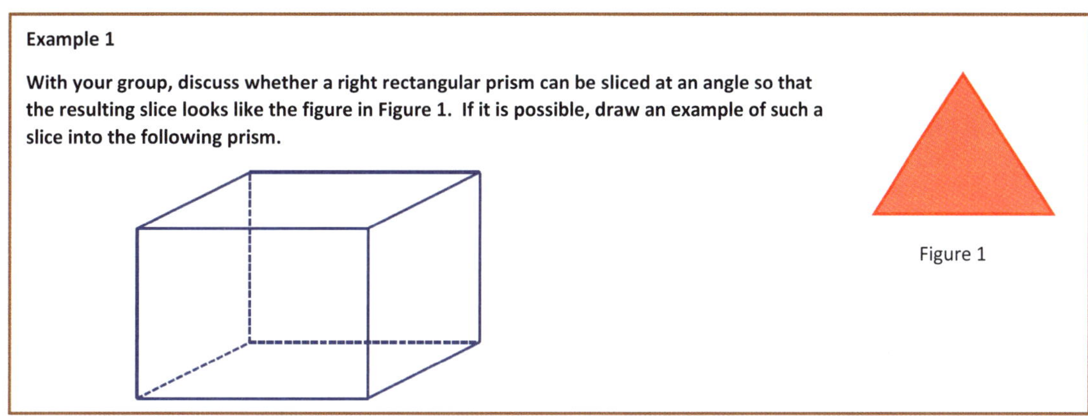

Example 1

With your group, discuss whether a right rectangular prism can be sliced at an angle so that the resulting slice looks like the figure in Figure 1. If it is possible, draw an example of such a slice into the following prism.

Figure 1

Once students have had some time to attempt Example 1, pose the following questions:

- Is it possible to make a triangular slice from this prism? Where would this slice have to be made? Students may choose their method of how to represent the slice (e.g., drawing a 2D or 3D sketch or describing the slice in words).
  - *Yes, it is possible. It can be done by slicing off a corner of the right rectangular prism.*

Here, and elsewhere, some students may be able to visualize this right away, while others may struggle.
Allow some time to see if any one group has a valid answer. If there is a valid answer, use it for the next item; otherwise, share the following triangular slice of a right rectangular prism.

- Here is a slice that results in a triangular region. Does this triangle have three equal sides, two equal sides, or no equal sides?

Encourage students to use a ruler to verify their answers.

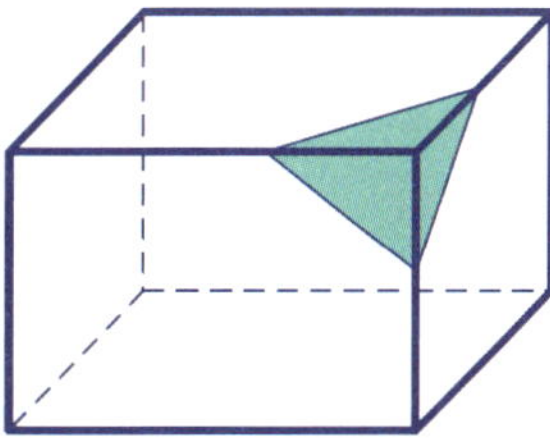

  - *The slice is a triangle with no equal sides.*
- At how many points does the slice meet an edge of the right rectangular prism? What makes these points important with respect to the triangle?
  - *The slice meets an edge of the right rectangular prism at three points, one on each of three edges; these three points are the vertices of the triangle.*
- Find another slice that will create another scalene triangular region. Mark the vertices on the edges of the prism.

Allow students a few moments to determine the slice before moving on.

Exercise 1 (7 minutes)

Students should experiment with their drawings to find the combination of lengths and positions of segments to form the shapes of the slices in parts (a) and (b).

MP.7

Exercise 1

a. **With your group, discuss how to slice a right rectangular prism so that the resulting slice looks like the figure in Figure 2. Justify your reasoning.**

I would use a ruler to measure two segments of equal length on two edges that meet at a common vertex. Then, I would join these two endpoints with a third segment.

Figure 2

b. **With your group, discuss how to slice a right rectangular prism so that the resulting slice looks like the figure in Figure 3. Justify your reasoning.**

I would use a ruler to measure three segments of equal length on three edges that meet at a common vertex.

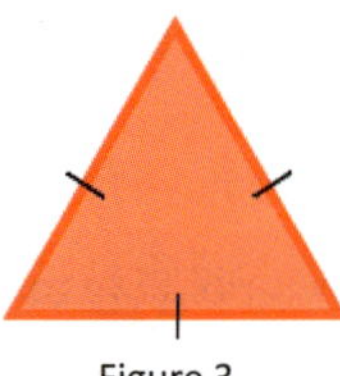

Figure 3

Example 2 (7 minutes)

MP.3

Example 2

With your group, discuss whether a right rectangular prism can be sliced at an angle so that the resulting slice looks like the figure in Figure 4. If it is possible, draw an example of such a slice into the following prism.

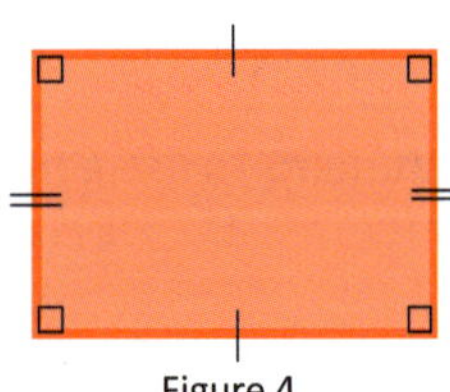

Figure 4

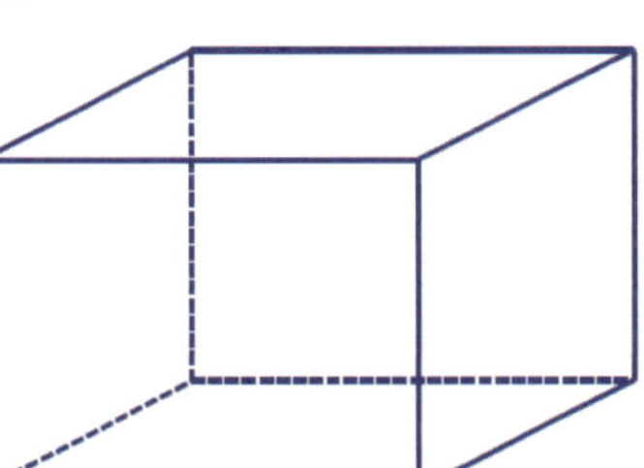

Once students have had some time to attempt Example 2, pose the following questions:

- Is it possible to slice this right rectangular prism to form a quadrilateral cross section? Remember, we are slicing at an angle now. Consider what you know about vertices and the edges they fall on from Example 1.

Again, some students may be able to visualize this right away, while some might not. Give students time to experiment with the solution. Use a valid response to the question to move forward, or simply share the image provided.

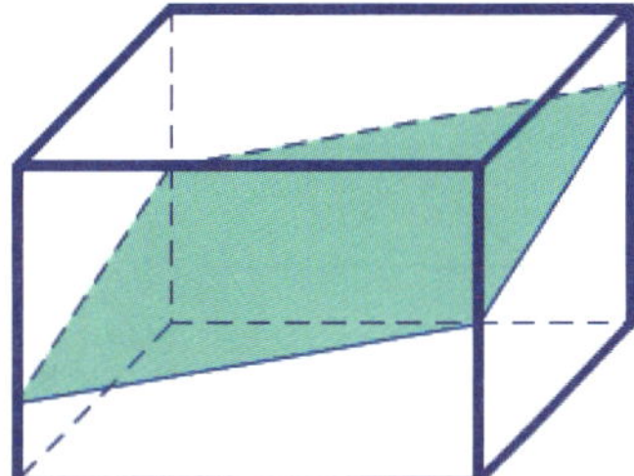

- Here is one possible slice in the shape of a quadrilateral. Notice that the slice is, in fact, made by a plane. A common error, especially when outlining a quadrilateral slice by its vertices, is to make a slice that is not a true slice because the figure could not be made by a single plane.
- What must be true about the opposite sides of the quadrilateral?
  - *The opposite sides are parallel.*

The conclusion that the opposite sides of the quadrilateral region are parallel is based on the above image. This conclusion likely comes from their understanding that the opposite faces of the right rectangular prism are parallel. Therefore, since the opposite sides lie in these faces, they too must be parallel.

This might be an appropriate time to show them an image like the following:

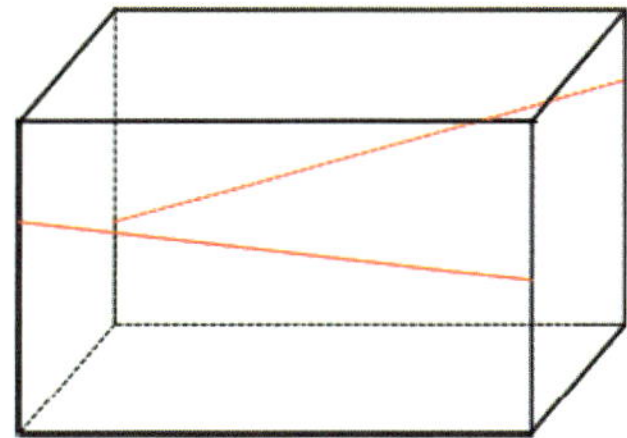

- Though the segments lie in planes that are a constant distance apart, the segments are not parallel.
- Because the earlier slice is made by one plane, the segments that form the sides of the quadrilateral-shaped slice both lie in the same plane as each other *and* in opposite faces that are an equal distance apart. Together, this means the segments are parallel.
- Not only is this slice a quadrilateral, but it is a special quadrilateral. What kind of special quadrilateral-shaped slice must it be?
  - *Since we have determined the opposite sides of the quadrilateral to be parallel, the quadrilateral must be a parallelogram.*

Have students draw another example of a slice through the right rectangular prism that results in a parallelogram shape.

Exercise 2 (5 minutes)

Exercise 2

In Example 2, we discovered how to slice a right rectangular prism to make the shapes of a rectangle and a parallelogram. Are there other ways to slice a right rectangular prism that result in other quadrilateral-shaped slices?

Allow students more time to experiment with other possible slices that might result in another kind of quadrilateral. If there is no valid response, share the figure below.

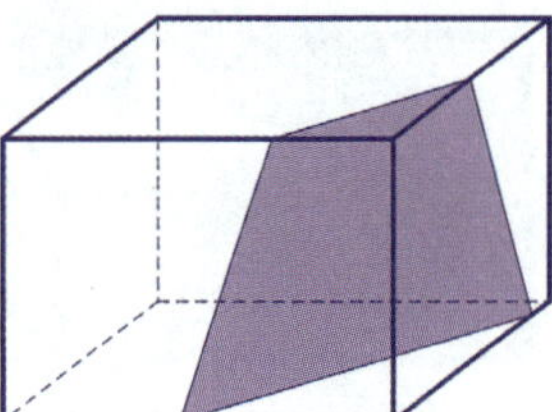

- A slice can be made to a right rectangular prism at an angle so that the resulting cross section is a trapezoid (as in the example shown above). In addition to slicing at an angle, it is also possible to slice perpendicular to a face or base to form a trapezoid-shaped slice.

Example 3 (5 minutes)

In Example 3, make sure students understand that the edges of a slice are determined by the number of faces the slicing plane meets. In other words, there is a correspondence between the sides of the polygonal region formed by the slice and the faces of the solid; the polygon cannot have more sides than there are faces of the solid.

MP.7

Example 3

a. **If slicing a plane through a right rectangular prism so that the slice meets the three faces of the prism, the resulting slice is in the shape of a triangle; if the slice meets four faces, the resulting slice is in the shape of a quadrilateral. Is it possible to slice the prism in a way that the region formed is a pentagon (as in Figure 5)? A hexagon (as in Figure 6)? An octagon (as in Figure 7)?**

Figure 5

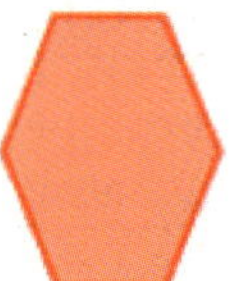

Figure 6

Figure 7

Yes, it is possible to slice a right rectangular prism with a plane so that the resulting cross section is a pentagon; the slice would have to meet five of the six faces of the prism. Similarly, it is possible for the slice to take the shape of a hexagon if the slice meets all six faces. It is impossible to create a slice in the shape of an octagon because a right rectangular prism has six faces, and it is not possible for the shape of a slice to have more sides than the number of faces of the solid.

b. **Draw an example of a slice in a pentagon shape and a slice in a hexagon shape.**

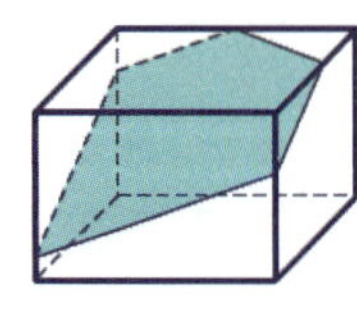

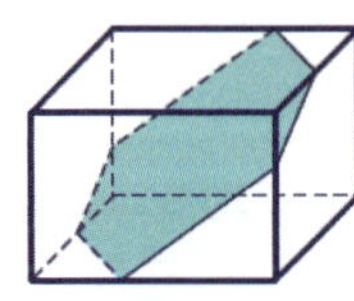

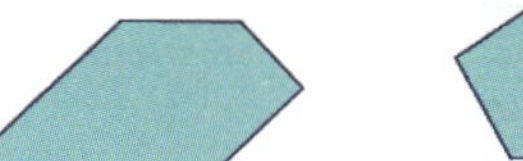

Remind students that marking the vertices of each slice on the edges of the prism facilitates the drawing of the slice.

Example 4 (5 minutes)

Students apply what they learned in Examples 1–3 to right rectangular pyramids.

- We have explored slices made parallel and perpendicular to the base of a right rectangular pyramid. What shapes did those slices have in common?
  - *Parallel slices yielded scale drawings of the base (with a scale factor less than* 1*), and slices made perpendicular to the base yielded slices in the shapes of triangles and trapezoids.*

MP.3

Example 4

a. **With your group, discuss whether a right rectangular pyramid can be sliced at an angle so that the resulting slice looks like the figure in Figure 8. If it is possible, draw an example of such a slice into the following pyramid.**

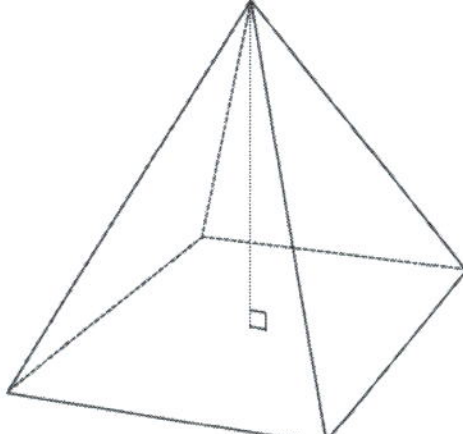

Figure 8

Allow students time to experiment with this slice. Remind students that marking the vertices of the slice on the edges of the pyramid facilitates the drawing of the slice.

If there is no valid response, share the figure below.

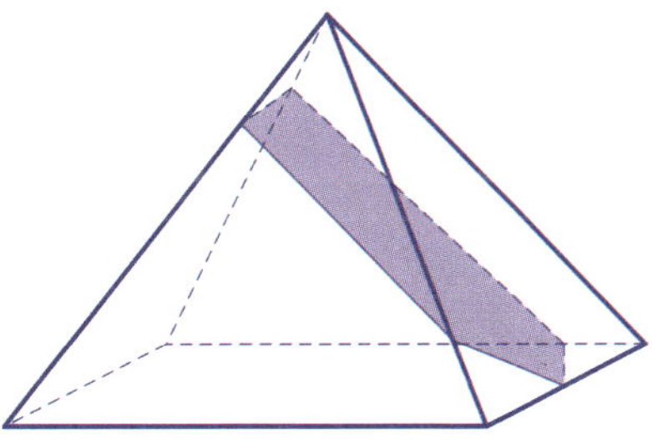

Ask students to find a second slice in the shape of a pentagon.

MP.7

b. **With your group, discuss whether a right rectangular pyramid can be sliced at an angle so that the resulting slice looks like the figure in Figure 9. If it is possible, draw an example of such a slice into the pyramid above.**

It is impossible to create a slice in the shape of a hexagon because a right rectangular pyramid has five faces, and it is not possible for the shape of a slice to have more sides than the number of faces of the solid.

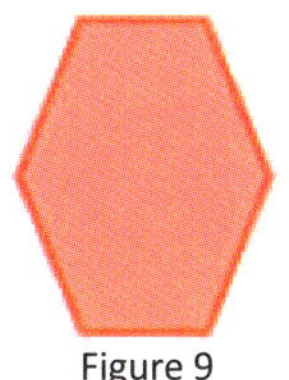

Figure 9

Closing (1 minute)

Refer students to an interactive experience of slicing solids at the following Annenberg Learner website:

http://www.learner.org/courses/learningmath/geometry/session9/part_c/

> **Lesson Summary**
>
> - **Slices made at an angle are neither parallel nor perpendicular to a base.**
> - **There cannot be more sides to the polygonal region of a slice than there are faces of the solid.**

Exit Ticket (5 minutes)

Name ______________________________ Date ______________

Lesson 18: Slicing on an Angle

Exit Ticket

Draw a slice that has the maximum possible number of sides for each solid. Explain how you got your answer.

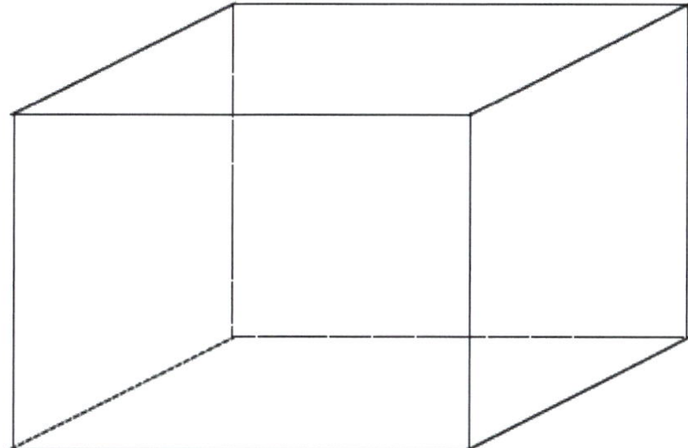

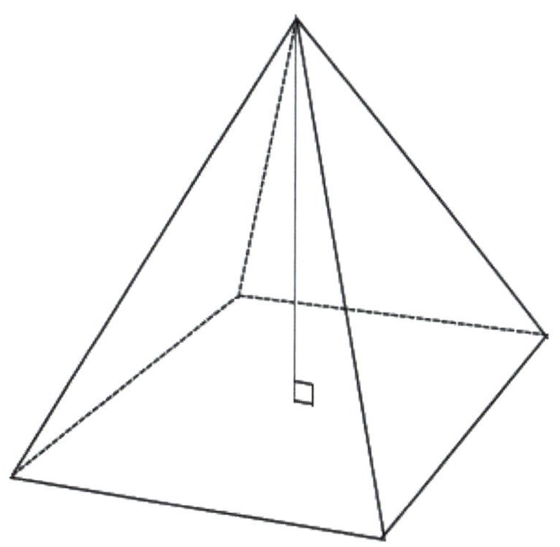

Exit Ticket Sample Solutions

Draw a slice that has the maximum possible number of sides for each solid. Explain how you got your answer.

The slice in the right rectangular prism should be hexagonal (diagrams will vary); the slice in the right rectangular pyramid should be pentagonal (again, diagrams will vary).

The edges of a slice are determined by the number of faces the slicing plane meets; there cannot be more sides to the polygon than there are faces of the solid.

Problem Set Sample Solutions

Note that though sample drawings have been provided in Problems 1 and 2, teachers should expect a variety of acceptable drawings from students.

1. Draw a slice into the right rectangular prism at an angle in the form of the provided shape, and draw each slice as a 2D shape.

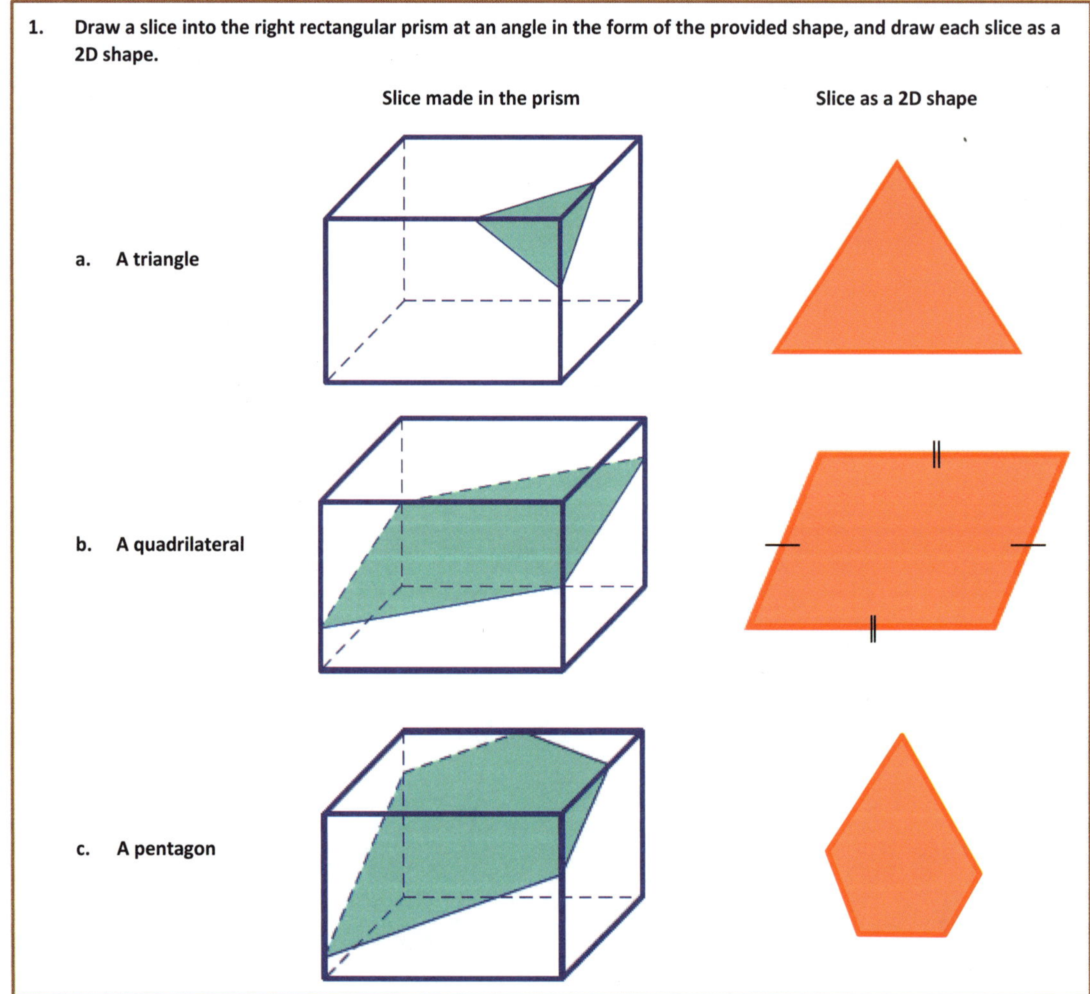

d. A hexagon

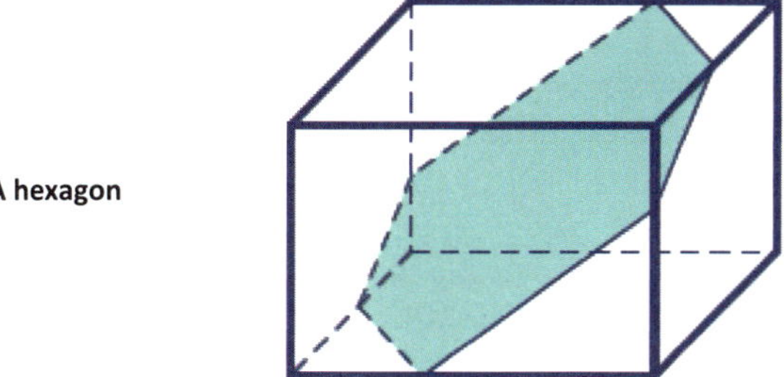

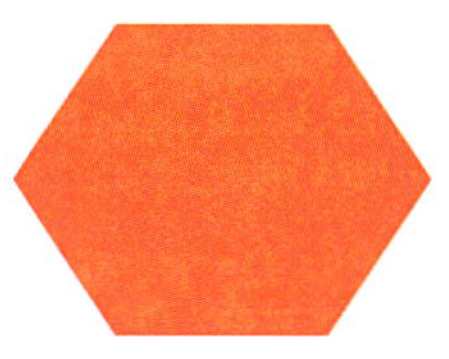

2. Draw slices at an angle in the form of each given shape into each right rectangular pyramid, and draw each slice as a 2D shape.

	Slice made in the pyramid	Slice as a 2D shape
a. A triangle	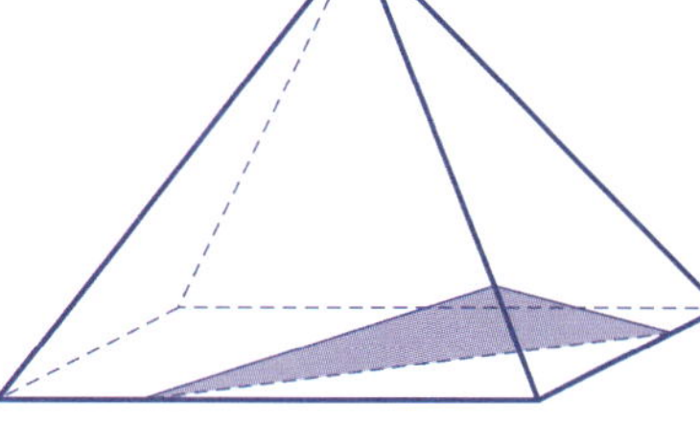	
b. A quadrilateral	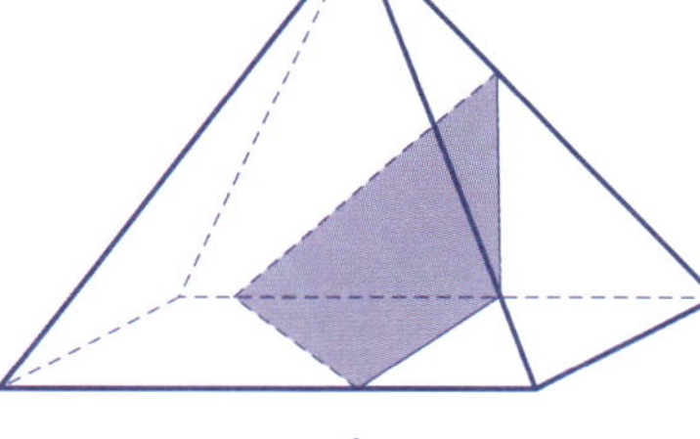	
c. A pentagon	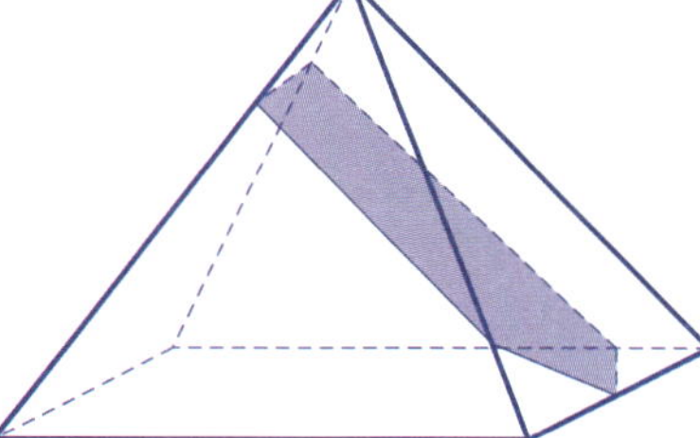	

3. Why is it not possible to draw a slice in the shape of a hexagon for a right rectangular pyramid?

   It is not possible for the shape of a slice to have more sides than the number of faces of the solid.

4. If the slicing plane meets every face of a right rectangular prism, then the slice is a hexagonal region. What can you say about opposite sides of the hexagon?

   The opposite sides of the hexagon lie in opposite faces; therefore, they are parallel.

5. Draw a right rectangular prism so that rectangles $ABCD$ and $A'B'C'D'$ are base faces. The line segments AA', BB', CC', and DD' are edges of the lateral faces.

   a. A slicing plane meets the prism so that vertices A, B, C, and D lie on one side of the plane, and vertices A', B', C', and D' lie on the other side. Based on the slice's position, what other information can be concluded about the slice?

      The slice misses the base faces $ABCD$ and $A'B'C'D'$ since all the vertices of each face lie on the same side of the plane. The slice meets each of the lateral faces in an interval since each lateral face has two vertices on each side. The slice is a quadrilateral. In fact, the slice is a parallelogram because opposite faces of a right rectangular prism lie in parallel planes.

   b. A slicing plane meets the prism so that vertices A, B, C, and B' are on one side of the plane, and vertices A', D', C', and D are on the other side. What other information can be concluded about the slice based on its position?

      The slice meets each face in line segments because, in each case, three of the vertices of the face are on one side of the plane and the remaining vertex lies in the opposite side. The slice is a hexagon because it has six edges. Opposite sides of the hexagon are parallel since they lie in parallel planes.

Lesson 19: Understanding Three-Dimensional Figures

Student Outcomes

- Students describe three-dimensional figures built from cubes by looking at horizontal slicing planes.

Lesson Notes

In high school Geometry, students study the link between the volume of a figure and its slices and, in doing so, consider a whole figure in slices versus any given slice as studied in Grade 7. In Lesson 19, students take an easy approach to thinking of a figure in slices. The approach is easy because each slice is made up of several cubes, and it is thereby not a stretch to visualize or to build. In this lesson, students examine figures built out of unit cubes. A one-unit grid is placed on a table. Cubes are fit into the squares on the grid and then stacked on top of each other to make a three-dimensional figure; the figure in Example 1 is one such example. Slices are made at each level of the figure so that each slice is actually between layers of cubes. Students learn to map the figure layer by layer, much like creating a blueprint for each floor of a building. Students are able to deconstruct a figure, mapping each slice on a grid to determine the number of cubes in the figure. Students are also able to do the reverse and construct a three-dimensional figure from a map of each horizontal slice of the figure. The use of unit cubes and a square unit grid is useful in this lesson; another strategy is to provide graph paper that students can use to draw the levels of each figure.

Classwork

Example 1 (10 minutes)

Students are to imagine each three-dimensional figure as a figure built on a tabletop. Each horizontal slice of the figure is to be mapped onto grid paper where each 1×1 cell represents the base of a unit cube. Level n means the slicing plane is n units above the tabletop. This means that Level 0 (0 units above the tabletop) is the level of the tabletop, while Level 1 is one unit above the tabletop. Recall that a slice is *the intersection of the solid with the slicing plane*. This means that the slice at Level 0 and the slice at Level 1 are always the same. This is also why there is a slice at Level 3, even though it is the *top* of the figure; a horizontal plane at that level would still intersect with the figure.

In the map of each slice, there should be a reference point that remains in the same position (i.e., the reference points are all exactly on top of or below each other), regardless of which slice the map is for. In Example 1, the reference point is marked for students. Reference points should also be marked in the image of the three-dimensional figure so that any reader can correctly compare the point of view of the three-dimensional figure to the slices.

> *Scaffolding:*
>
> Consider using unit cubes and grid paper throughout the examples. As an alternative, have students build the figures in the examples using the net for a cube that is provided at the end of the module.

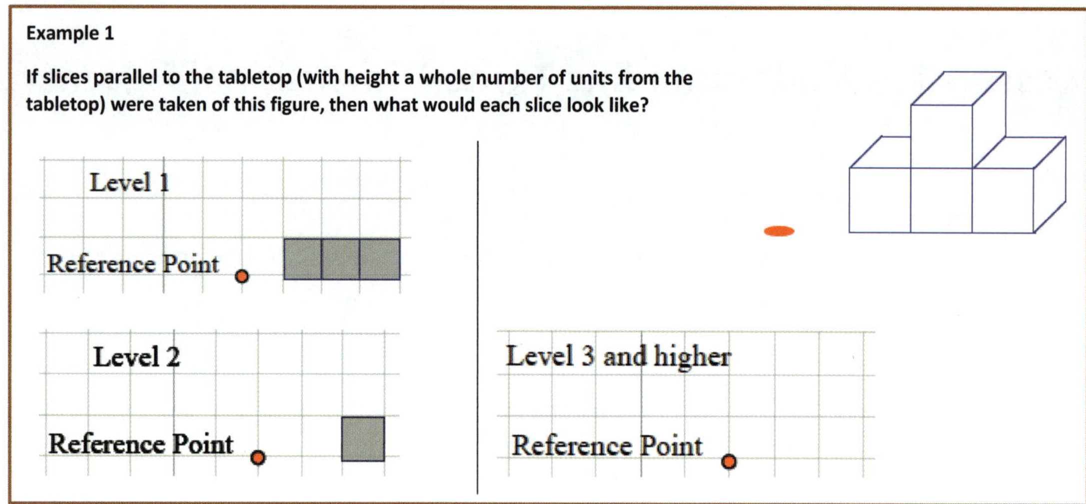

Example 2 (7 minutes)

MP.7

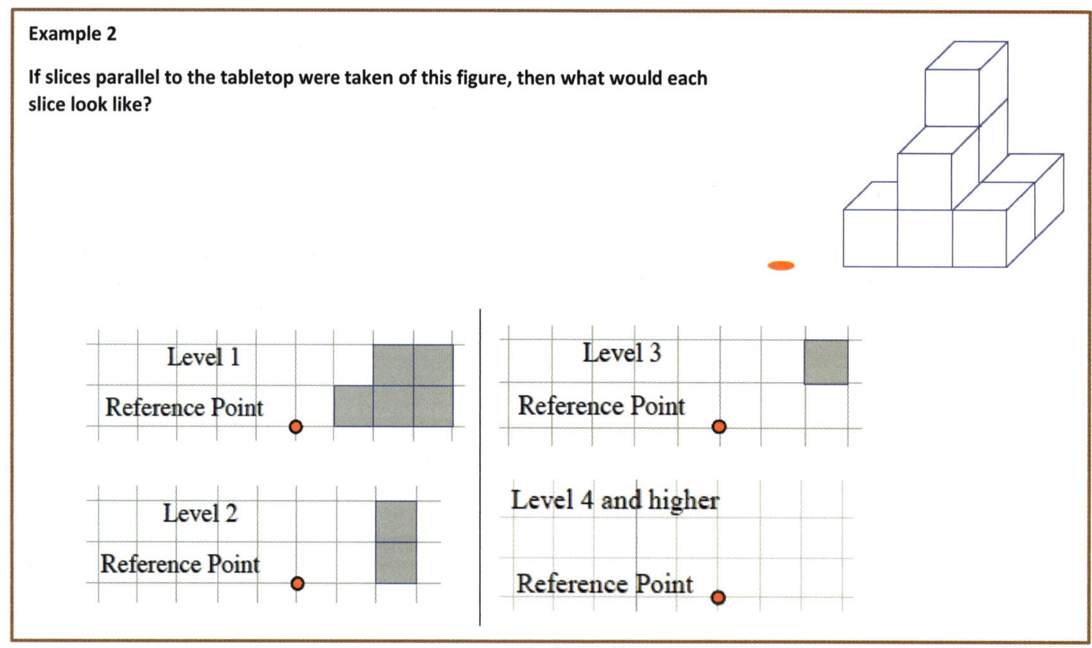

Check in with students for each level in the example. Pull the whole class together if discussion is needed. Remind students that this perspective of the three-dimensional solid allows a full view of two of four *sides* of the figure.

Exercise 1 (5 minutes)

Exercise 1

Based on the level slices you determined in Example 2, how many unit cubes are in the figure?

The number of unit cubes can be determined by counting the shaded squares in Levels 1–3.

Level 1: Five shaded squares; there are 5 cubes between Level 0 and Level 1.

Level 2: Two shaded squares; there are 2 cubes between Level 1 and Level 2.

Level 3: One shaded square; there is 1 cube between Level 2 and Level 3.

The total number of cubes in the solid is 8.

Exercise 2 (7 minutes)

Exercise 2

a. If slices parallel to the tabletop were taken of this figure, then what would each slice look like?

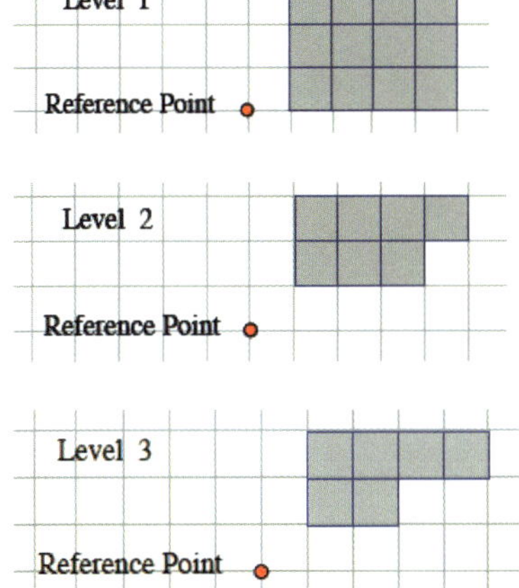

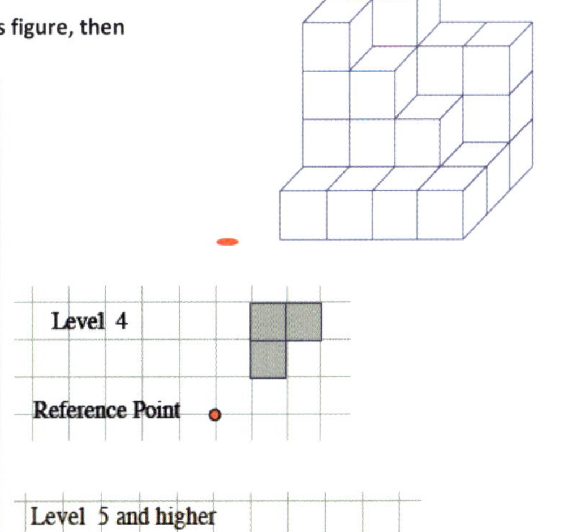

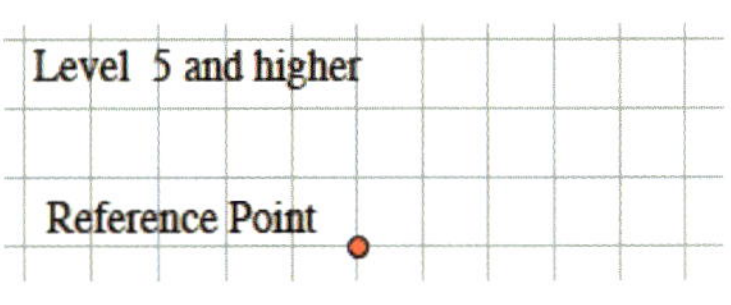

b. Given the level slices in the figure, how many unit cubes are in the figure?

The number of unit cubes can be determined by counting the shaded squares in Levels 1–4.

Level 1: There are 12 cubes between Level 0 and Level 1.

Level 2: There are 7 cubes between Level 1 and Level 2.

Level 3: There are 6 cubes between Level 2 and Level 3.

Level 4: There are 3 cubes between Level 3 and Level 4.

The total number of cubes in the solid is 28.

Example 3 (7 minutes)

Example 3

Given the level slices in the figure, how many unit cubes are in the figure?

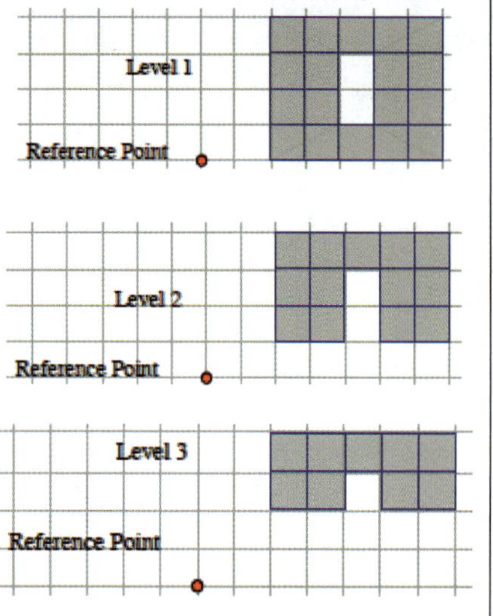

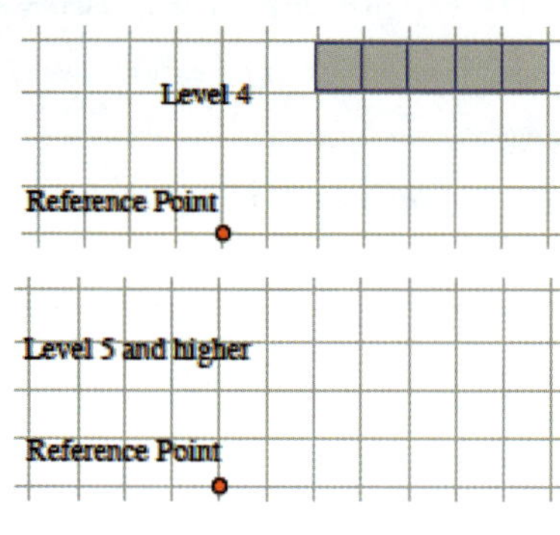

The number of unit cubes can be determined by counting the shaded squares in Levels 1–4.

Level 1: There are 18 *cubes between Level 0 and Level 1.*

Level 2: There are 13 *cubes between Level 1 and Level 2.*

Level 3: There are 9 *cubes between Level 2 and Level 3.*

Level 4: There are 5 *cubes between Level 3 and Level 4.*

The total number of cubes in the solid is 45.

Exercise 3 (optional)

Exercise 3

Sketch your own three-dimensional figure made from cubes and the slices of your figure. Explain how the slices relate to the figure.

Responses will vary.

Closing (3 minutes)

We take a different perspective of three-dimensional figures built from unit cubes by examining the horizontal whole-unit slices. The slices allow a way to count the number of unit cubes in the figure, which is particularly useful when the figure is layered in a way so that many cubes are hidden from view.

> **Lesson Summary**
>
> **We can examine the horizontal whole-unit scales to look at three-dimensional figures. These slices allow a way to count the number of unit cubes in the figure, which is useful when the figure is layered in a way so that many cubes are hidden from view.**

Exit Ticket (6 minutes)

Name ______________________________ Date ______________

Lesson 19: Understanding Three-Dimensional Figures

Exit Ticket

1. The following three-dimensional figure is built on a tabletop. If slices parallel to the tabletop are taken of this figure, then what would each slice look like?

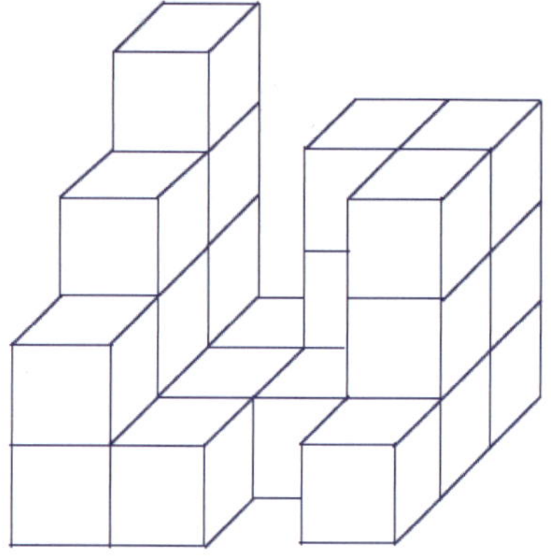

2. Given the level slices in the figure, how many cubes are in the figure?

Exit Ticket Sample Solutions

1. The following three-dimensional figure is built on a tabletop. If slices parallel to the tabletop are taken of this figure, then what would each slice look like?

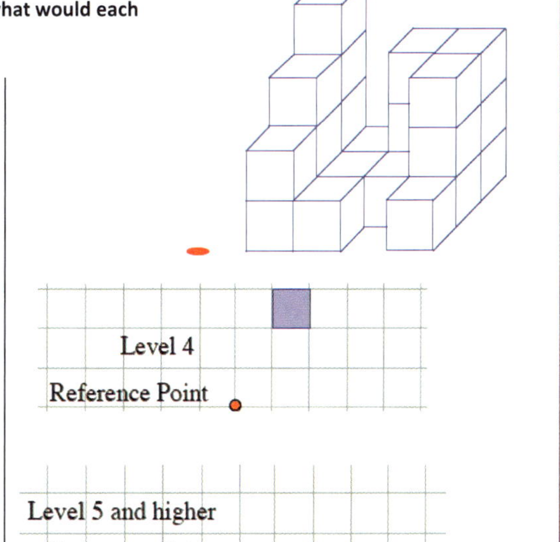

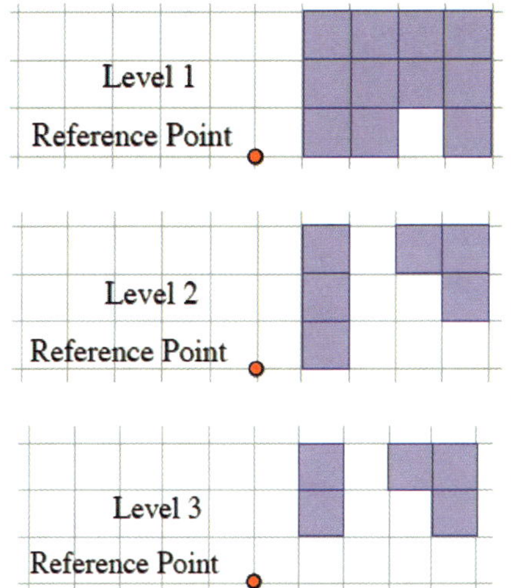

2. Given the level slices in the figure, how many cubes are in the figure?

   The number of unit cubes can be determined by counting the shaded squares in Levels 1–4.

   Level 1: There are 11 cubes between Level 0 and Level 1.

   Level 2: There are 6 cubes between Level 1 and Level 2.

   Level 3: There are 5 cubes between Level 2 and Level 3.

   Level 4: There is 1 cube between Level 3 and Level 4.

   The total number of cubes in the solid is 23.

Problem Set Sample Solutions

In the given three-dimensional figures, unit cubes are stacked exactly on top of each other on a tabletop. Each block is either visible or below a visible block.

1.

a. The following three-dimensional figure is built on a tabletop. If slices parallel to the tabletop are taken of this figure, then what would each slice look like?

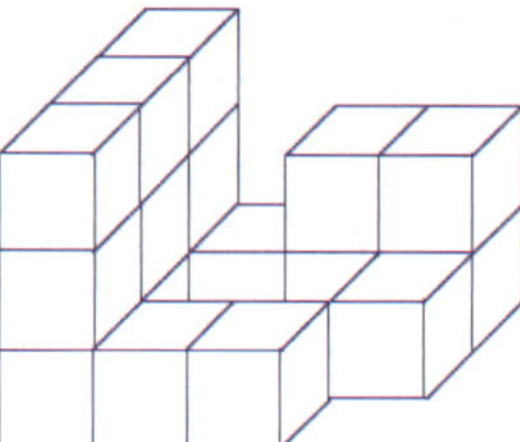

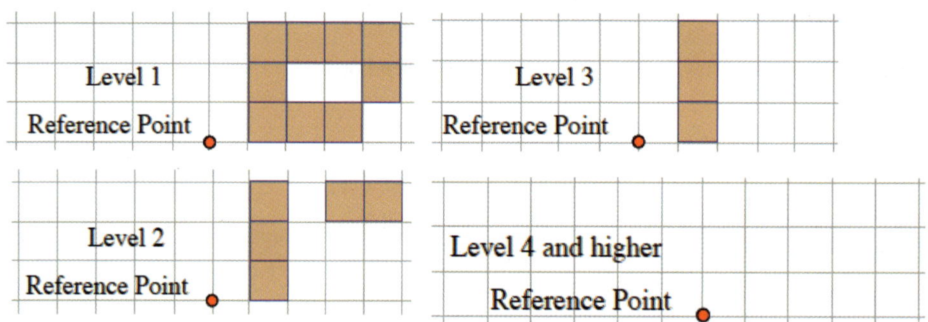

b. Given the level slices in the figure, how many cubes are in the figure?

The number of unit cubes can be determined by counting the shaded squares in Levels 1–3.

Level 1: There are 9 *cubes between Level 0 and Level 1.*

Level 2: There are 5 *cubes between Level 1 and Level 2.*

Level 3: There are 3 *cubes between Level 2 and Level 3.*

The total number of cubes in the solid is 17.

EUREKA MATH

2.

a. The following three-dimensional figure is built on a tabletop. If slices parallel to the tabletop are taken of this figure, then what would each slice look like?

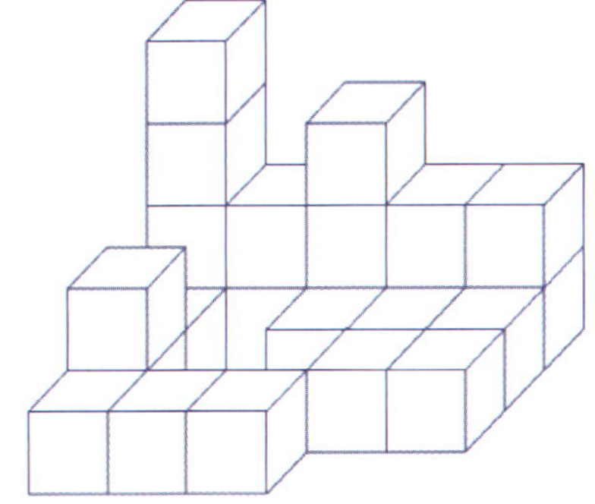

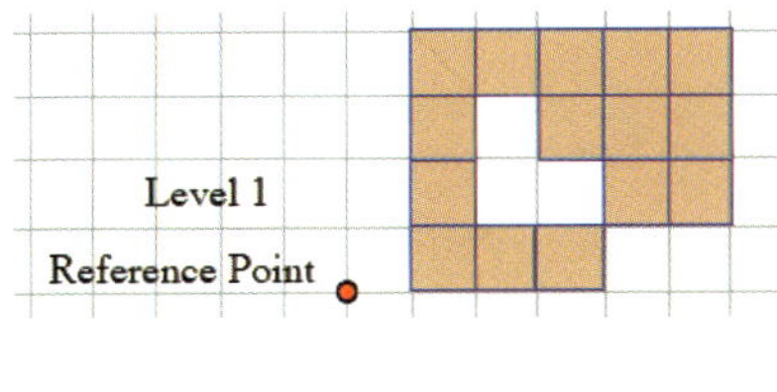

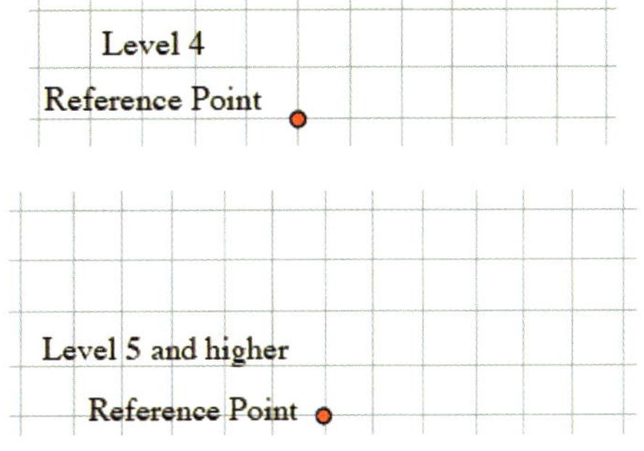

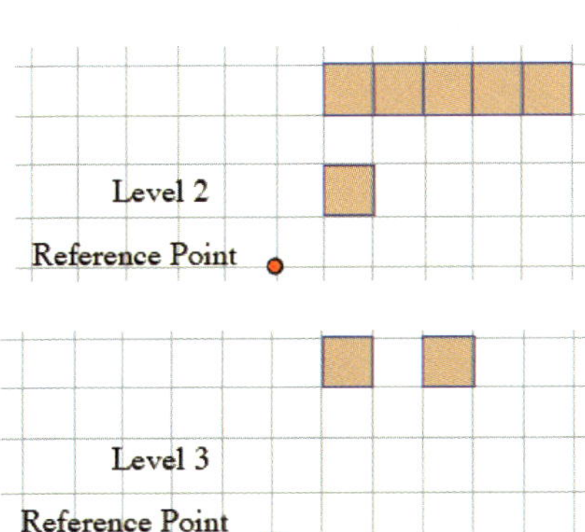

Level 5 and higher

Reference Point

b. Given the level slices in the figure, how many cubes are in the figure?

The number of unit cubes can be determined by counting the shaded squares in Levels 1–4.

Level 1: There are 15 *cubes between Level 0 and Level 1.*

Level 2: There are 6 *cubes between Level 1 and Level 2.*

Level 3: There are 2 *cubes between Level 2 and Level 3.*

Level 4: There is 1 *cube between Level 3 and Level 4.*

The total number of cubes in the solid is 24.

3.

a. **The following three-dimensional figure is built on a tabletop. If slices parallel to the tabletop are taken of this figure, then what would each slice look like?**

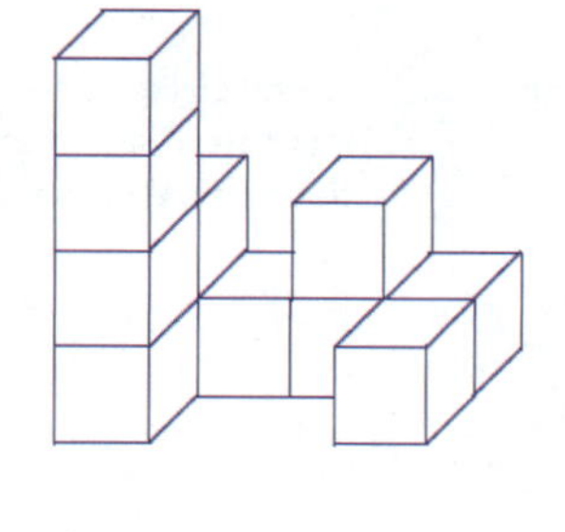

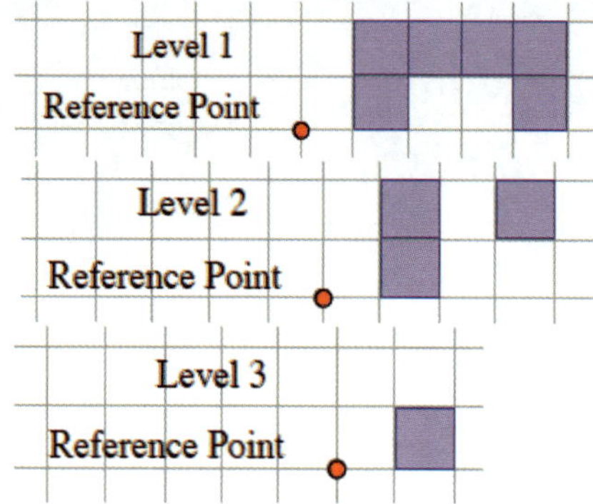

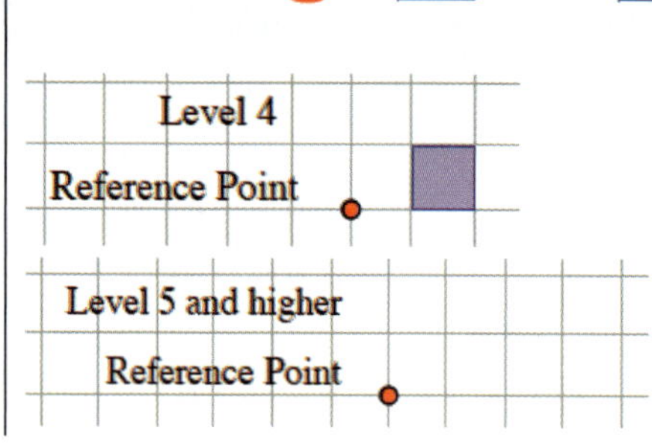

b. **Given the level slices in the figure, how many cubes are in the figure?**

The number of unit cubes can be determined by counting the shaded squares in Levels 1–4.

Level 1: There are 6 *cubes between Level 0 and Level 1.*

Level 2: There are 3 *cubes between Level 1 and Level 2.*

Level 3: There is 1 *cube between Level 2 and Level 3.*

Level 4: There is 1 *cube between Level 3 and Level 4.*

The total number of cubes in the solid is 11.

4. **John says that we should be including the Level 0 slice when mapping slices. Naya disagrees, saying it is correct to start counting cubes from the Level 1 slice. Who is right?**

Naya is right because the Level 0 slice and Level 1 slice are the tops and bottoms of the same set of cubes; counting cubes in both slices would be double counting cubes.

5. **Draw a three-dimensional figure made from cubes so that each successive layer farther away from the tabletop has one less cube than the layer below it. Use a minimum of three layers. Then draw the slices, and explain the connection between the two.**

Responses will vary.

EUREKA MATH

Topic D

Problems Involving Area and Surface Area

7.G.B.6

Focus Standard:	7.G.B.6	Solve real-world and mathematical problems involving area, volume, and surface area of two- and three-dimensional objects composed of triangles, quadrilaterals, polygons, cubes, and right prisms.
Instructional Days:	5	
Lesson 20:	Real-World Area Problems (S)[1]	
Lesson 21:	Mathematical Area Problems (S)	
Lesson 22:	Area Problems with Circular Regions (P)	
Lessons 23–24:	Surface Area (P, P)	

Students enter Grade 7 having studied area in several grade levels. Most recently, they found the areas of both basic geometric shapes and of more complex shapes by decomposition, and they applied these skills in a real-world context (**6.G.A.1**). Lesson 20 reintroduces students to these concepts with area problems embedded in a real-world context (e.g., finding the cost of carpeting a home based on a floor plan, calculating the cost of seeding a lawn, and determining how many stop signs can be painted with a given amount of paint). In Lesson 21, students use the area properties to justify the repeated use of the distributive property. Students apply their knowledge of finding the area of both polygons and circles to find the area of composite figures made of both categories of shapes in Lesson 22. The figures in this lesson are similar to those in Module 3 in that they are composite figures, some of which have "holes" or missing sections in the form of geometric shapes. However, the figures in Lesson 22 are more complex; therefore, their areas are more challenging to determine. In Lessons 23 and 24, the content transitions from area to surface area, asking students to find the surface area of basic and composite three-dimensional figures. As with the topic of area, the figures are missing sections. These missing sections are, of course, now three-dimensional, so students must take this into account when calculating surface area.

[1]Lesson Structure Key: **P**-Problem Set Lesson, **M**-Modeling Cycle Lesson, **E**-Exploration Lesson, **S**-Socratic Lesson

Lesson 20: Real-World Area Problems

Student Outcomes

- Students determine the area of composite figures in real-life contextual situations using composition and decomposition of polygons and circular regions.

Lesson Notes

Students apply their understanding of the area of polygons and circular regions to real-world contexts.

Classwork

Opening Exercise (8 minutes)

Opening Exercise

Find the area of each shape based on the provided measurements. Explain how you found each area.

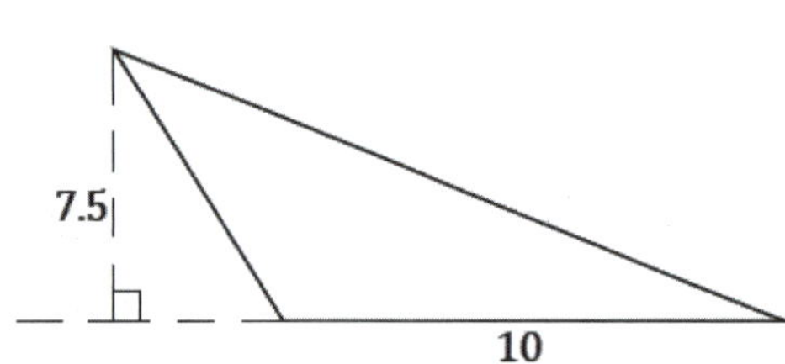

Triangular region: half base times height

$$\text{area} = \frac{1}{2} \cdot 10 \text{ units} \cdot 7.5 \text{ units}$$
$$= 37.5 \text{ units}^2$$

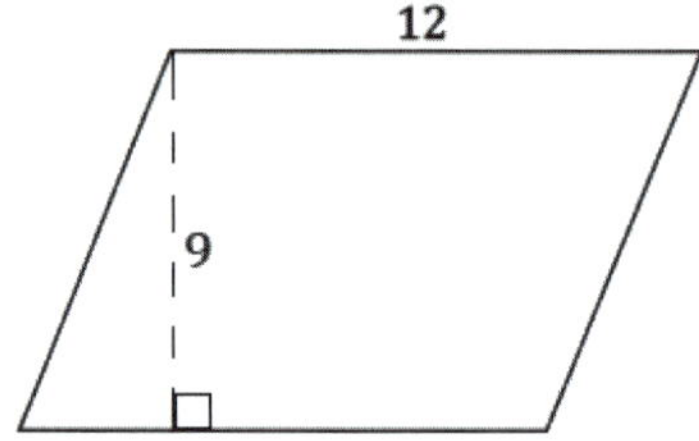

Parallelogram: base times height

$$\text{area} = 12 \text{ units} \cdot 9 \text{ units}$$
$$= 108 \text{ units}^2$$

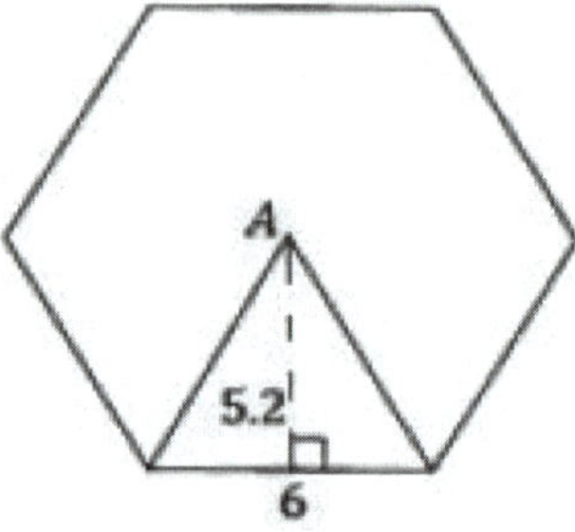

Regular hexagon: area of the shown triangle times six for the six triangles that fit into the hexagon

$$\text{area} = 6\left(\frac{1}{2} \cdot 6 \text{ units} \cdot 5.2 \text{ units}\right)$$
$$= 93.6 \text{ units}^2$$

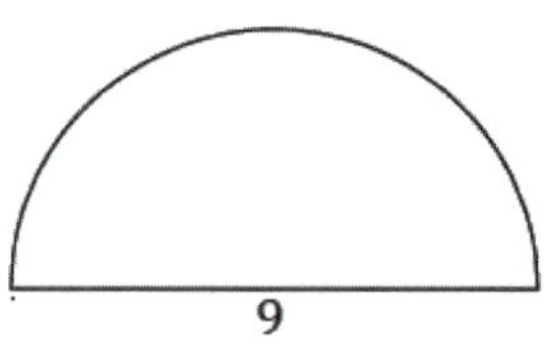

Semicircle: half the area of a circle with the same radius

$$\text{area} = \frac{\pi}{2}(4.5 \text{ unit})^2$$
$$= 10.125\pi \text{ units}^2$$
$$\approx 31.81 \text{ units}^2$$

Example 1 (10 minutes)

Students should first attempt this question without assistance. Once students have had time to work on their own, lead the class through a discussion using the following prompts according to student need.

Example 1

A landscape company wants to plant lawn seed. A 20 lb. bag of lawn seed will cover up to 420 sq. ft. of grass and costs $\$49.98$ plus the 8% sales tax. A scale drawing of a rectangular yard is given. The length of the longest side is 100 ft. The house, driveway, sidewalk, garden areas, and utility pad are shaded. The unshaded area has been prepared for planting grass. How many 20 lb. bags of lawn seed should be ordered, and what is the cost?

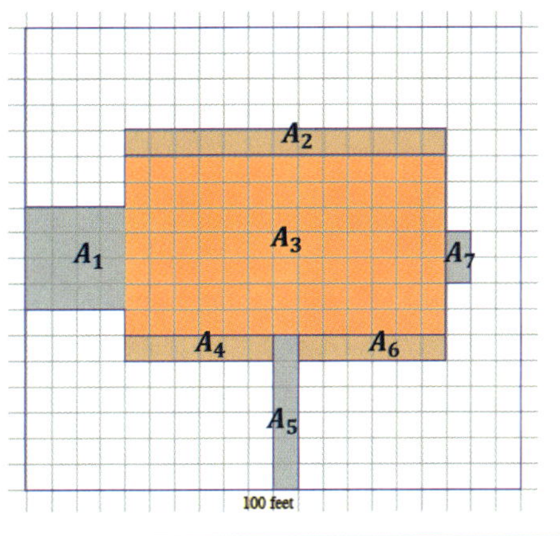

The following calculations demonstrate how to find the area of the lawn by subtracting the area of the home from the area of the entire yard.

- Find the non-grassy sections in the map of the yard and their areas.
  - $A_1 = 4 \text{ units} \cdot 4 \text{ units} = 16 \text{ units}^2$
  - $A_2 = 1 \text{ units} \cdot 13 \text{ units} = 13 \text{ units}^2$
  - $A_3 = 7 \text{ units} \cdot 13 \text{ units} = 91 \text{ units}^2$
  - $A_4 = 1 \text{ units} \cdot 6 \text{ units} = 6 \text{ units}^2$
  - $A_5 = 6 \text{ units} \cdot 1 \text{ units} = 6 \text{ units}^2$
  - $A_6 = 1 \text{ units} \cdot 6 \text{ units} = 6 \text{ units}^2$
  - $A_7 = 2 \text{ units} \cdot 1 \text{ units} = 2 \text{ units}^2$

> *Scaffolding:*
> An alternative image with fewer regions is provided after the initial solution.

- What is the total area of the non-grassy sections?
  - $A_1 + A_2 + A_3 + A_4 + A_5 + A_6 + A_7 = 16 \text{ units}^2 + 13 \text{ units}^2 + 91 \text{ units}^2 + 6 \text{ units}^2 + 6 \text{ units}^2 + 6 \text{ units}^2 + 2 \text{ units}^2 = 140 \text{ units}^2$
- What is the area of the grassy section of the yard?
  - *Subtract the area of the non-grassy sections from the area of the yard.*
    $A = (20 \text{ units} \cdot 18 \text{ units}) - 140 \text{ units}^2 = 220 \text{ units}^2$
- What is the scale of the map of the yard?
  - *The scale of the map is* 5 ft.
- What is the grassy area in square feet?
  - $220 \text{ units}^2 \cdot 25 \dfrac{\text{ft}^2}{\text{units}^2} = 5{,}500 \text{ ft}^2$
    The area of the grassy space is $5{,}500 \text{ ft}^2$.

MP.2

- If one 20 lb. bag covers 420 square feet, write a numerical expression for the number of bags needed to cover the grass in the yard. Explain your expression.
  - *Grassy area ÷ area that one bag of seed covers*

    $5{,}500 \div 420$
- How many bags are needed to cover the grass in the yard?
  - $5{,}500 \div 420 \approx 13.1$

    It will take 14 *bags to seed the yard.*
- What is the final cost of seeding the yard?
  - $1.08 \cdot 14 \cdot \$49.98 \approx \$755.70$

    The final cost with sales tax is $\$755.70$.

Encourage students to write or state an explanation for how they solved the problem.

Alternative image of property:

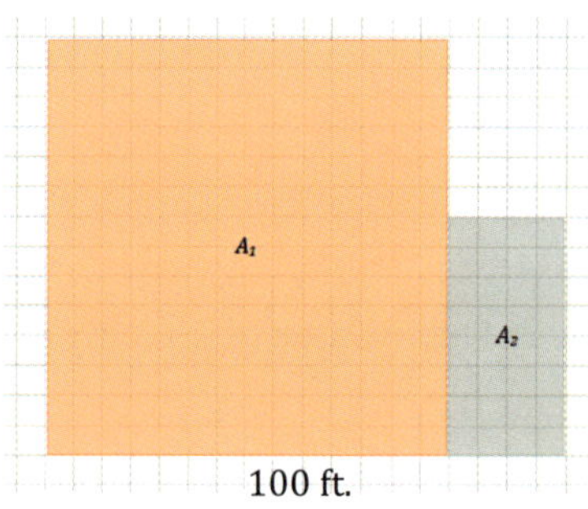

- Find the non-grassy sections in the map of the yard and their areas.
  - $A_1 = 14 \text{ units} \cdot 14 \text{ units} = 196 \text{ units}^2$
  - $A_2 = 4 \text{ units} \cdot 8 \text{ units} = 32 \text{ units}^2$
- What is the total area of the non-grassy sections?
  - $A_1 + A_2 = 196 \text{ units}^2 + 32 \text{ units}^2 = 228 \text{ units}^2$
- What is the area of the grassy section of the yard?
  - *Subtract the area of the non-grassy sections from the area of the yard.*

    $A = (20 \text{ units} \cdot 15 \text{ units}) - 228 \text{ units}^2 = 72 \text{ units}^2$
- What is the scale of the map of the yard?
  - *The scale of the map is* 5 ft.
- What is the grassy area in square feet?
  - $72 \cdot 25 = 1{,}800$

    The area of the grassy space is $1{,}800 \text{ ft}^2$.

- If one 20 lb. bag covers 420 square feet, write a numerical expression for the number of bags needed to cover the grass in the yard. Explain your expression.
  - *Grassy area ÷ area that one bag of seed covers*

    $1{,}800 \div 420$

- How many bags are needed to cover the grass in the yard?
  - $1{,}800 \div 420 \approx 4.3$

    It will take 5 *bags to seed the yard.*
- What is the final cost of seeding the yard?
  - $1.08 \cdot 5 \cdot \$49.98 \approx \$269.89$

    The final cost with sales tax is $\$269.89$.

Exercise 1 (6 minutes)

Exercise 1

A landscape contractor looks at a scale drawing of a yard and estimates that the area of the home and garage is the same as the area of a rectangle that is 100 ft. $\times$ 35 ft. The contractor comes up with $5,500\ \text{ft}^2$. How close is this estimate?

The entire yard (home and garage) has an area of 100 ft. $\times$ 35 ft. $= 3,500\ \text{ft}^2$. The contractor's estimate is $5,500\ \text{ft}^2$. He is $2,000\ \text{ft}^2$ over the actual area, which is quite a bit more ($2,000\ \text{ft}^2$ is roughly 57% of the actual area).

Example 2 (10 minutes)

Example 2

Ten dartboard targets are being painted as shown in the following figure. The radius of the smallest circle is 3 in., and each successive larger circle is 3 in. more in radius than the circle before it. A can of red paint and a can of white paint are purchased to paint the target. Each 8 oz. can of paint covers $16\ \text{ft}^2$. Is there enough paint of each color to create all ten targets?

Let each circle be labeled as in the diagram.

> ***Radius of C_1 is 3 in.; area of C_1 is $9\pi\ \text{in}^2$.***
>
> ***Radius of C_2 is 6 in.; area of C_2 is $36\pi\ \text{in}^2$.***
>
> ***Radius of C_3 is 9 in.; area of C_3 is $81\pi\ \text{in}^2$.***
>
> ***Radius of C_4 is 12 in.; area of C_4 is $144\pi\ \text{in}^2$.***

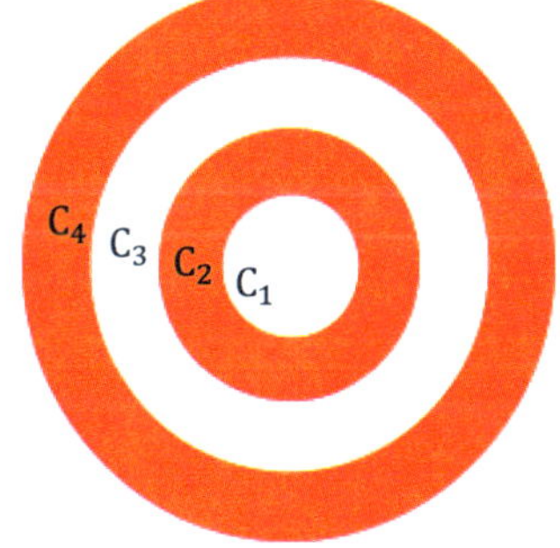

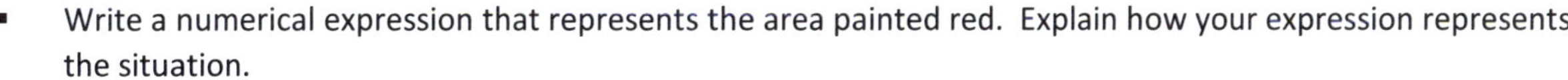

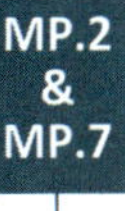

- Write a numerical expression that represents the area painted red. Explain how your expression represents the situation.
  - *The area of red and white paint in square inches is found by finding the area between circles of the target board.*

    Red paint: $(144\pi\ \text{in}^2 - 81\pi\ \text{in}^2) + (36\pi\ \text{in}^2 - 9\pi\ \text{in}^2)$

    White paint: $(81\pi\ \text{in}^2 - 36\pi\ \text{in}^2) + 9\pi\ \text{in}^2$

The following calculations demonstrate how to find the area of red and white paint in the target.

Target area painted red

The area between C_4 and C_3:	$144\pi \text{ in}^2 - 81\pi \text{ in}^2 = 63\pi \text{ in}^2$
The area between C_2 and C_1:	$36\pi \text{ in}^2 - 9\pi \text{ in}^2 = 27\pi \text{ in}^2$
Area painted red in one target:	$63\pi \text{ in}^2 + 27\pi \text{ in}^2 = 90\pi \text{ in}^2$; *approximately* 282.7 in^2
Area of red paint for one target in square feet:	$282.7 \text{ in}^2 \left(\frac{1 \text{ ft}^2}{144 \text{ in}^2}\right) \approx 1.96 \text{ ft}^2$
Area to be painted red for ten targets in square feet:	$1.96 \text{ ft}^2 \times 10 = 19.6 \text{ ft}^2$

Target area painted white

The area between C_3 and C_2:	$81\pi \text{ in}^2 - 36\pi \text{ in}^2 = 45\pi \text{ in}^2$
The area of C_1:	$9\pi \text{ in}^2$
Area painted white in one target:	$45\pi \text{ in}^2 + 9\pi \text{ in}^2 = 54\pi \text{ in}^2$; *approximately* 169.6 in^2
Area of white paint for one target in square feet:	$169.6 \text{ in}^2 \left(\frac{1 \text{ ft}^2}{144 \text{ in}^2}\right) \approx 1.18 \text{ ft}^2$
Area of white paint for ten targets in square feet:	$1.18 \text{ ft}^2 \times 10 = 11.8 \text{ ft}^2$

There is not enough red paint in one 8 oz. can of paint to complete all ten targets; however, there is enough white paint in one 8 oz. can of paint for all ten targets.

Closing (2 minutes)

- What is a useful strategy when tackling area problems with real-world context?
  - *Decompose drawings into familiar polygons and circular regions, and identify all relevant measurements.*
  - *Pay attention to the unit needed in a response to each question.*

Lesson Summary

- **One strategy to use when solving area problems with real-world context is to decompose drawings into familiar polygons and circular regions while identifying all relevant measurements.**
- **Since the area problems involve real-world context, it is important to pay attention to the units needed in each response.**

Exit Ticket (9 minutes)

Name ______________________________ Date ____________

Lesson 20: Real-World Area Problems

Exit Ticket

A homeowner called in a painter to paint the bedroom walls and ceiling. The bedroom is 18 ft. long, 12 ft. wide, and 8 ft. high. The room has two doors each 3 ft. by 7 ft. and three windows each 3 ft. by 5 ft. The doors and windows do not have to be painted. A gallon of paint can cover $300\ \text{ft}^2$. A hired painter claims he will need 4 gallons. Show that the estimate is too high.

Exit Ticket Sample Solutions

A homeowner called in a painter to paint the bedroom walls and ceiling. The bedroom is 18 ft. long, 12 ft. wide, and 8 ft. high. The room has two doors each 3 ft. by 7 ft. and three windows each 3 ft. by 5 ft. The doors and windows do not have to be painted. A gallon of paint can cover $300\ \text{ft}^2$. A hired painter claims he will need 4 gallons. Show that the estimate is too high.

Area of 2 walls: $2(18\ \text{ft.} \cdot 8\ \text{ft.}) = 288\ \text{ft}^2$

Area of remaining 2 walls: $2(12\ \text{ft.} \cdot 8\ \text{ft.}) = 192\ \text{ft}^2$

Area of ceiling: $18\ \text{ft.} \cdot 12\ \text{ft.} = 216\ \text{ft}^2$

Area of 2 doors: $2(3\ \text{ft.} \cdot 7\ \text{ft.}) = 42\ \text{ft}^2$

Area of 3 windows: $3(3\ \text{ft.} \cdot 5\ \text{ft.}) = 45\ \text{ft}^2$

Area to be painted: $(288\ \text{ft}^2 + 192\ \text{ft}^2 + 216\ \text{ft}^2) - (42\ \text{ft}^2 + 45\ \text{ft}^2) = 609\ \text{ft}^2$

Gallons of paint needed: $609 \div 300 = 2.03$; *The painter will need a little more than* 2 gal.

The painter's estimate for how much paint is necessary is too high.

Problem Set Sample Solutions

1. A farmer has four pieces of unfenced land as shown to the right in the scale drawing where the dimensions of one side are given. The farmer trades all of the land and $\$10,000$ for 8 acres of similar land that is fenced. If one acre is equal to $43,560\text{ ft}^2$, how much per square foot for the extra land did the farmer pay rounded to the nearest cent?

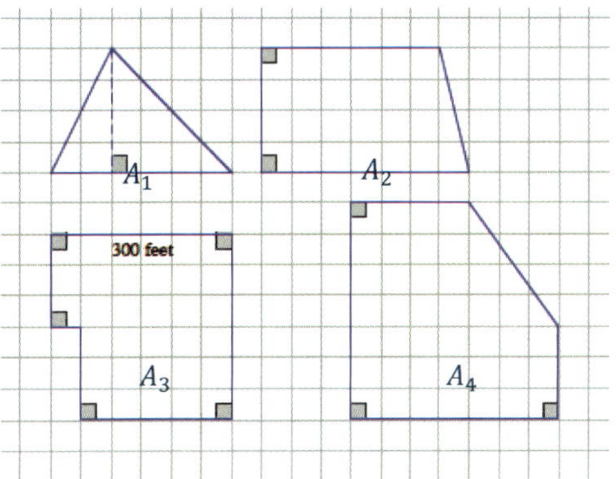

$A_1 = \frac{1}{2}(6\text{ units} \cdot 4\text{ units}) = 12\text{ units}^2$

$A_2 = \frac{1}{2}(6\text{ units} + 7\text{ units})(4\text{ units}) = 26\text{ units}^2$

$A_3 = (3\text{ units} \cdot 6\text{ units}) + (3\text{ units} \cdot 5\text{ units}) = 33\text{ units}^2$

$A_4 = (4\text{ units} \cdot 7\text{ units}) + (3\text{ units} \cdot 3\text{ units}) + \frac{1}{2}(3\text{ units} \cdot 4\text{ units}) = 43\text{ units}^2$

The sum of the farmer's four pieces of land:

$A_1 + A_2 + A_3 + A_4 = 12\text{ units}^2 + 26\text{ units}^2 + 33\text{ units}^2 + 43\text{ units}^2 = 114\text{ units}^2$

The sum of the farmer's four pieces of land in square feet:

$6\text{ units} = 300\text{ ft.}$; *divide each side by* 6.

$1\text{ unit} = 50\text{ ft.}$ *and* $1\text{ unit}^2 = 2,500\text{ ft}^2$

$114 \cdot 2,500 = 285,000$

The total area of the farmer's four pieces of land: $285,000\text{ ft}^2$.

The sum of the farmer's four pieces of land in acres:

$285,000 \div 43,560 \approx 6.54$

The farmer's four pieces of land total about 6.54 *acres.*

Extra land purchased with $\$10,000$: $8\text{ acres} - 6.54\text{ acres} = 1.46\text{ acres}$

Extra land in square feet:

$$1.46\text{ acres}\left(\frac{43,560\text{ ft}^2}{1\text{ acre}}\right) \approx 63597.6\text{ ft}^2$$

Price per square foot for extra land:

$$\left(\frac{\$10,000}{63,597.6\text{ ft}^2}\right) \approx \$0.16$$

2. An ordinance was passed that required farmers to put a fence around their property. The least expensive fences cost $\$10$ for each foot. Did the farmer save money by moving the farm?

At $\$10$ for each foot, $\$10,000$ would purchase $1,000$ feet of fencing. The perimeter of the third piece of land (labeled A_3) has a perimeter of $1,200$ ft. So, it would have cost over $\$10,000$ just to fence that piece of property. The farmer did save money by moving the farm.

3. A stop sign is an octagon (i.e., a polygon with eight sides) with eight equal sides and eight equal angles. The dimensions of the octagon are given. One side of the stop sign is to be painted red. If Timmy has enough paint to cover 500 ft^2, can he paint 100 stop signs? Explain your answer.

Area of top trapezoid $= \frac{1}{2}(12 \text{ in.} + 29 \text{ in.})(8.5 \text{ in.}) = 174.25 \text{ in}^2$

Area of middle rectangle $= 12 \text{ in.} \cdot 29 \text{ in.} = 348 \text{ in}^2$

Area of bottom trapezoid $= \frac{1}{2}(12 \text{ in.} + 29 \text{ in.})(8.5 \text{ in.}) = 174.25 \text{ in}^2$

Total area of stop sign in square inches:

$A_1 + A_2 + A_3 = 174.25 \text{ in}^2 + 348 \text{ in}^2 + 174.25 \text{ in}^2 = 696.5 \text{ in}^2$

Total area of stop sign in square feet:

$696.5 \text{ in}^2 \left(\frac{1 \text{ ft}^2}{144 \text{ in}^2}\right) \approx 4.84 \text{ ft}^2$

Yes, the area of one stop sign is less than 5 ft^2 (approximately 4.84 ft^2). Therefore, 100 stop signs would be less than 500 ft^2.

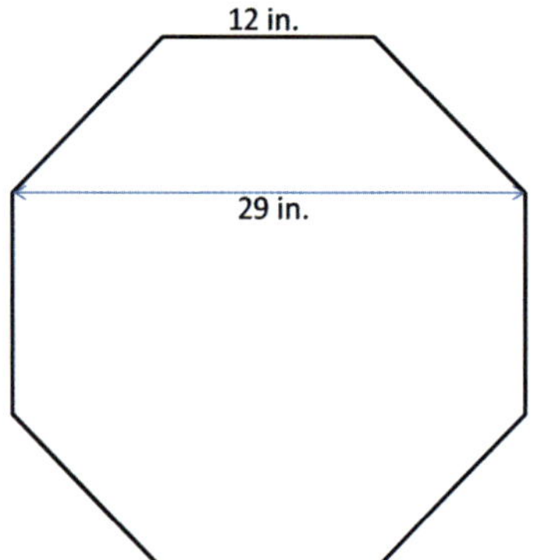

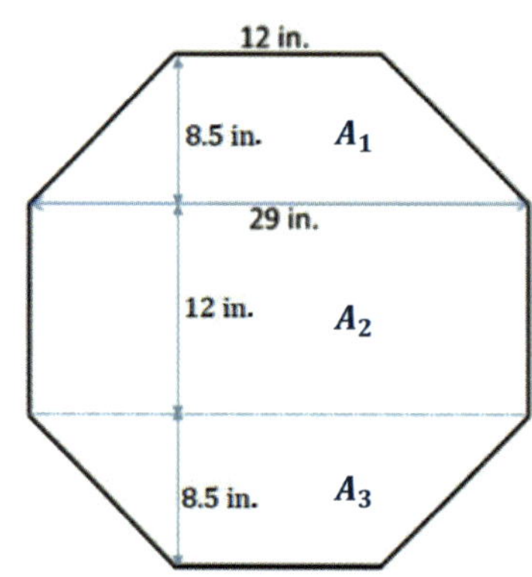

EUREKA MATH™

4. The Smith family is renovating a few aspects of their home. The following diagram is of a new kitchen countertop. Approximately how many square feet of counter space is there?

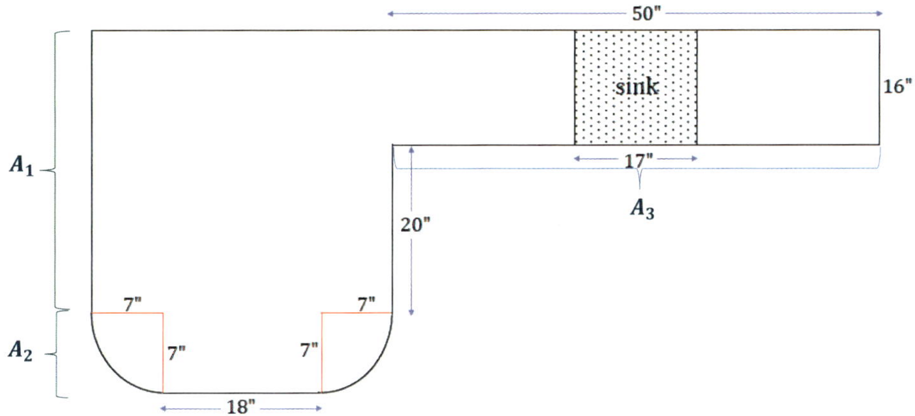

$A_1 = (20 \text{ in.} + 16 \text{ in.})(18 \text{ in.} + 14 \text{ in.}) = 1,152 \text{ in}^2$

$A_2 = (18 \text{ in.} \cdot 7 \text{ in.}) + \frac{1}{2}(49\pi \text{ in}^2)$

$\approx (126 \text{ in}^2 + 77 \text{ in}^2)$

$\approx 203 \text{ in}^2$

$A_3 = (50 \text{ in.} \cdot 16 \text{ in.}) - (17 \text{ in.} \cdot 16 \text{ in.}) = 528 \text{ in}^2$

Total area of counter space in square inches:

$$A_1 + A_2 + A_3 \approx 1,152 \text{ in}^2 + 203 \text{ in}^2 + 528 \text{ in}^2$$
$$A_1 + A_2 + A_3 \approx 1,883 \text{ in}^2$$

Total area of counter space in square feet:

$$1,883 \text{ in}^2 \left(\frac{1 \text{ ft}^2}{144 \text{ in}^2}\right) \approx 13.1 \text{ ft}^2$$

There is approximately 13.1 ft^2 *of counter space.*

5. In addition to the kitchen renovation, the Smiths are laying down new carpet. Everything but closets, bathrooms, and the kitchen will have new carpet. How much carpeting must be purchased for the home?

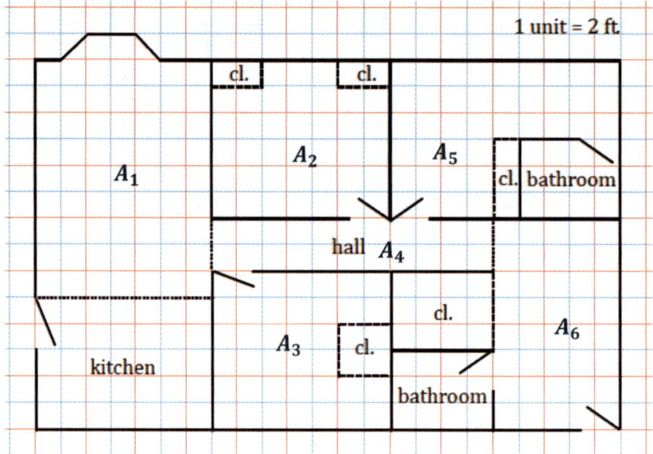

$A_1 = (9 \text{ units} \cdot 7 \text{ units}) + 3 \text{ units}^2 = 66 \text{ units}^2$

$A_2 = (6 \text{ units} \cdot 7 \text{ units}) - 4 \text{ units}^2 = 38 \text{ units}^2$

$A_3 = (6 \text{ units} \cdot 7 \text{ units}) - 4 \text{ units}^2 = 38 \text{ units}^2$

$A_4 = 2 \text{ units} \cdot 11 \text{ units} = 22 \text{ units}^2$

$A_5 = (5 \text{ units} \cdot 3 \text{ units}) + (4 \text{ units} \cdot 6 \text{ units}) = 39 \text{ units}^2$

$A_6 = 5 \text{ units} \cdot 8 \text{ units} = 40 \text{ units}^2$

Total area that needs carpeting:

$$\begin{aligned} A_1 + A_2 + A_3 + A_4 + A_5 + A_6 &= 66 \text{ units}^2 + 38 \text{ units}^2 + 38 \text{ units}^2 + 22 \text{ units}^2 + 39 \text{ units}^2 + 40 \text{ units}^2 \\ &= 243 \text{ units}^2 \end{aligned}$$

Scale: $1 \text{ unit} = 2 \text{ ft.}$; $1 \text{ unit}^2 = 4 \text{ ft}^2$

Total area that needs carpeting in square feet:

$$243 \text{ units}^2 \left(\frac{4 \text{ ft}^2}{1 \text{ unit}^2}\right) = 972 \text{ ft}^2$$

6. Jamie wants to wrap a rectangular sheet of paper completely around cans that are $8\frac{1}{2}$ in. high and 4 in. in diameter. She can buy a roll of paper that is $8\frac{1}{2}$ in. wide and 60 ft. long. How many cans will this much paper wrap?

A can with a 4*-inch diameter has a circumference of* 4π in. *(i.e., approximately* 12.57 in.*).* 60 ft. *is the same as* 720 in.; $720 \text{ in.} \div 12.57 \text{ in.}$ *is approximately* 57.3 in.*, so this paper will cover* 57 *cans.*

Lesson 21: Mathematical Area Problems

Student Outcomes

- Students use the area properties to justify the repeated use of the distributive property to expand the product of linear expressions.

Lesson Notes

In Lesson 21, students use area models to make an explicit connection between area and algebraic expressions. The lesson opens with a numeric example of a rectangle (a garden) expanding in both dimensions. Students calculate the area of the garden as if it expands in a single dimension, once just in length and then in width, and observe how the areas change with each change in dimension. Similar changes are then made to a square. Students record the areas of several squares with expanded side lengths, eventually generalizing the pattern of areas to the expansion of $(a + b)^2$ (MP.2 and MP.8). This generalization is reinforced through the repeated use of the distributive property, which clarifies the link between the terms of the algebraic expression and the sections of area in each expanded figure.

Classwork

Opening Exercise (7 minutes)

The objective of the lesson is to generalize a formula for the area of rectangles that result from adding to the length and width. Using visuals and concrete (numerical) examples throughout the lesson helps students make this generalization.

Opening Exercise

Patty is interested in expanding her backyard garden. Currently, the garden plot has a length of 4 ft. and a width of 3 ft.

a. **What is the current area of the garden?**

The garden has an area of 12 ft^2.

Patty plans on extending the length of the plot by 3 ft. and the width by 2 ft.

b. **What will the new dimensions of the garden be? What will the new area of the garden be?**

The new dimensions of the garden will be 7 ft. *by* 5 ft.*, and it will have an area of* 35 ft^2.

Scaffolding:

Use the following visual as needed.

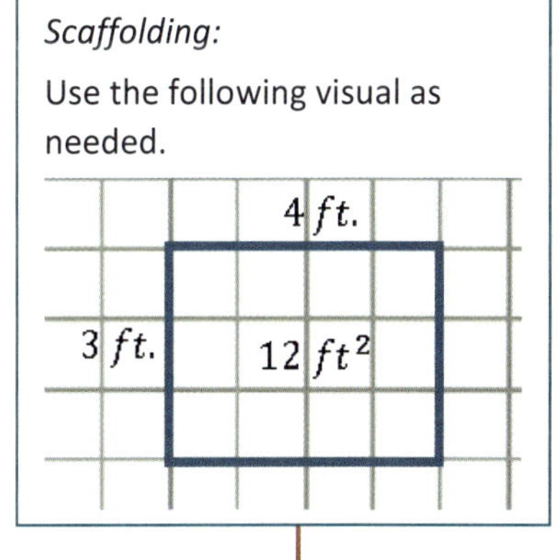

Part (c) asks students to draw a plan of the expanded garden and quantify how the area increases. Allow students time to find a way to show this. Share out student responses; if none are able to produce a valid response, share the diagram shown on the next page.

c. Draw a diagram that shows the change in dimension and area of Patty's garden as she expands it. The diagram should show the original garden as well as the expanded garden.

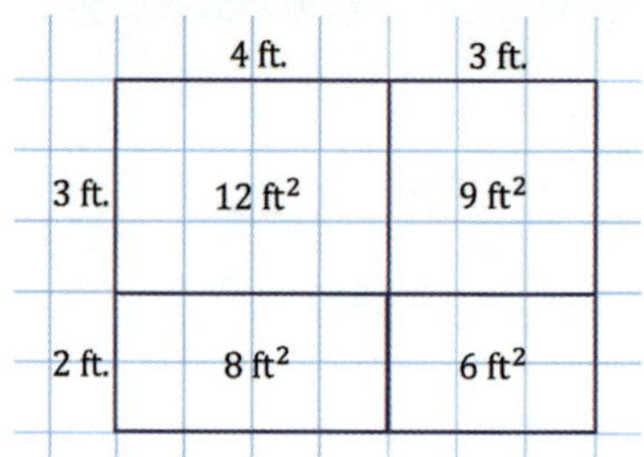

d. Based on your diagram, can the area of the garden be found in a way other than by multiplying the length by the width?

The area can be found by taking the sum of the smaller sections of areas.

e. Based on your diagram, how would the area of the original garden change if only the length increased by 3 ft.? By how much would the area increase?

The area of the garden would increase by 9 ft^2.

f. How would the area of the original garden change if only the width increased by 2 ft.? By how much would the area increase?

The area of the garden would increase by 8 ft^2.

g. Complete the following table with the numeric expression, area, and increase in area for each change in the dimensions of the garden.

Dimensions of the Garden	Numeric Expression for the Area of the Garden	Area of the Garden	Increase in Area of the Garden
The original garden with length of 4 ft. and width of 3 ft.	$4 \text{ ft.} \cdot 3 \text{ ft.}$	12 ft^2	–
The original garden with length extended by 3 ft. and width extended by 2 ft.	$(4+3) \text{ ft.} \cdot (3+2) \text{ ft.}$	35 ft^2	23 ft^2
The original garden with only the length extended by 3 ft.	$(4+3) \text{ ft.} \cdot 3 \text{ ft.}$	21 ft^2	9 ft^2
The original garden with only the width extended by 2 ft.	$4 \text{ ft.} \cdot (3+2) \text{ ft.}$	20 ft^2	8 ft^2

h. Will the increase in both the length and width by 3 ft. and 2 ft., respectively, mean that the original area will increase strictly by the areas found in parts (e) and (f)? If the area is increasing by more than the areas found in parts (e) and (f), explain what accounts for the additional increase.

The area of the garden increases not only by 9 ft^2 *and* 8 ft^2*, but also by an additional* 6 ft^2*. This additional* 6 ft^2 *is the corresponding area formed by the 3 ft. and 2 ft. extensions in both dimensions (length and width); that is, this area results from not just an extension in the length or just in the width but because the extensions occurred in both length and width.*

Example 1 (8 minutes)

Students increase the dimensions of several squares and observe the pattern that emerges from each area model.

Example 1

Examine the change in dimension and area of the following square as it increases by 2 units from a side length of 4 units to a new side length of 6 units. Observe the way the area is calculated for the new square. The lengths are given in units, and the areas of the rectangles and squares are given in units squared.

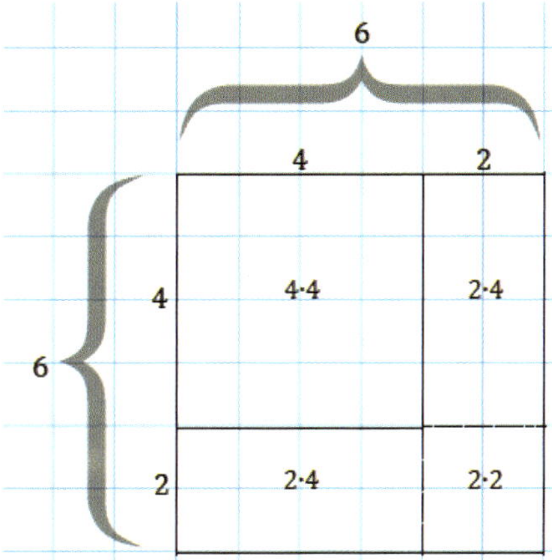

Area of the 6 by 6 square $= (4+2)^2$ units^2 $= 4^2$ units^2 $+ 2(2\cdot 4)$ units^2 $+ 2^2$ units^2 $= 36$ units^2

The area of the 6 by 6 square can be calculated either by multiplying its sides represented by $(4+2)(4+2)$ or by adding the areas of the subsections, represented by $(4\cdot 4)$, $2(2\cdot 4)$, and $(2\cdot 2)$.

a. **Based on the example above, draw a diagram for a square with a side length of 3 units that is increasing by 2 units. Show the area calculation for the larger square in the same way as in the example.**

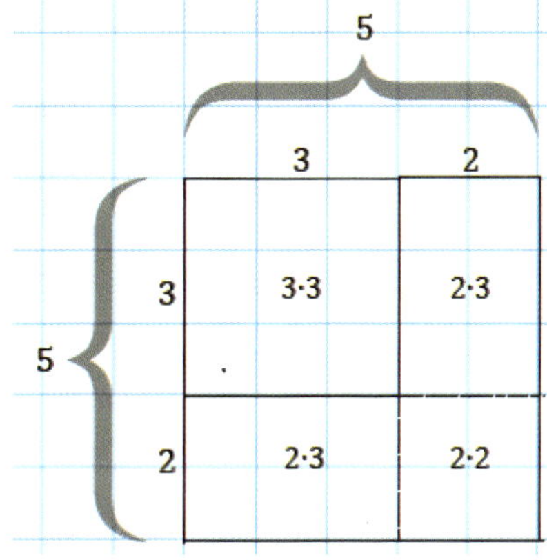

Area of the 5 by 5 square $= (3+2)^2$ units^2 $= 3^2$ units^2 $+ 2(2\cdot 3)$ units^2 $+ 2^2$ units^2 $= 25$ units^2

The area of the 5 by 5 square can be calculated either by multiplying its sides represented by $(3+2)(3+2)$ or by adding the areas of the subsections, represented by $(3\cdot 3)$, $2(2\cdot 3)$, and $(2\cdot 2)$.

b. **Draw a diagram for a square with a side length of 5 units that is increased by 3 units. Show the area calculation for the larger square in the same way as in the example.**

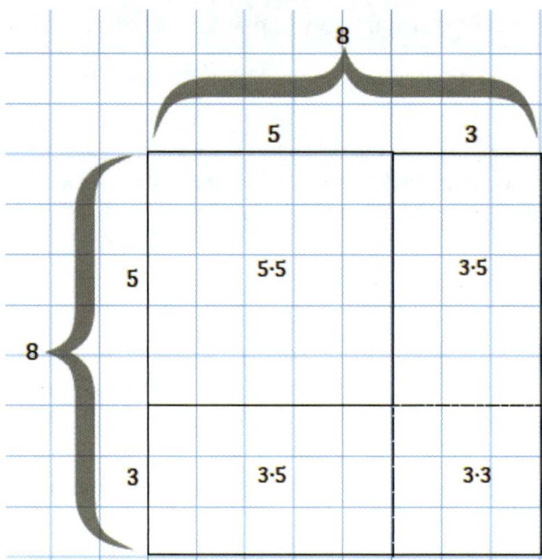

Area of the 8 by 8 square $= (5+3)^2 \text{ units}^2 = 5^2 \text{ units}^2 + 2(3 \cdot 5) \text{ units}^2 + 3^2 \text{ units}^2 = 64 \text{ units}^2$

The area of the 8 by 8 square can be calculated either by multiplying its sides represented by $(5+3)(5+3)$ or by adding the areas of the subsections, represented by $(5 \cdot 5)$, $2(3 \cdot 5)$, and $(3 \cdot 3)$.

MP.2 & MP.8

c. **Generalize the pattern for the area calculation of a square that has an increase in dimension. Let the length of the original square be a units and the increase in length be b units. Use the diagram below to guide your work.**

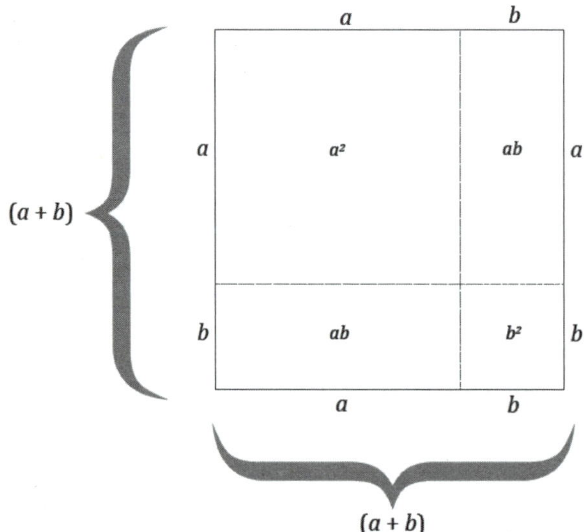

Area of the $(a+b)$ by $(a+b)$ square $= (a+b)^2 \text{ units}^2 = (a^2 + 2ab + b^2) \text{ units}^2$

The area of the square with side length $(a+b)$ is equal to the sum of the areas of the subsections. Describing the area as $(a+b)^2$ units^2 is a way to state the area in terms of the dimensions of the figure; whereas describing the area as $(a^2 + 2ab + b^2)$ units^2 is a way to state the area in terms of the sum of the areas of the sections formed by the extension of b units in each dimension.

Show how the distributive property can be used to rewrite the expression $(a+b)^2$:

$$\boxed{a+b}(a+b) = \boxed{a+b}\cdot a + \boxed{a+b}\cdot b$$

$$(a+b)\cdot a + (a+b)\cdot b = (a\cdot a + b\cdot a) + (a\cdot b + b\cdot b)$$

$$(a\cdot a + b\cdot a) + (a\cdot b + b\cdot b) = a^2 + 2ab + b^2$$

Example 2 (5 minutes)

Students model an increase of one dimension of a square by an unknown amount. Students may hesitate with how to draw this. Instruct them to select an extension length of their choice and label it x so that the reader recognizes the length is an unknown quantity.

Example 2

Bobby draws a square that is 10 units by 10 units. He increases the length by x units and the width by 2 units.

a. **Draw a diagram that models this scenario.**

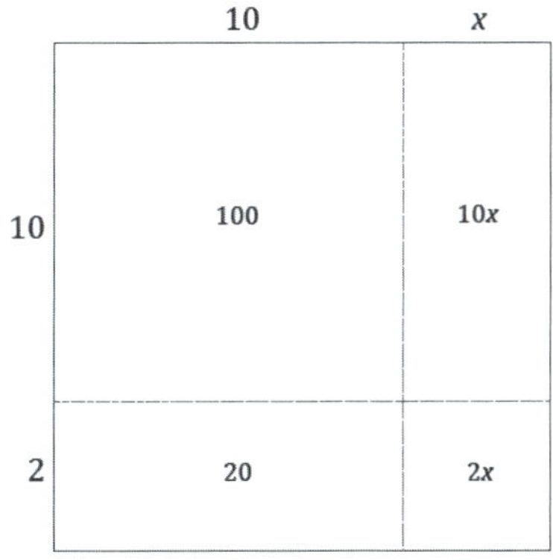

b. **Assume the area of the large rectangle is 156 units^2. Find the value of x.**

$$156 = 100 + 20 + 10x + 2x$$
$$156 = 120 + 12x$$
$$156 - 120 = 120 + 12x - 120$$
$$36 = 12x$$
$$\left(\frac{1}{12}\right)36 = \left(\frac{1}{12}\right)12x$$
$$x = 3$$

Example 3 (7 minutes)

In Example 3, students model an increase in dimensions of a square with a side length of x units, where the increase in the length is different than the increase in the width.

Example 3

The dimensions of a square with a side length of x units are increased. In this figure, the indicated lengths are given in units, and the indicated areas are given in units^2.

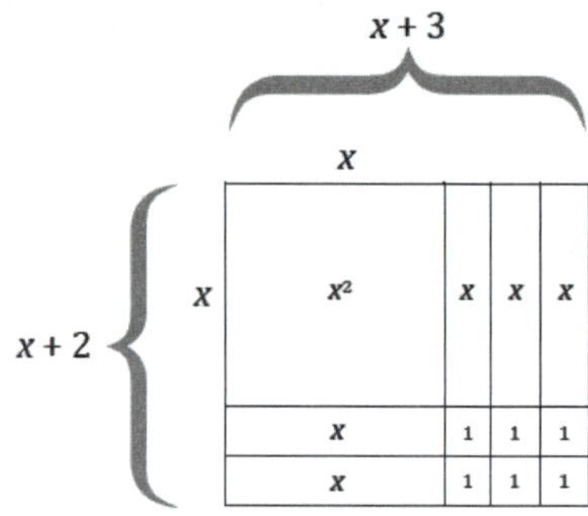

a. **What are the dimensions of the large rectangle in the figure?**

The length (or width) is $(x+3)$ units, and the width (or length) is $(x+2)$ units.

b. **Use the expressions in your response from part (a) to write an equation for the area of the large rectangle, where A represents area.**

$A = (x+3)(x+2)$ units^2

c. **Use the areas of the sections within the diagram to express the area of the large rectangle.**

$A = (x^2 + 3x + 2x + 6)$ units^2

d. **What can be concluded from parts (b) and (c)?**

$(x+2)(x+3) = x^2 + 3x + 2x + 6$

e. **Explain how the expressions $(x+2)(x+3)$ and $x^2 + 3x + 2x + 6$ differ within the context of the area of the figure.**

The expression $(x+2)(x+3)$ shows the area is equal to the quantity of the width increased by 2 units times the quantity of the length increased by 3 units. The expression $x^2 + 3x + 2x + 6$ shows the area as the sum of four sections of the expanded rectangle.

Discussion (12 minutes)

- Even though the context of the area calculation only makes sense for positive values of x, what happens if you substitute a negative number into the equation you stated in part (d) of Example 3? (Hint: Try some values.) Is the resulting number sentence true? Why?

Give students time to try a few negative values for x in the equation $(x+2)(x+3) = x^2 + 5x + 6$. Encourage students to try small numbers (less than 1) and large numbers (greater than 100). Conclusion: The equation becomes a true number sentence for all values in the equation.

- The resulting number sentence is true for negative values of x because of the distributive property. The area properties explain why the equation is true for positive values of x, but the distributive property holds for both positive and negative values of x.
- Show how the distributive property can be used to rewrite the expression from Example 3, $(x+2)(x+3)$, as $x^2 + 5x + 6$. Explain how each step relates back to the area calculation you did above when x is positive.
- Think of $(x+2)$ as a single number, and distribute it over $(x+3)$.

$$\boxed{x+2}(x+3) = \boxed{x+2} \cdot x + \boxed{x+2} \cdot 3$$

(This step is equivalent to relating the area of the entire rectangle to the areas of each of the two corresponding rectangles in the diagram above.)

- Distribute the x over $(x+2)$, and distribute the 3 over $(x+2)$:

$$(x+2) \cdot x + (x+2) \cdot 3 = (x \cdot x + 2 \cdot x) + (x \cdot 3 + 2 \cdot 3)$$

(This step is equivalent to relating the area of the entire rectangle to the areas of each of the two corresponding rectangles in the diagram above.)

- Collecting like terms gives us the right-hand side of the equation displayed in Example 2(d), showing that the two expressions are equivalent both by area properties (when x is positive) and by the properties of operations.

$$(x \cdot x + 2 \cdot x) + (x \cdot 3 + 2 \cdot 3) = x^2 + 5x + 6$$

Closing (1 minute)

The properties of area, because they are limited to positive numbers for lengths and areas, are not as robust as properties of operations, but the area properties do support why the properties of operations are true.

> **Lesson Summary**
>
> - **The properties of area are limited to positive numbers for lengths and areas.**
> - **The properties of area do support why the properties of operations are true.**

Exit Ticket (5 minutes)

Name ______________________________ Date ______________

Lesson 21: Mathematical Area Problems

Exit Ticket

1. Create an area model to represent this product: $(x+4)(x+2)$.

2. Write two different expressions that represent the area.

3. Explain how each expression represents different information about the situation.

4. Show that the two expressions are equal using the distributive property.

Exit Ticket Sample Solutions

1. Create an area model to represent this product: $(x+4)(x+2)$.

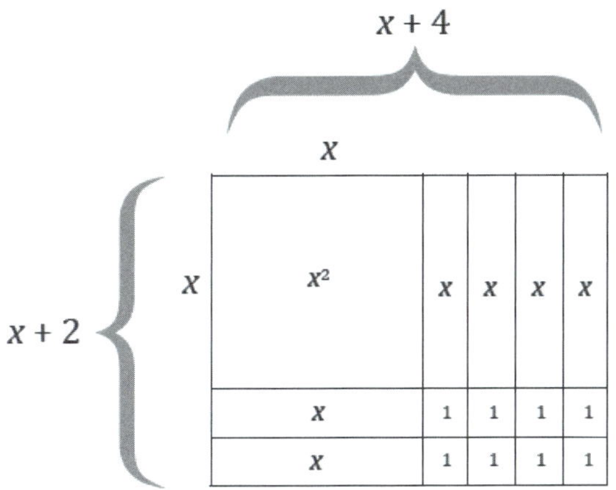

2. Write two different expressions that represent the area.

   $(x+2)(x+4)$ and x^2+6x+8

3. Explain how each expression represents different information about the situation.

   The expression $(x+2)(x+4)$ shows the area is equal to the quantity of the length increased by 4 times the quantity of the width increased by 2. The expression $x^2+4x+2x+8$ shows the area as the sum of four sections of the expanded rectangle.

4. Show that the two expressions are equal using the distributive property.

$$(x+4)(x+2) = (x+4)\cdot x + (x+4)\cdot 2$$

$$(x+4)\cdot x + (x+4)\cdot 2 = (x\cdot x + 4\cdot x) + (x\cdot 2 + 4\cdot 2)$$

$$(x\cdot x + 4\cdot x) + (x\cdot 2 + 4\cdot 2) = x^2 + 6x + 8$$

Problem Set Sample Solutions

1. A square with a side length of a units is decreased by b units in both length and width.

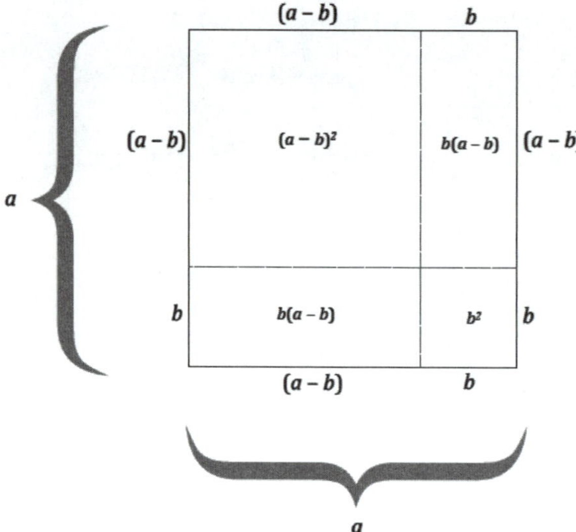

Use the diagram to express $(a-b)^2$ in terms of the other a^2, ab, and b^2 by filling in the blanks below:

$(a-b)^2 = a^2 - b(a-b) - b(a-b) - b^2$

$= a^2 -$ ____ $+$ ____ $-$ ____ $+$ ____ $-b^2$

$= a^2 - 2ab +$ ____ $-b^2$

$=$ ______________________

$(a-b)^2 = a^2 - b(a-b) - b(a-b) - b^2$

$= a^2 - \underline{ba + b^2 - ba + b^2} - b^2$

$= a^2 - 2ab + \underline{2b^2} - b^2$

$= \underline{a^2 - 2ab + b^2}$

2. In Example 3, part (c), we generalized that $(a+b)^2 = a^2 + 2ab + b^2$. Use these results to evaluate the following expressions by writing $1,001 = 1,000 + 1$.

   a. Evaluate 101^2.

   $(100+1)^2 = 100^2 + 2(100 \cdot 1) + 1^2$
   $= 10,000 + 200 + 1$
   $= 10,201$

   b. Evaluate $1,001^2$.

   $(1,000+1)^2 = 1,000^2 + 2(1,000 \cdot 1) + 1^2$
   $= 1,000,000 + 2,000 + 1$
   $= 1,002,001$

c. Evaluate 21^2.

$$(20+1)^2 = 20^2 + 2(20 \cdot 1) + 1^2$$
$$= 400 + 40 + 1$$
$$= 441$$

3. Use the results of Problem 1 to evaluate 999^2 by writing $999 = 1,000 - 1$.

$$(1,000-1)^2 = 1,000^2 - 2(1,000 \cdot 1) + 1^2$$
$$= 1,000,000 - 2,000 + 1$$
$$= 998,001$$

4. The figures below show that $8^2 - 5^2$ is equal to $(8-5)(8+5)$.

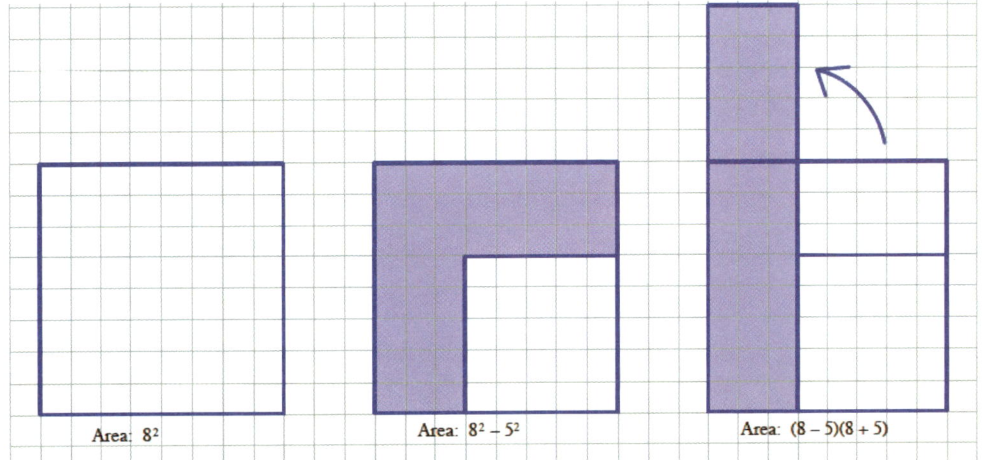

a. Create a drawing to show that $a^2 - b^2 = (a-b)(a+b)$.

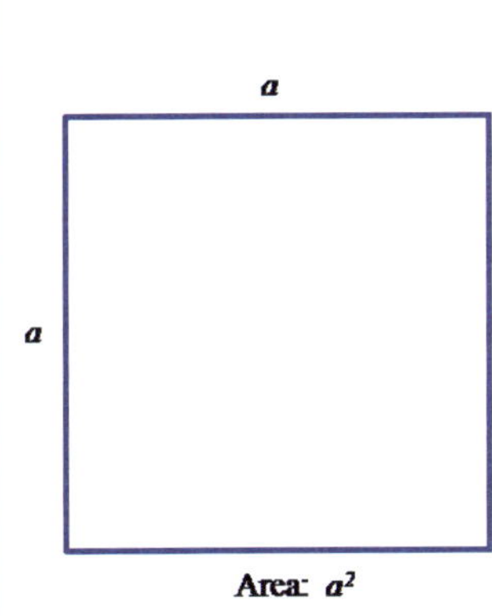

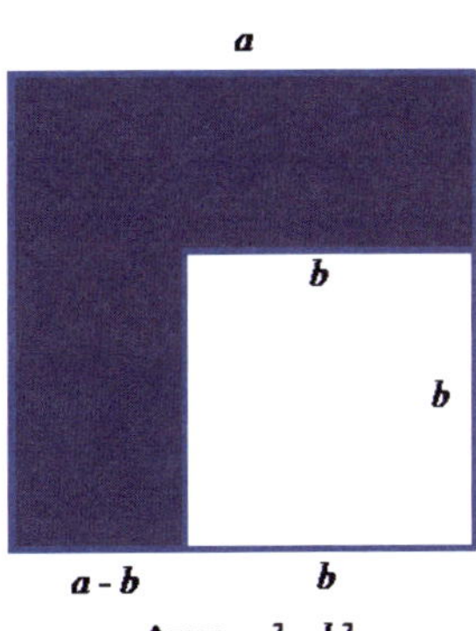

b

a

a

b

b

a - b

b

Area: $(a-b)(a+b)$

$a^2 - b^2 = (a-b)(a+b)$

b. Use the result in part (a), $a^2 - b^2 = (a - b)(a + b)$, to explain why:

i. $35^2 - 5^2 = (30)(40)$.

$35^2 - 5^2 = (35 - 5)(35 + 5) = (30)(40)$

ii. $21^2 - 18^2 = (3)(39)$.

$21^2 - 18^2 = (21 - 18)(21 + 18) = (3)(39)$

iii. $104^2 - 63^2 = (41)(167)$.

$104^2 - 63^2 = (104 - 63)(104 + 63) = (41)(167)$

c. Use the fact that $35^2 = (30)(40) + 5^2 = 1,225$ to create a way to mentally square any two-digit number ending in 5.

$15^2 = (10)(20) + 25 = 225$

$25^2 = (20)(30) + 25 = 625$

$35^2 = (30)(40) + 25 = 1,225$

In general, if the first digit is n, then the first digit(s) are $n(n + 1)$, and the last two digits are 25.

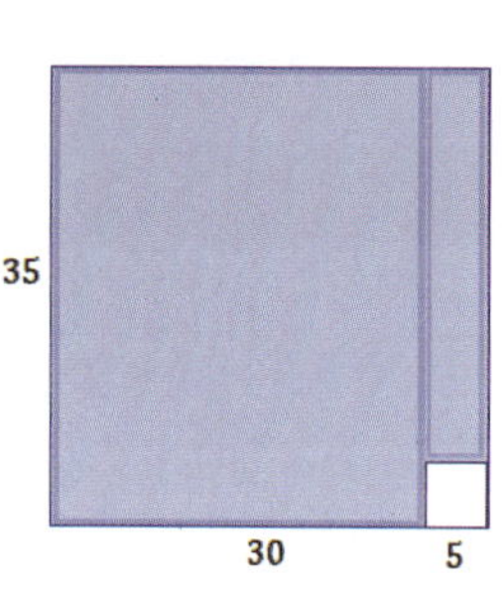

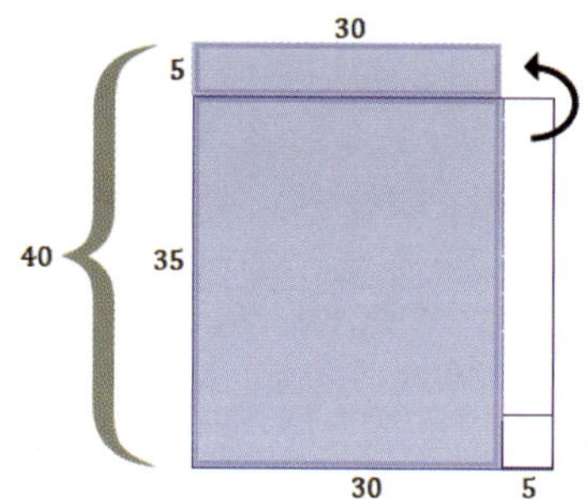

5. Create an area model for each product. Use the area model to write an equivalent expression that represents the area.

a. $(x + 1)(x + 4) = x^2 + x + 4x + 4$

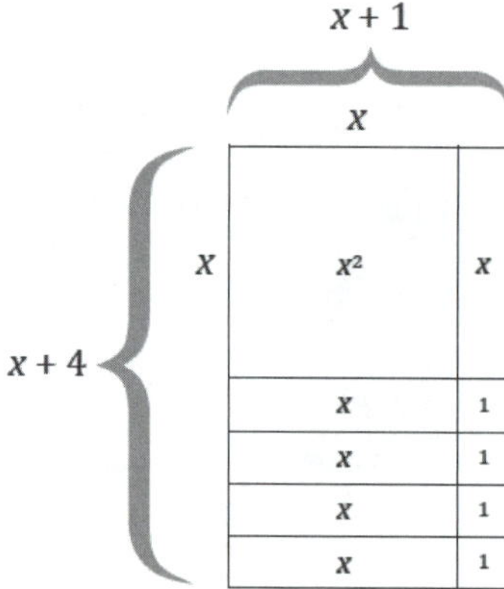

b. $(x+5)(x+2) = x^2 + 5x + 2x + 10$

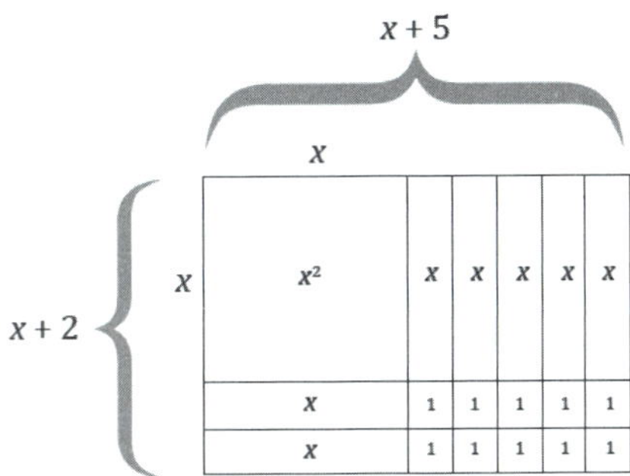

c. Based on the context of the area model, how do the expressions provided in parts (a) and (b) differ from the equivalent expression answers you found for each?

The expression provided in each question shows the area as a length times width product or as the product of two expanded lengths. Each equivalent expression shows the area as the sum of the sections of the expanded rectangle.

6. Use the distributive property to multiply the following expressions.

a. $(2+6)(2+4)$

$$\begin{aligned}(2+6)(2+4) &= (2+6)\cdot 2 + (2+6)\cdot 4\\ &= (2\cdot 2 + 6\cdot 2) + (2\cdot 4 + 6\cdot 4)\\ &= 2^2 + 10(2) + 24\\ &= 48\end{aligned}$$

b. $(x+6)(x+4)$; draw a figure that models this multiplication problem.

$$\begin{aligned}(x+6)(x+4) &= (x+6)\cdot x + (x+6)\cdot 4\\ &= (x\cdot x + 6\cdot x) + (x\cdot 4 + 6\cdot 4)\\ &= x^2 + 10x + 24\end{aligned}$$

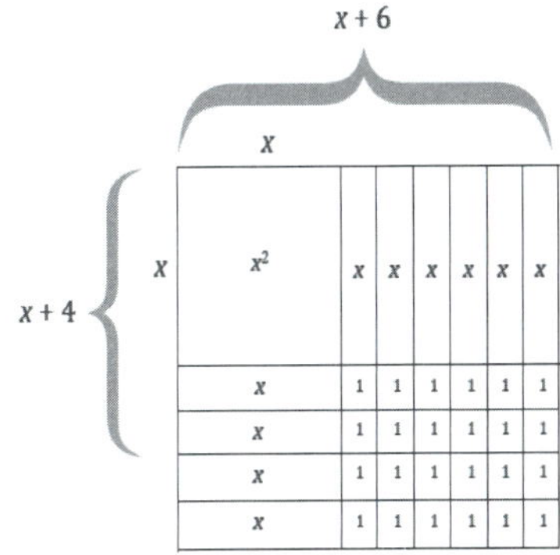

c. $(10+7)(10+7)$

$$\begin{aligned}(10+7)(10+7) &= (10+7)\cdot 10 + (10+7)\cdot 7\\ &= (10\cdot 10 + 7\cdot 10) + (7\cdot 10 + 7\cdot 7)\\ &= 10^2 + 2(7\cdot 10) + 49\\ &= 10^2 + 140 + 49\\ &= 289\end{aligned}$$

d. $(a+7)(a+7)$

$$\begin{aligned}(a+7)(a+7) &= (a+7)\cdot a+(a+7)\cdot 7\\ &= (a\cdot a+7\cdot a)+(a\cdot 7+7\cdot 7)\\ &= a^2+2(7\cdot a)+49\\ &= a^2+14a+49\end{aligned}$$

e. $(5-3)(5+3)$

$$\begin{aligned}(5-3)(5+3) &= (5-3)\cdot 5+(5-3)\cdot 3\\ &= (5\cdot 5-3\cdot 5)+(5\cdot 3-3\cdot 3)\\ &= 5^2-(3\cdot 5)+(5\cdot 3)-3^2\\ &= 5^2-3^2\\ &= 16\end{aligned}$$

f. $(x-3)(x+3)$

$$\begin{aligned}(x-3)(x+3) &= (x-3)\cdot x+(x-3)\cdot 3\\ &= (x\cdot x-3\cdot x)+(x\cdot 3-3\cdot 3)\\ &= x^2-(3\cdot x)+(x\cdot 3)-3^2\\ &= x^2-9\end{aligned}$$

Lesson 22: Area Problems with Circular Regions

Student Outcomes

- Students determine the area of composite figures and of missing regions using composition and decomposition of polygons.

Lesson Notes

Students learned how to calculate the area of circles previously in Grade 7. Throughout this lesson, they apply this knowledge in order to calculate the area of composite shapes. The problems become progressively more challenging. It is important to remind students that they have all the necessary knowledge and skills for every one of these problems.

Classwork

Example 1 (5 minutes)

Allow students time to struggle through the following question. Use the bullet points to lead a discussion to review the problem.

Example 1

a. **The circle to the right has a diameter of 12 cm. Calculate the area of the shaded region.**

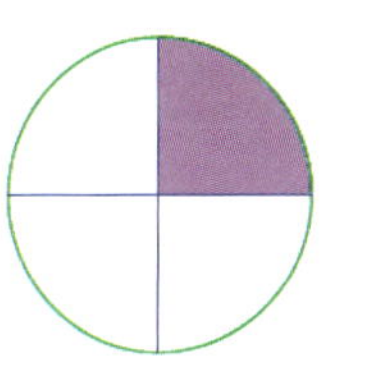

MP.1

- What information do we need to calculate the area?
  - *We need the radius of the circle and the shaded fraction of the circle.*
- What is the radius of the circle? How do you know?
  - *The radius of the circle is 6 cm since the length of the radius is half the length of the diameter.*
- Now that we know the radius, how can we find the area of the shaded region?
  - *Find the area of one quarter of the circle because the circle is divided into four identical parts, and only one part is shaded.*
- Choose a method discussed, and calculate the area of the shaded region.
  - $A_{\text{quarter circle}} = \frac{1}{4}\pi r^2$
  - *The area in squared centimeters:*

$$\frac{1}{4}\pi(6)^2 = 9\pi \approx 28.27$$

  - *The area of the shaded region is about 28.27 cm^2.*

Scaffolding:

Place a prominent visual display in the classroom of a circle and related key relationships (formulas for circumference and area).

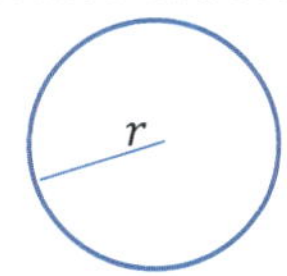

$$\text{Circumference} = 2\pi r = \pi d$$

$$\text{Area} = \pi r^2$$

b. Sasha, Barry, and Kyra wrote three different expressions for the area of the shaded region. Describe what each student was thinking about the problem based on his or her expression.

Sasha's expression: $\frac{1}{4}\pi(6^2)$

Sasha's expression gets directly to the shaded area as it represents a quarter of the area of the whole circle.

Barry's expression: $\pi(6^2) - \frac{3}{4}\pi(6^2)$

Barry's expression shows the area of the whole circle minus the unshaded area of the circle or three-quarters of the area of the circle.

Kyra's expression: $\frac{1}{2}\left(\frac{1}{2}\pi(6^2)\right)$

Kyra's expression arrives at the shaded area by taking half the area of the whole circle, which is half of the circle, and taking half of that area, which leaves a quarter of the area of the whole circle.

Exercise 1 (5 minutes)

Exercise 1

a. Find the area of the shaded region of the circle to the right.

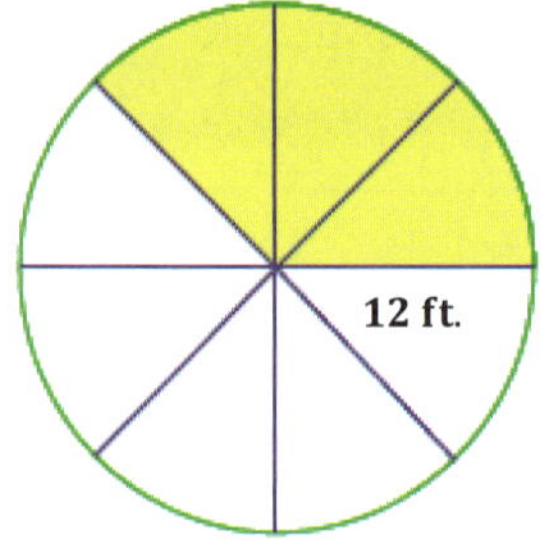

$A = \frac{3}{8}\pi r^2$

$A = \frac{3}{8}(\pi)(12\text{ ft})^2$

$A = 54\pi\text{ ft}^2$

$A \approx 169.65\text{ ft}^2$

The shaded area of the circle is approximately 169.65 ft^2.

b. Explain how the expression you used represents the area of the shaded region.

The expression $\frac{3}{8}(\pi)(12)^2$ *takes the area of a whole circle with a radius of* 12 ft., *which is just the portion that reads* $(\pi)(12)^2$, *and then multiplies that by* $\frac{3}{8}$. *The shaded region is just three out of eight equal pieces of the circle.*

Exercise 2 (7 minutes)

Exercise 2

Calculate the area of the figure below that consists of a rectangle and two quarter circles, each with the same radius. Leave your answer in terms of pi.

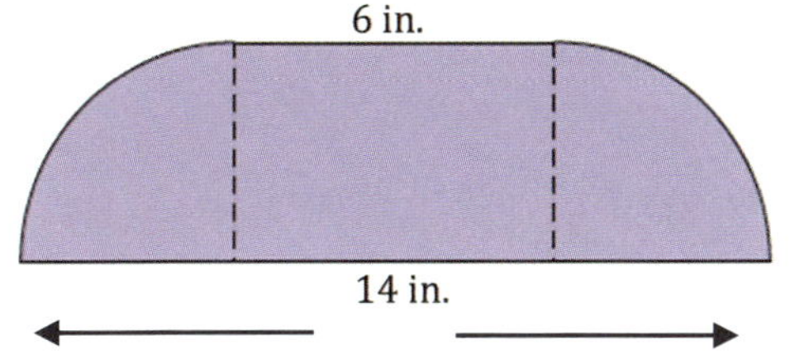

6 in.

4 in. 4 in.

4 in. 6 in. 4 in.

$A_{\text{rectangle}} = l \cdot w$
$A = 6 \text{ in.} \cdot 4 \text{ in.}$
$A = 24 \text{ in}^2$

The area of the rectangle is 24 in^2.

$A_{\text{half circle}} = \frac{1}{2}\pi r^2$
$A = \frac{1}{2}(\pi)(4 \text{ in.})^2$
$A = 8\pi \text{ in}^2$

The area of the two quarter circles, or one semicircle, is $8\pi \text{ in}^2$.

The area of the entire figure is $A = (24 + 8\pi) \text{ in}^2$.

Example 2 (7 minutes)

Example 2

The square in this figure has a side length of 14 inches. The radius of the quarter circle is 7 inches.

a. Estimate the shaded area.

b. What is the exact area of the shaded region?

c. What is the approximate area using $\pi \approx \frac{22}{7}$?

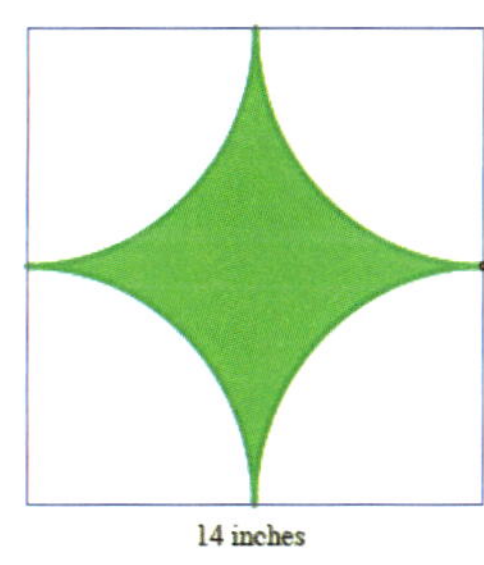

MP.1

- Describe a strategy to find the area of the shaded region.
  - *Find the area of the entire square, and subtract the area of the unshaded region because the unshaded region is four quarter circles of equal radius or a whole circle.*
- What is the difference between parts (b) and (c) in Example 2?
  - *Part (b) asks for the exact area, which means the answer must be left in terms of pi; part (c) asks for the approximate area, which means the answer must be rounded off.*
- How would you estimate the shaded area?
  - *Responses will vary. One possible response might be that the shaded area looks to be approximately one-quarter the area of the entire square, roughly* $\frac{1}{4}(14 \text{ in.})^2 = 49 \text{ in}^2$.

- What is the area of the square?
  - $A = 14 \text{ in.} \cdot 14 \text{ in.} = 196 \text{ in}^2$
    The area of the square is 196 in^2.
- What is the area of each quarter circle?
  - $A = \frac{\pi}{4} r^2 = \frac{\pi}{4}(7 \text{ in.})^2 = \frac{49\pi}{4} \text{ in}^2$
    The area of each quarter circle is $\frac{49\pi}{4} \text{ in}^2$.
    The area of the entire unshaded region, or all four quarter circles, is $49\pi \text{ in}^2$.

MP.1

- What is the exact area of the shaded region?
  - $A = 196 \text{ in}^2 - 49\pi \text{ in}^2$
    The area of the shaded region is $(196 - 49\pi) \text{ in}^2$.
- What is the approximate area using $\pi \approx \frac{22}{7}$?
  - $A \approx 196 \text{ in}^2 - \left(\frac{22}{7}\right)(7 \text{ in.})^2$
    $\approx 196 \text{ in}^2 - 154 \text{ in}^2$
    $\approx 42 \text{ in}^2$
    The area of the shaded region is approximately 42 in^2.

Exercise 3 (7 minutes)

Exercise 3

The vertices A and B of rectangle $ABCD$ are centers of circles each with a radius of 5 inches.

a. **Find the exact area of the shaded region.**

$A_{\text{rectangle}} = 10 \text{ in.} \cdot 5 \text{ in.} = 50 \text{ in}^2$

$A_{\text{semicircle}} = \frac{1}{2}\pi(5 \text{ in.})^2$

$A_{\text{semicircle}} = \frac{25\pi}{2} \text{ in}^2$

$A_{\text{shaded area}} = \left(50 - \frac{25\pi}{2}\right) \text{ in}^2$

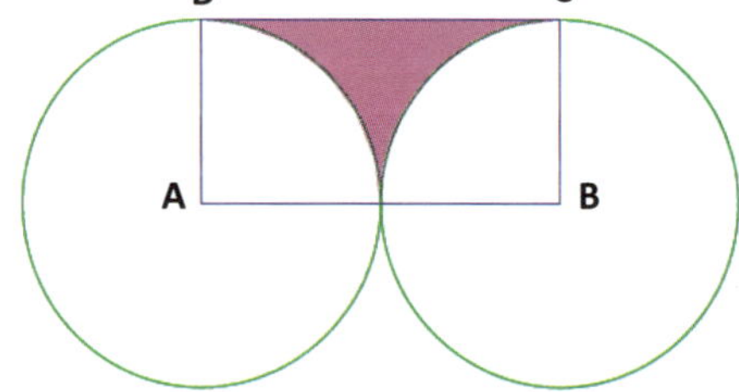

b. **Find the approximate area using $\pi \approx \frac{22}{7}$.**

$A_{\text{shaded area}} = \left(50 - \frac{25\pi}{2}\right) \text{ in}^2$

$A_{\text{shaded area}} \approx \left(50 - \frac{25}{2}\left(\frac{22}{7}\right)\right) \text{ in}^2$

$A_{\text{shaded area}} \approx \left(50 - \frac{275}{7}\right) \text{ in}^2$

$A_{\text{shaded area}} \approx 10\frac{5}{7} \text{ in}^2$

The area of the shaded region in the figure is approximately $10\frac{5}{7} \text{ in}^2$.

EUREKA MATH™

c. Find the area to the nearest hundredth using the π key on your calculator.

$$A_{\text{shaded area}} = \left(50 - \frac{25\pi}{2}\right) \text{ in}^2$$

$$A_{\text{shaded area}} \approx 10.73 \text{ in}^2$$

The area of the shaded region in the figure is approximately 10.73 in^2.

Exercise 4 (5 minutes)

Exercise 4

The diameter of the circle is 12 in. Write and explain a numerical expression that represents the area of the shaded region.

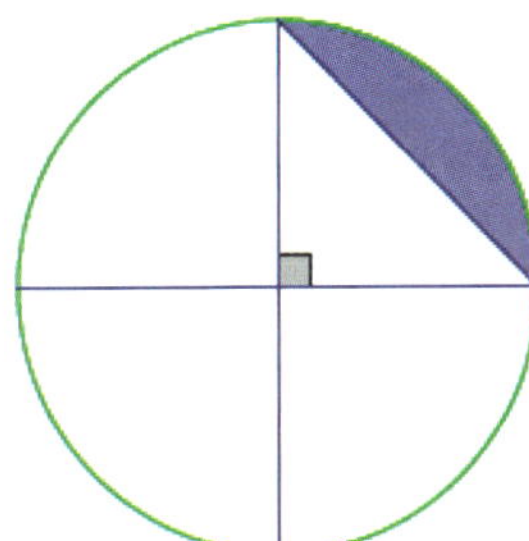

$$A_{\text{shaded}} = A_{\text{quarter circle}} - A_{\text{triangle}}$$

$$A_{\text{shaded}} = \frac{\pi}{4}(6 \text{ in.})^2 - \frac{1}{2}(6 \text{ in.} \cdot 6 \text{ in.})$$

$$A_{\text{shaded}} = (9\pi - 18) \text{ in}^2$$

The expression represents the area of one quarter of the entire circle less the area of the right triangle, whose legs are formed by radii of the circle.

Closing (2 minutes)

Lesson Summary

To calculate composite figures with circular regions:

- Identify relevant geometric areas (such as rectangles or squares) that are part of a figure with a circular region.
- Determine which areas should be subtracted or added based on their positions in the diagram.
- Answer the question, noting if the exact or approximate area is to be found.

Exit Ticket (7 minutes)

Name ______________________________ Date ______________

Lesson 22: Area Problems with Circular Regions

Exit Ticket

A circle with a 10 cm radius is cut into a half circle and two quarter circles. The three circular arcs bound the region below.

a. Write and explain a numerical expression that represents the area.

b. Then, find the area of the figure.

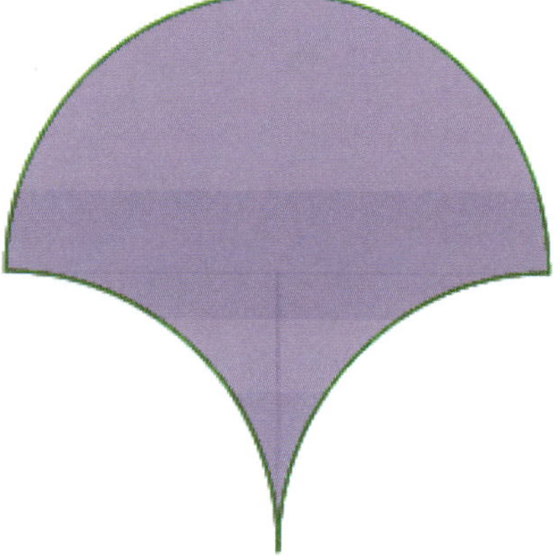

Exit Ticket Sample Solutions

A circle with a 10 cm radius is cut into a half circle and two quarter circles. The three circular arcs bound the region below.

a. Write and explain a numerical expression that represents the area.

b. Then, find the area of the figure.

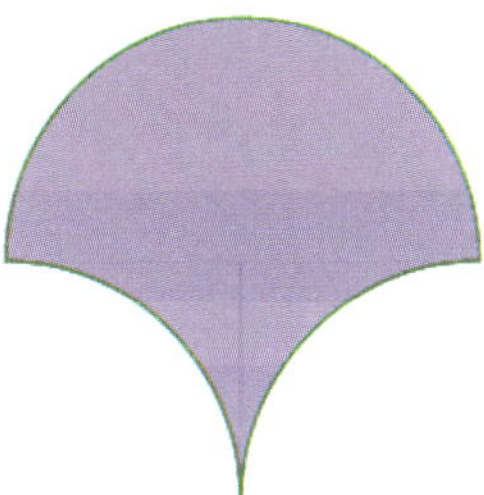

a. Numeric expression 1 for the area:

$10 \text{ cm} \cdot 20 \text{ cm}$

The expression for the area represents the region when it is cut into three pieces and rearranged to make a complete rectangle as shown.

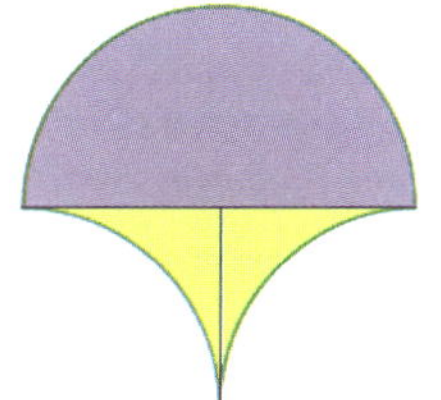
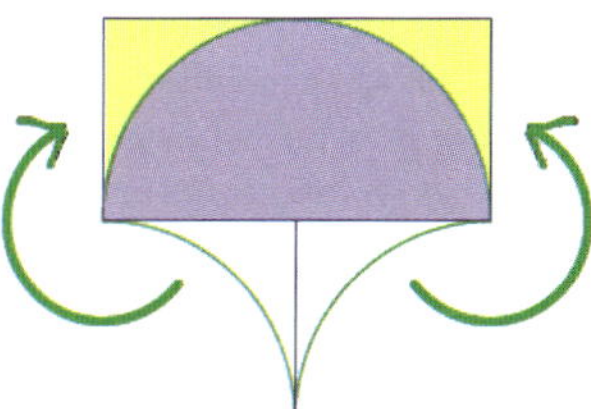

Numeric expression 2 for the area: $(50\pi) \text{ cm}^2 + (200 - 50\pi) \text{ cm}^2$

The expression for the area is calculated as is; in other words, by finding the area of the semicircle in the top portion of the figure and then the area of the carved-out regions in the bottom portion of the figure.

b. The area of the figure is 200 cm^2.

Problem Set Sample Solutions

1. A circle with center O has an area of 96 in^2. Find the area of the shaded region.

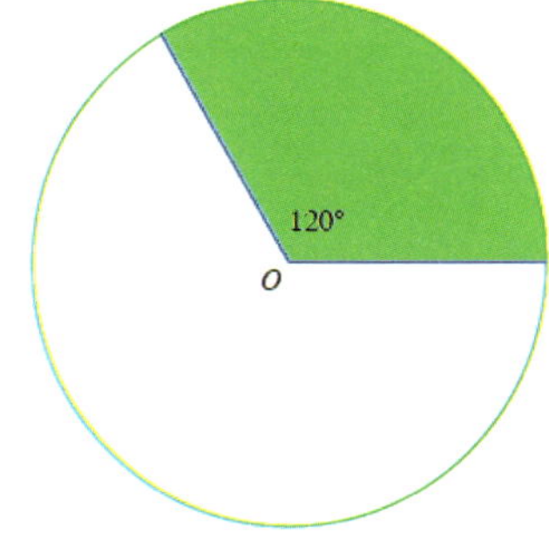

Peyton's Solution

$$A = \frac{1}{3}(96 \text{ in}^2) = 32 \text{ in}^2$$

Monte's Solution

$$A = \frac{96}{120} \text{ in}^2 = 0.8 \text{ in}^2$$

Which person solved the problem correctly? Explain your reasoning.

Peyton solved the problem correctly because he correctly identified the shaded region as one-third of the area of the entire circle. The shaded region represents $\frac{1}{3}$ *of the circle because* $120°$ *is one third of* $360°$*. To find the area of the shaded region, one-third of the area of the entire circle,* 96 in^2*, must be calculated, which is what Peyton did to get his solution.*

2. The following region is bounded by the arcs of two quarter circles, each with a radius of 4 cm, and by line segments 6 cm in length. The region on the right shows a rectangle with dimensions 4 cm by 6 cm. Show that both shaded regions have equal areas.

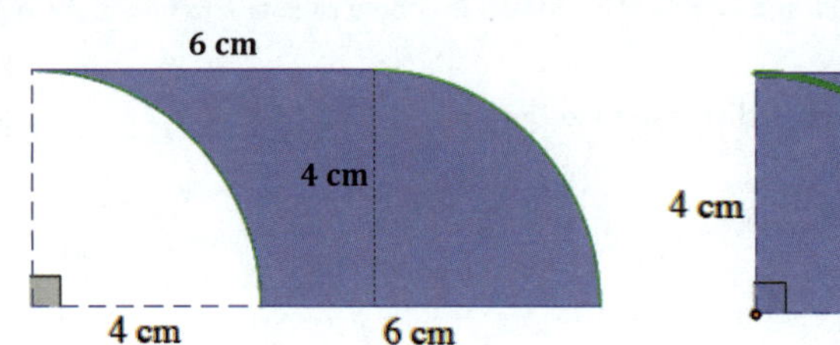

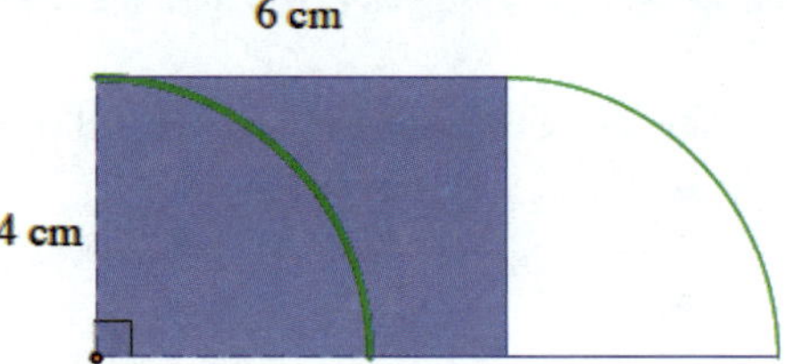

$A = \left((4\text{ cm} \cdot 6\text{ cm}) - \frac{1}{4}\pi(4\text{ cm})^2\right) + \frac{1}{4}\pi(4\text{ cm})^2$

$A = 24\text{ cm}^2$

$A = 4\text{ cm} \cdot 6\text{ cm}$

$A = 24\text{ cm}^2$

3. A square is inscribed in a paper disc (i.e., a circular piece of paper) with a radius of 8 cm. The paper disc is red on the front and white on the back. Two edges of the circle are folded over. Write and explain a numerical expression that represents the area of the figure. Then, find the area of the figure.

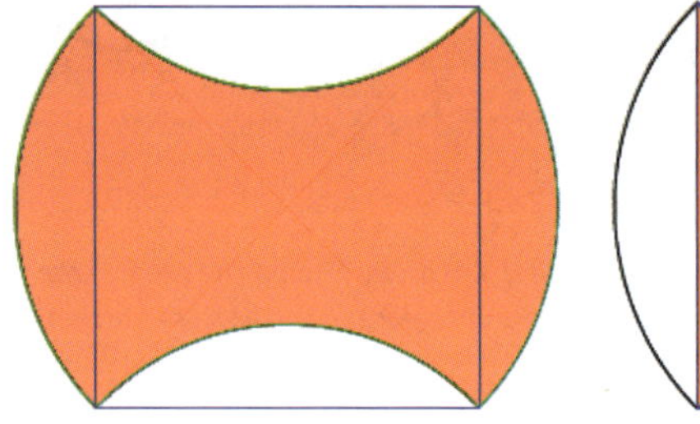

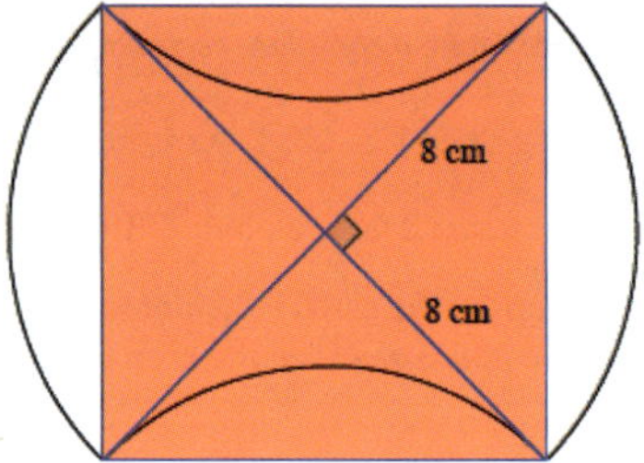

Numeric expression for the area: $4\left(\frac{1}{2} \cdot 8\text{ cm} \cdot 8\text{ cm}\right)$

The shaded (red) area is the same as the area of the square. The radius is 8 cm, which is the length of one leg of each of the four equal-sized right triangles within the square. Thus, we find the area of one triangle and multiply by 4.

The area of the shaded region is 128 cm^2.

4. The diameters of four half circles are sides of a square with a side length of 7 cm.

Figure 1

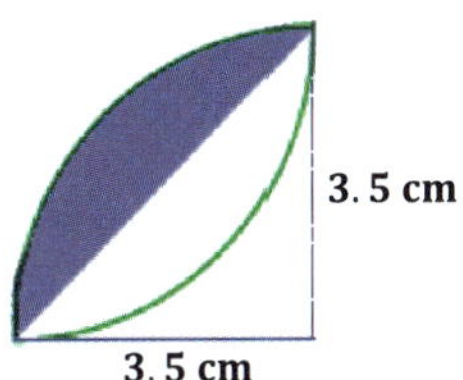

Figure 2
(Not drawn to scale)

a. Find the exact area of the shaded region.

Figure 2 isolates one quarter of Figure 1. The shaded area in Figure 2 can be found as follows:

Shaded area = Area of the quarter circle − Area of the isosceles right triangle.

Shaded area:

$$\left(\frac{\pi}{4}\left(\frac{7}{2}\text{ cm}\right)^2\right) - \left(\frac{1}{2}\cdot\frac{7}{2}\text{ cm}\cdot\frac{7}{2}\text{ cm}\right) = \frac{49\pi}{16}\text{ cm}^2 - \frac{49}{8}\text{ cm}^2$$
$$= \frac{49}{16}(\pi - 2)\text{ cm}^2$$

The area of the shaded region is $\frac{49}{16}(\pi - 2)\text{ cm}^2$. *There are* 8 *such regions in the figure, so we multiply this answer by* 8.

Total shaded area:

$$8\left(\frac{49}{16}(\pi - 2)\right) = \frac{49}{2}(\pi - 2) = \frac{49\pi}{2} - 49$$

The exact area of the shaded region is $\left(\frac{49\pi}{2} - 49\right)\text{ cm}^2$.

b. Find the approximate area using $\pi \approx \frac{22}{7}$.

$$A_{\text{total shaded}} \approx \frac{49}{2}\left(\frac{22}{7} - 2\right)\text{ cm}^2$$
$$A_{\text{total shaded}} \approx (77 - 49)\text{ cm}^2$$
$$A_{\text{total shaded}} \approx 28\text{ cm}^2$$

The approximate area of the shaded region is 28 cm^2.

c. Find the area using the π button on your calculator and rounding to the nearest thousandth.

$$A_{\text{total shaded}} = \frac{49}{2}(\pi - 2)\text{ cm}^2$$
$$A_{\text{total shaded}} \approx 27.969\text{ cm}^2$$

The approximate area of the shaded region is 27.969 cm^2.

5. **A square with a side length of 14 inches is shown below, along with a quarter circle (with a side of the square as its radius) and two half circles (with diameters that are sides of the square). Write and explain a numerical expression that represents the area of the figure.**

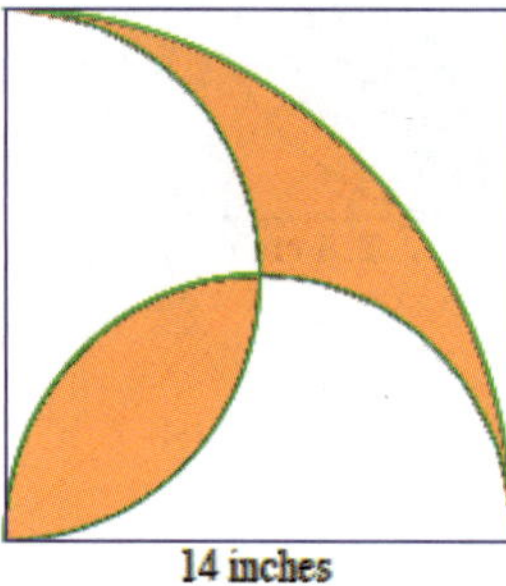

Figure 1

14 inches

Figure 2

Numeric expression for the area: $\frac{1}{4}\pi(14\text{ in.})^2 - \left(\frac{1}{2} \cdot 14\text{ in.} \cdot 14\text{ in.}\right)$

The shaded area in Figure 1 is the same as the shaded area in Figure 2. This area can be found by subtracting the area of the right triangle with leg lengths of 14 in. *from the area of the quarter circle with a radius of* 14 in.

$$\frac{1}{4}\pi(14\text{ in.})^2 - \left(\frac{1}{2} \cdot 14\text{ in.} \cdot 14\text{ in.}\right) = (49\pi - 98)\text{ in}^2$$

6. **Three circles have centers on segment AB. The diameters of the circles are in the ratio $3:2:1$. If the area of the largest circle is 36 ft^2, find the area inside the largest circle but outside the smaller two circles.**

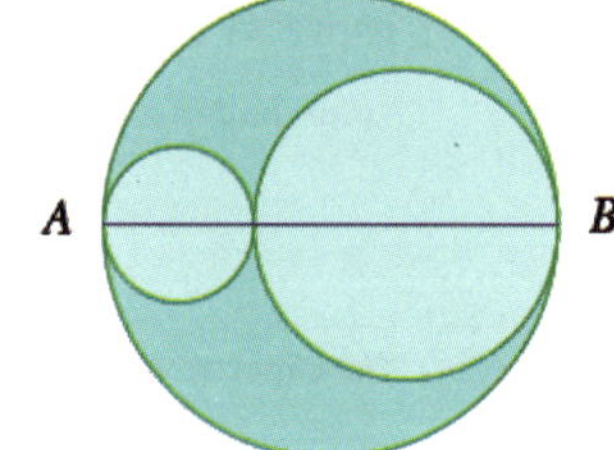

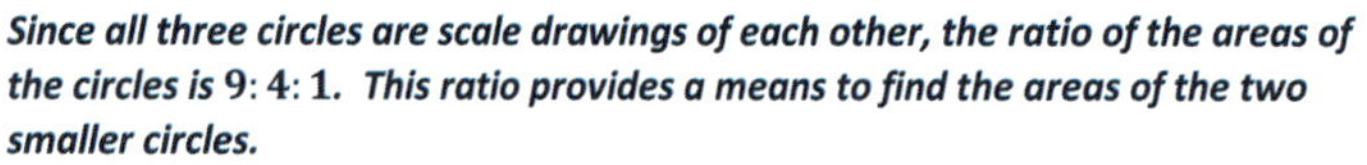

Since all three circles are scale drawings of each other, the ratio of the areas of the circles is $9:4:1$. *This ratio provides a means to find the areas of the two smaller circles.*

Area of medium-sized circle in ft^2:

$$\frac{9}{4} = \frac{36}{x}$$
$$x = 16$$

The area of the medium-sized circle is 16 ft^2.

Area of small-sized circle in ft^2:

$$\frac{9}{1} = \frac{36}{y}$$
$$y = 4$$

The area of the small-sized circle is 4 ft^2.

The area inside the largest circle but outside the smaller two circles is

$$A = 36\text{ ft}^2 - 16\text{ ft}^2 - 4\text{ ft}^2$$
$$A = 16\text{ ft}^2.$$

The area inside the largest circle but outside the smaller two circles is 16 ft^2.

7. A square with a side length of 4 ft. is shown, along with a diagonal, a quarter circle (with a side of the square as its radius), and a half circle (with a side of the square as its diameter). Find the exact, combined area of regions I and II.

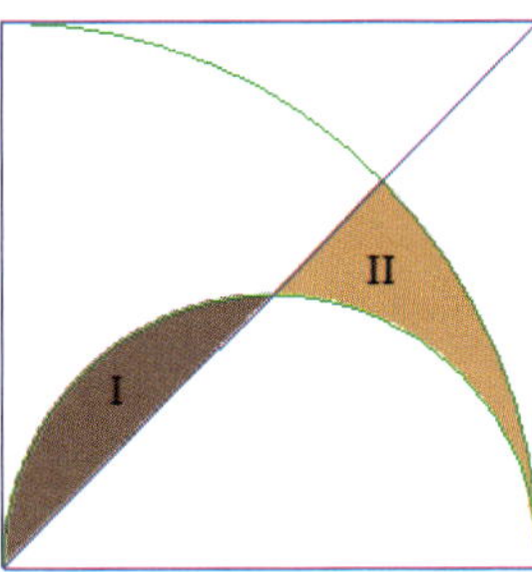

The area of I is the same as the area of III in the following diagram.

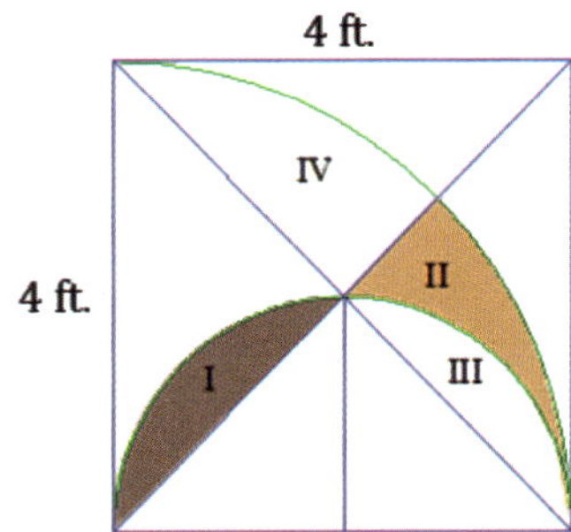

Since the area of I is the same as the area of III, we need to find the combined area of II and III. The combined area of II and III is half the area of II, III, and IV. The area of II, III, and IV is the area of the quarter circle minus the area of the triangle.

$$A_{\text{II and III}} = \frac{1}{2}\left(\frac{1}{4}\pi(4\text{ ft.})^2 - \frac{1}{2}\cdot 4\text{ ft}\cdot 4\text{ ft.}\right)$$

$$A_{\text{II and III}} = \frac{1}{2}\left(\frac{16\pi}{4}\text{ ft}^2 - \frac{4\text{ ft.}\cdot 4\text{ ft.}}{2}\right)$$

$$A_{\text{II and III}} = (2\pi - 4)\text{ ft}^2$$

The combined area of I and II is $(2\pi - 4)\text{ ft}^2$.

Lesson 23: Surface Area

Student Outcomes

- Students determine the surface area of three-dimensional figures, including both composite figures and those missing sections.

Lesson Notes

This lesson is an extension of the work done on surface area in Module 5 of Grade 6 (Lessons 15–19) as well as Module 3 of Grade 7 (Lessons 21–22).

Classwork

Opening Exercise (5 minutes)

> *Scaffolding:*
>
> Encourage students that are struggling to draw a net of the figure to help them determine the surface area.

Opening Exercise

Calculate the surface area of the square pyramid.

5 cm
5 cm
8 cm
8 cm

Area of the square base:

$$s^2 = (8 \text{ cm})^2$$
$$= 64 \text{ cm}^2$$

Area of the triangular lateral sides:

$$\frac{1}{2}bh = \frac{1}{2}(8 \text{ cm})(5 \text{ cm})$$
$$= 20 \text{ cm}^2$$

***There are four lateral sides. So, the area of all 4 triangles is** 80 cm^2.*

MP.1

Surface area:

$$(8 \text{ cm} \cdot 8 \text{ cm}) + 4\left(\frac{1}{2}(8 \text{ cm} \cdot 5 \text{ cm})\right) = 64 \text{ cm}^2 + 80 \text{ cm}^2 = 144 \text{ cm}^2$$

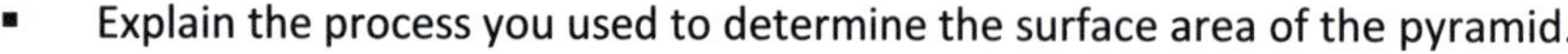

- Explain the process you used to determine the surface area of the pyramid.
  - *Answers will vary. I drew a net to determine the area of each side and added the areas together.*
- The surface area of a pyramid is the union of its base region and all its lateral faces.
- Explain how $(8 \text{ cm} \cdot 8 \text{ cm}) + 4\left(\frac{1}{2}(8 \text{ cm} \cdot 5 \text{ cm})\right)$ represents the surface area.
  - *The area of the square with side lengths of* 8 cm *is represented by* $(8 \text{ cm} \cdot 8 \text{ cm})$*, and the area of the four lateral faces with base lengths of* 8 cm *and heights of* 5 cm *is represented by* $4\left(\frac{1}{2}(8 \text{ cm} \cdot 5 \text{ cm})\right)$.

- Would this method work for any prism or pyramid?
  - *Answers will vary. Calculating the area of each face, including bases, will determine the surface area even if the areas are determined in different orders, by using a formula or net, or any other method.*

Example 1 (10 minutes)

Students find the surface area of the rectangular prism. Then, students determine the surface area of the rectangular prism when it is broken into two separate pieces. Finally, students compare the surface areas before and after the split.

> *Scaffolding:*
>
> Students may benefit from a physical demonstration of this, perhaps using base ten blocks.

Example 1

a. **Calculate the surface area of the rectangular prism.**

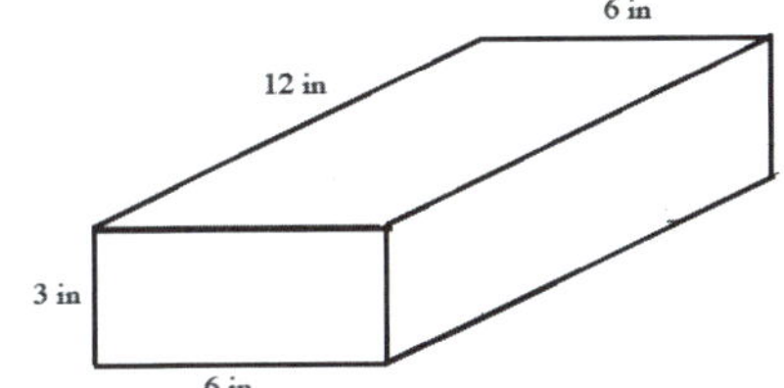

Surface area:

$2(3\text{ in.} \times 6\text{ in.}) + 2(3\text{ in.} \times 12\text{ in.}) + 2(6\text{ in.} \times 12\text{ in.})$
$= 2(18\text{ in}^2) + 2(36\text{ in}^2) + 2(72\text{ in}^2)$
$= 36\text{ in}^2 + 72\text{ in}^2 + 144\text{ in}^2$
$= 252\text{ in}^2$

MP.3

Have students predict in writing or in discussion with a partner whether or not the sum of the two surface areas in part (b) will be the same as the surface area in part (a).

b. **Imagine that a piece of the rectangular prism is removed. Determine the surface area of both pieces.**

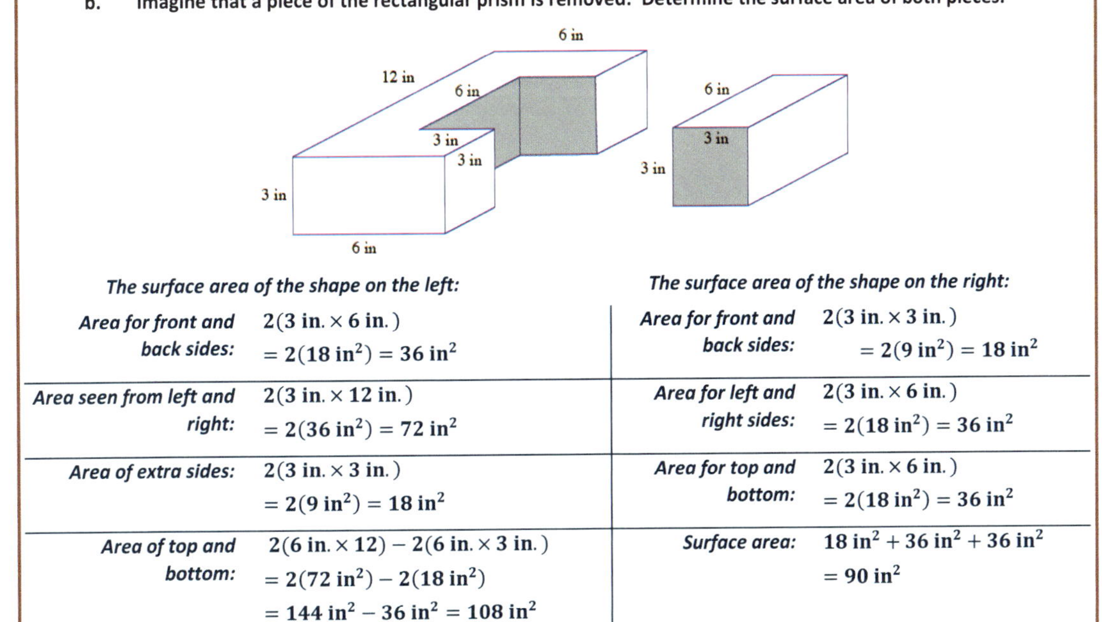

The surface area of the shape on the left:		*The surface area of the shape on the right:*	
Area for front and back sides:	$2(3\text{ in.} \times 6\text{ in.})$ $= 2(18\text{ in}^2) = 36\text{ in}^2$	*Area for front and back sides:*	$2(3\text{ in.} \times 3\text{ in.})$ $= 2(9\text{ in}^2) = 18\text{ in}^2$
Area seen from left and right:	$2(3\text{ in.} \times 12\text{ in.})$ $= 2(36\text{ in}^2) = 72\text{ in}^2$	*Area for left and right sides:*	$2(3\text{ in.} \times 6\text{ in.})$ $= 2(18\text{ in}^2) = 36\text{ in}^2$
Area of extra sides:	$2(3\text{ in.} \times 3\text{ in.})$ $= 2(9\text{ in}^2) = 18\text{ in}^2$	*Area for top and bottom:*	$2(3\text{ in.} \times 6\text{ in.})$ $= 2(18\text{ in}^2) = 36\text{ in}^2$
Area of top and bottom:	$2(6\text{ in.} \times 12) - 2(6\text{ in.} \times 3\text{ in.})$ $= 2(72\text{ in}^2) - 2(18\text{ in}^2)$ $= 144\text{ in}^2 - 36\text{ in}^2 = 108\text{ in}^2$	*Surface area:*	$18\text{ in}^2 + 36\text{ in}^2 + 36\text{ in}^2$ $= 90\text{ in}^2$
Surface area:	$36\text{ in}^2 + 72\text{ in}^2 + 18\text{ in}^2 + 108\text{ in}^2$ $= 234\text{ in}^2$		

- How did you determine the surface area of the shape on the left?
  - *I was able to calculate the area of the sides that are rectangles using length times width. For the two bases that are C-shaped, I used the area of the original top and bottom and subtracted the piece that was taken off.*

> c. **How is the surface area in part (a) related to the surface area in part (b)?**
>
> ***If I add the surface area of both figures, I will get more than the surface area of the original shape.***
>
> $$234 \text{ in}^2 + 90 \text{ in}^2 = 324 \text{ in}^2$$
>
> $$324 \text{ in}^2 - 252 \text{ in}^2 = 72 \text{ in}^2$$
>
> ***72 in^2 is twice the area of the region where the two pieces fit together.***
>
> ***There are 72 more square inches when the prisms are separated.***

Exercises 1–5 (18 minutes)

Exercises

Determine the surface area of the right prisms.

1.

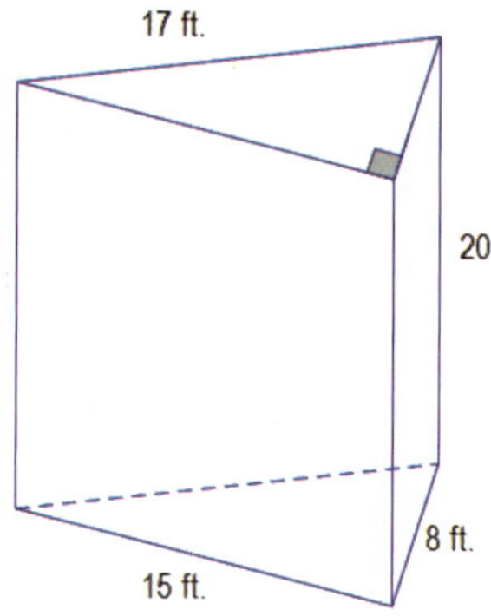

Area of top and bottom: $2\left(\frac{1}{2}(15 \text{ ft.} \times 8 \text{ ft.})\right)$

$= 15 \text{ ft.} \times 8 \text{ ft.} = 120 \text{ ft}^2$

Area of front: $15 \text{ ft.} \times 20 \text{ ft.} = 300 \text{ ft}^2$

Area that can be seen from left: $17 \text{ ft.} \times 20 \text{ ft.} = 340 \text{ ft}^2$

Area that can be seen from right: $8 \text{ ft.} \times 20 \text{ ft.} = 160 \text{ ft}^2$

Surface area: $120 \text{ ft}^2 + 300 \text{ ft}^2 + 340 \text{ ft}^2 + 160 \text{ ft}^2 = 920 \text{ ft}^2$

2.

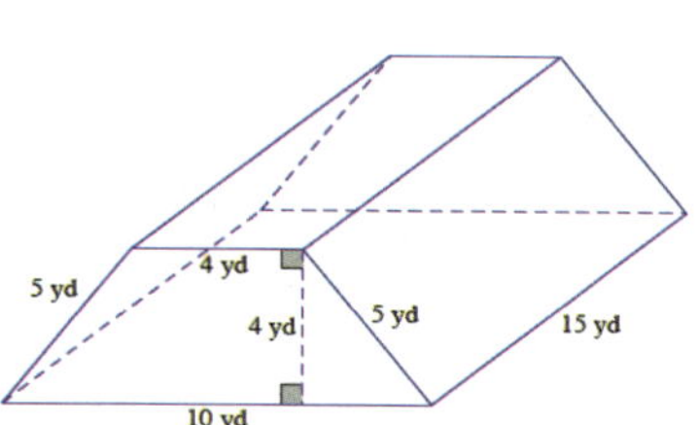

Area of front and back: $2\left(\frac{1}{2}(10 \text{ yd.} + 4 \text{ yd.})4 \text{ yd.}\right)$

$= 14 \text{ yd.} \times 4 \text{ yd.} = 56 \text{ yd}^2$

Area of top: $4 \text{ yd.} \times 15 \text{ yd.} = 60 \text{ yd}^2$

Area that can be seen from left and right: $2(5 \text{ yd.} \times 15 \text{ yd.})$

$= 2(75 \text{ yd}^2) = 150 \text{ yd}^2$

Area of bottom: $10 \text{ yd.} \times 15 \text{ yd.} = 150 \text{ yd}^2$

Surface area: $56 \text{ yd}^2 + 60 \text{ yd}^2 + 150 \text{ yd}^2 + 150 \text{ yd}^2 = 416 \text{ yd}^2$

3.

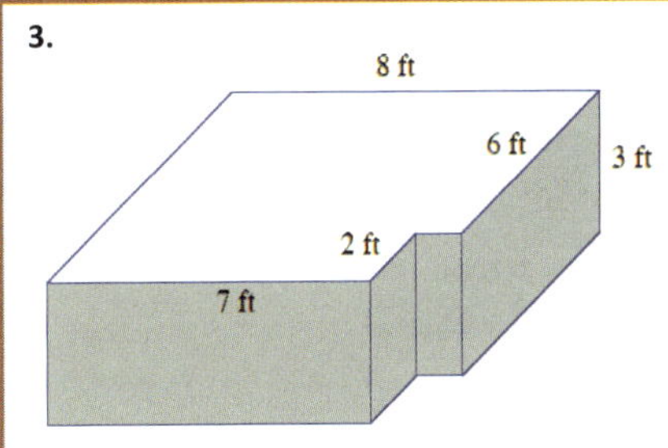

Area of top and bottom: $2((8 \text{ ft.} \times 6 \text{ ft.}) + (7 \text{ ft.} \times 2 \text{ ft.}))$
$= 2(48 \text{ ft}^2 + 14 \text{ ft}^2)$
$= 2(62 \text{ ft}^2) = 124 \text{ ft}^2$

Area for back: $8 \text{ ft.} \times 3 \text{ ft.} = 24 \text{ ft}^2$

Area for front: $7 \text{ ft.} \times 3 \text{ ft.} = 21 \text{ ft}^2$

Area of corner cutout: $(2 \text{ ft.} \times 3 \text{ ft.}) + (1 \text{ ft.} \times 3 \text{ ft.}) = 9 \text{ ft}^2$

Area of right side: $6 \text{ ft.} \times 3 \text{ ft.} = 18 \text{ ft}^2$

Area of left side: $8 \text{ ft.} \times 3 \text{ ft.} = 24 \text{ ft}^2$

Surface area: $124 \text{ ft}^2 + 24 \text{ ft}^2 + 21 \text{ ft}^2 + 9 \text{ ft}^2 + 18 \text{ ft}^2 + 24 \text{ ft}^2 = 220 \text{ ft}^2$

4.

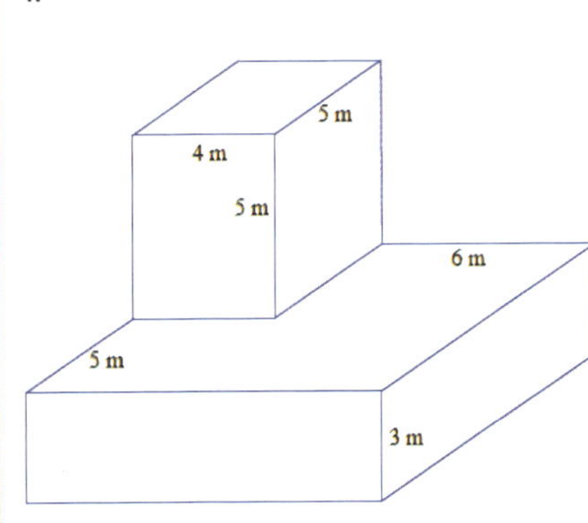

Surface area of top prism:

Area of top: $4 \text{ m} \times 5 \text{ m} = 20 \text{ m}^2$

Area of front and back sides: $2(4 \text{ m} \times 5 \text{ m}) = 40 \text{ m}^2$

Area of left and right sides: $2(5 \text{ m} \times 5 \text{ m}) = 50 \text{ m}^2$

Total surface area of top prism: $20 \text{ m}^2 + 40 \text{ m}^2 + 50 \text{ m}^2 = 110 \text{ m}^2$

Surface area of bottom prism:

Area of top: $10 \text{ m} \times 10 \text{ m} - 20 \text{ m}^2 = 80 \text{ m}^2$

Area of bottom: $10 \text{ m} \times 10 \text{ m} = 100 \text{ m}^2$

Area of front and back sides: $2(10 \text{ m} \times 3 \text{ m}) = 60 \text{ m}^2$

Area of left and right sides: $2(10 \text{ m} \times 3 \text{ m}) = 60 \text{ m}^2$

Total surface area of bottom prism: $80 \text{ m}^2 + 100 \text{ m}^2 + 60 \text{ m}^2 + 60 \text{ m}^2$
$= 300 \text{ m}^2$

Surface area: $110 \text{ m}^2 + 300 \text{ m}^2 = 410 \text{ m}^2$

5.

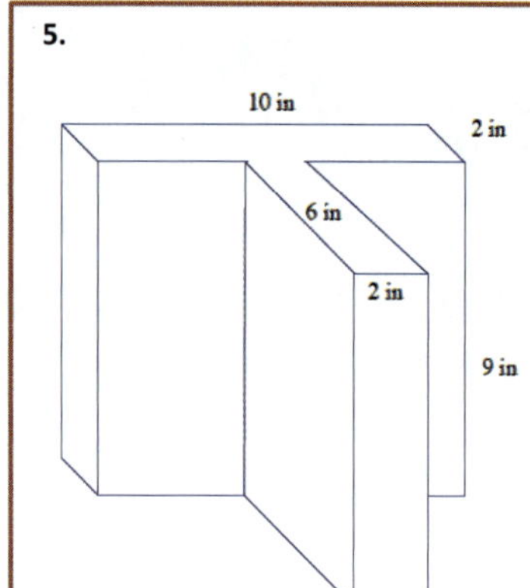

Area of top and bottom faces: $2(10 \text{ in.} \times 2 \text{ in.}) + 2(6 \text{ in.} \times 2 \text{ in.})$
$= 40 \text{ in}^2 + 24 \text{ in}^2$
$= 64 \text{ in}^2$

Area of lateral faces: $2(9 \text{ in.} \times 2 \text{ in.}) + 2(6 \text{ in.} \times 9 \text{ in.}) + 2(9 \text{ in.} \times 10 \text{ in.})$
$= 36 \text{ in}^2 + 108 \text{ in}^2 + 180 \text{ in}^2$
$= 324 \text{ in}^2$

Surface area: $64 \text{ in}^2 + 324 \text{ in}^2 = 388 \text{ in}^2$

Closing (2 minutes)

- Describe the process you use to find the surface area of right prisms that are composite figures or that are missing sections.
  - *To determine the surface area of a right prism, find the area of each lateral face and the two base faces, and then add the areas of all the faces together.*

Lesson Summary

To determine the surface area of right prisms that are composite figures or missing sections, determine the area of each lateral face and the two base faces, and then add the areas of all the faces together.

Exit Ticket (10 minutes)

Name ______________________________ Date ______________

Lesson 23: Surface Area

Exit Ticket

Determine and explain how to find the surface area of the following right prisms.

1.

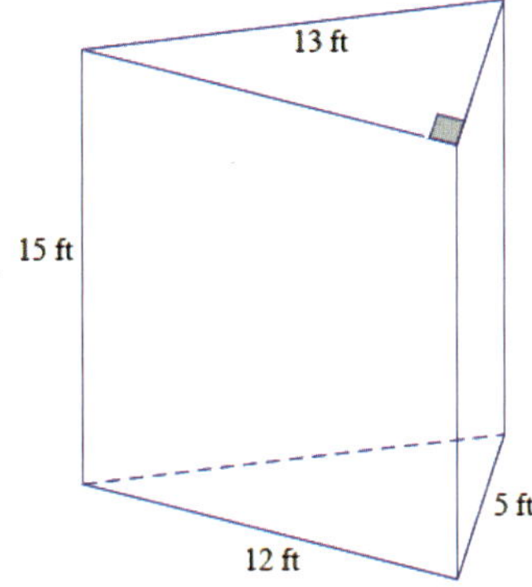

2.

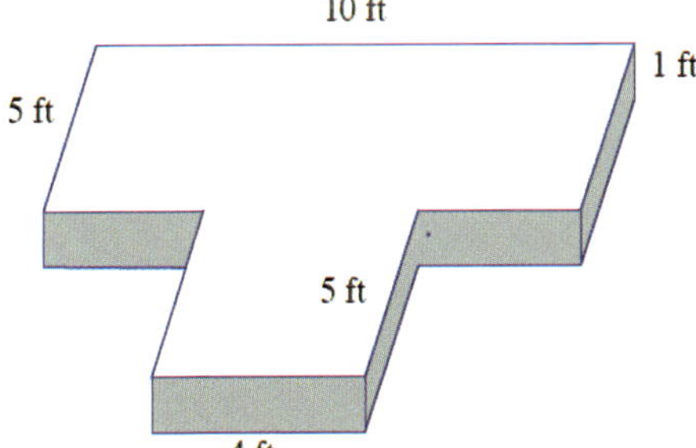

Exit Ticket Sample Solutions

Determine and explain how to find the surface area of the following right prisms.

1.

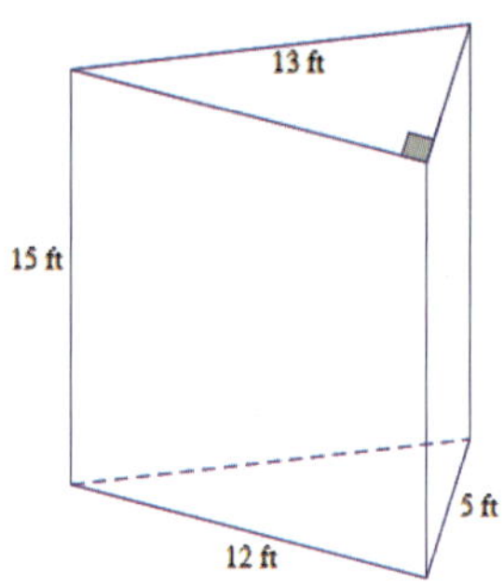

Area of top and bottom: $2\left(\frac{1}{2}(12\text{ ft.} \times 5\text{ ft.})\right)$
$= 12\text{ ft.} \times 5\text{ ft.}$
$= 60\text{ ft}^2$

Area of front: $12\text{ ft.} \times 15\text{ ft.}$
$= 180\text{ ft}^2$

Area seen from left: $13\text{ ft.} \times 15\text{ ft.}$
$= 195\text{ ft}^2$

Area seen from right: $5\text{ ft.} \times 15\text{ ft.}$
$= 75\text{ ft}^2$

Surface area: $60\text{ ft}^2 + 180\text{ ft}^2 + 195\text{ ft}^2 + 75\text{ ft}^2$
$= 510\text{ ft}^2$

To find the surface area of the triangular prism, I must sum the areas of two triangles (the bases that are equal in area) and the areas of three different-sized rectangles.

2.

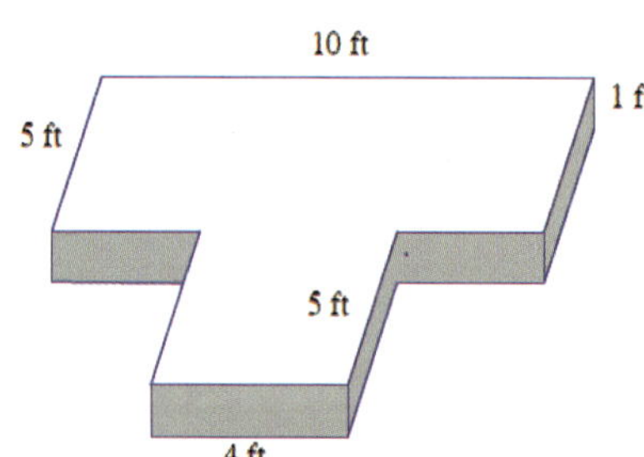

Area of front and back: $2(10\text{ ft.} \times 1\text{ ft.}) = 20\text{ ft}^2$

Area of sides: $2(10\text{ ft.} \times 1\text{ ft.}) = 20\text{ ft}^2$

Area of top and bottom: $2(10\text{ ft.} \times 5\text{ ft.}) + 2(4\text{ ft.} \times 5\text{ ft.})$
$= 100\text{ ft}^2 + 40\text{ ft}^2$
$= 140\text{ ft}^2$

Surface area: $20\text{ ft}^2 + 20\text{ ft}^2 + 140\text{ ft}^2$
$= 180\text{ ft}^2$

To find the surface area of the prism, I must sum the composite area of the bases with rectangular areas of the sides of the prism.

Problem Set Sample Solutions

Determine the surface area of the figures.

1.

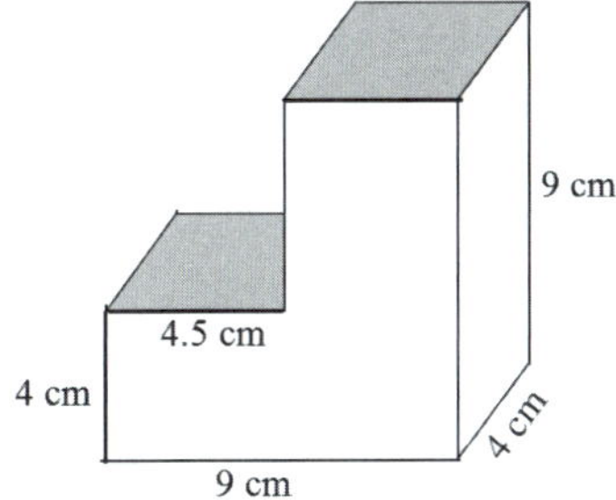

Area of top and bottom: $2(9\text{ cm} \times 4\text{ cm}) = 72\text{ cm}^2$

Area of left and right sides: $2(4\text{ cm} \times 9\text{ cm}) = 72\text{ cm}^2$

Area of front and back: $2(9\text{ cm} \times 4\text{ cm}) + 2(4.5\text{ cm} \times 5\text{ cm})$
$= 117\text{ cm}^2$

Surface area: $72\text{ cm}^2 + 72\text{ cm}^2 + 117\text{ cm}^2 = 261\text{ cm}^2$

2.

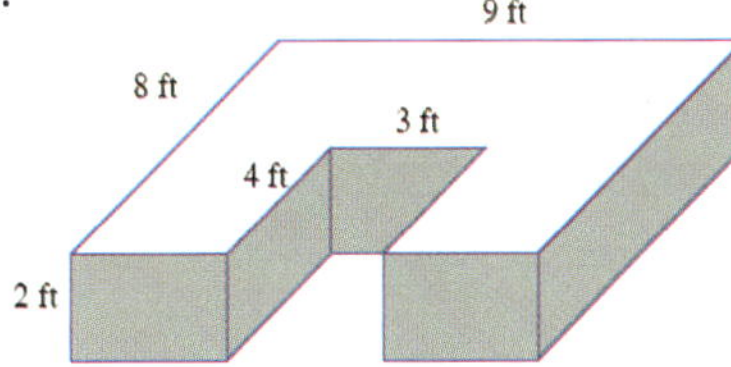

Area of front and back: $2(9\text{ ft.} \times 2\text{ ft.}) = 36\text{ ft}^2$

Area of sides: $2(8\text{ ft.} \times 2\text{ ft.}) = 32\text{ ft}^2$

Area of top and bottom: $2(9\text{ ft.} \times 8\text{ ft.}) - 2(4\text{ ft.} \times 3\text{ ft.})$
$= 144\text{ ft}^2 - 24\text{ ft}^2$
$= 120\text{ ft}^2$

Area of interior sides: $2(4\text{ ft.} \times 2\text{ ft.}) = 16\text{ ft}^2$

Surface area: $36\text{ ft}^2 + 32\text{ ft}^2 + 120\text{ ft}^2 + 16\text{ ft}^2 = 204\text{ ft}^2$

3.

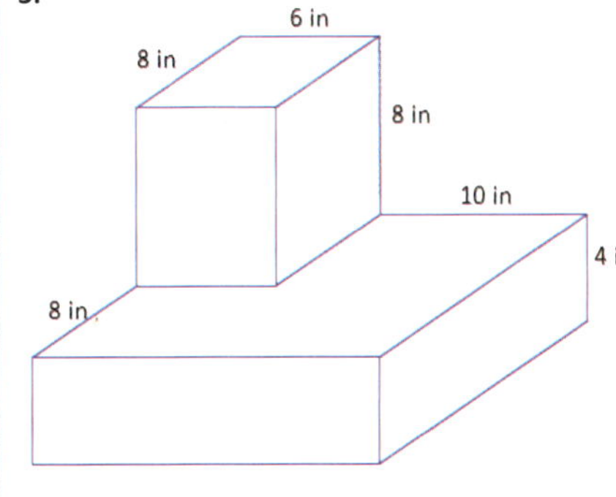

Surface area of top prism:

Area of top: $8\text{ in.} \times 6\text{ in.} = 48\text{ in}^2$

Area of front and back sides: $2(6\text{ in.} \times 8\text{ in.}) = 96\text{ in}^2$

Area of left and right sides: $2(8\text{ in.} \times 8\text{ in.}) = 128\text{ in}^2$

Total surface area of top prism: $48\text{ in}^2 + 96\text{ in}^2 + 128\text{ in}^2 = 272\text{ in}^2$

Surface area of bottom prism:

Area of top: $16\text{ in.} \times 16\text{ in.} - 48\text{ in}^2 = 208\text{ in}^2$

Area of bottom: $16\text{ in.} \times 16\text{ in.} = 256\text{ in}^2$

Area of front and back sides: $2(16\text{ in.} \times 4\text{ in.}) = 128\text{ in}^2$

Area of left and right sides: $2(16\text{ in.} \times 4\text{ in.}) = 128\text{ in}^2$

Total surface area of bottom prism: $208\text{ in}^2 + 256\text{ in}^2 + 128\text{ in}^2 + 128\text{ in}^2$
$= 720\text{ in}^2$

Surface area: $272\text{ in}^2 + 720\text{ in}^2 = 992\text{ in}^2$

4.

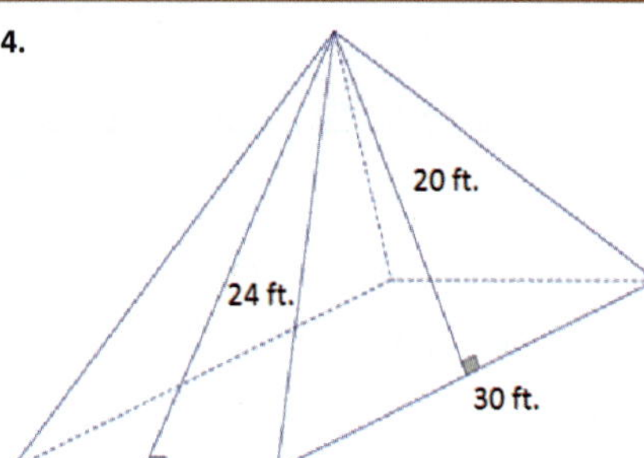

Area of the rectangle base: $14 \text{ ft.} \times 30 \text{ ft.} = 420 \text{ ft}^2$

Area of the triangular lateral sides:

Area of front and back: $\frac{1}{2}bh$

$= 2\left(\frac{1}{2}(14 \text{ ft.})(24 \text{ ft.})\right)$

$= 336 \text{ ft}^2$

Area that can be seen from left and right: $\frac{1}{2}bh$

$= 2\left(\frac{1}{2}(30 \text{ ft.})(20 \text{ ft.})\right)$

$= 600 \text{ ft}^2$

Surface area: $420 \text{ ft}^2 + 336 \text{ ft}^2 + 600 \text{ ft}^2 = 1,356 \text{ ft}^2$

5.

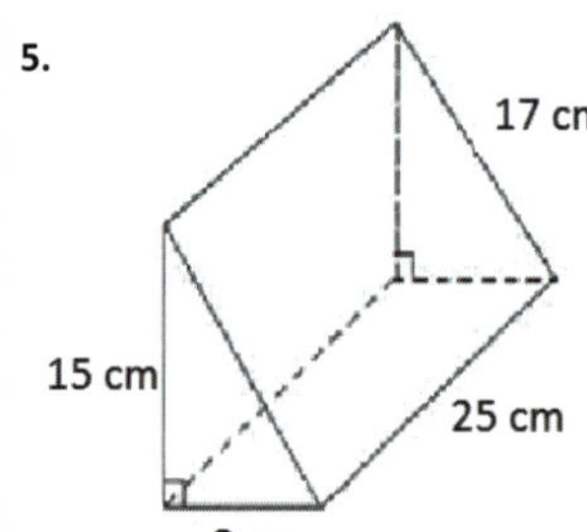

Area of front and back: $2\left(\frac{1}{2}(8 \text{ cm} \times 15 \text{ cm})\right)$

$= 8 \text{ cm} \times 15 \text{ cm}$

$= 120 \text{ cm}^2$

Area of bottom: $8 \text{ cm} \times 25 \text{ cm}$

$= 200 \text{ cm}^2$

Area that can be seen from left side: $25 \text{ cm} \times 15 \text{ cm}$

$= 375 \text{ cm}^2$

Area that can be seen from right side: $25 \text{ cm} \times 17 \text{ cm}$

$= 425 \text{ cm}^2$

Surface area: $120 \text{ cm}^2 + 200 \text{ cm}^2 + 375 \text{ cm}^2 + 425 \text{ cm}^2 = 1,120 \text{ cm}^2$

Lesson 24: Surface Area

Student Outcomes

- Students determine the surface area of three-dimensional figures—those that are composite figures and those that have missing sections.

Lesson Notes

This lesson is a continuation of Lesson 23. Students continue to work on surface area advancing to figures with missing sections.

Classwork

Example 1 (8 minutes)

Scaffolding:

As in Lesson 23, students can draw nets of the figures to help them visualize the area of the faces. They could determine the area of these without the holes first and subtract the surface area of the holes.

Students should solve this problem on their own.

Example 1

Determine the surface area of the image.

Surface area of top or bottom prism:

Lateral sides $= 4(12\text{ in.} \times 3\text{ in.}) = 144\text{ in}^2$

Base face $= 12\text{ in.} \times 12\text{ in.} = 144\text{ in}^2$

Base face with hole $= 12\text{ in.} \times 12\text{ in.} - 4\text{ in.} \times 4\text{ in.}$
$= 128\text{ in}^2$

There are two of these, making up 832 in^2.

Surface area of middle prism:

Lateral sides: $= 4(4\text{ in.} \times 8\text{ in.}) = 128\text{ in}^2$

Surface area: $832\text{ in}^2 + 128\text{ in}^2 = 960\text{ in}^2$

12 in
12 in
3 in
8 in
4 in
4 in
12 in
12 in
3 in

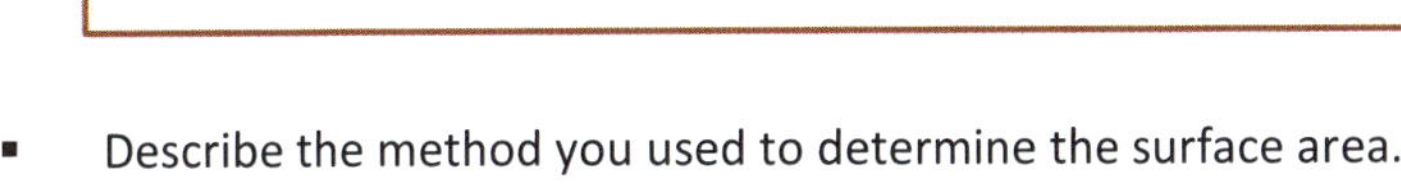

- Describe the method you used to determine the surface area.
  - *Answers will vary. I determined the surface area of each prism separately and added them together. Then, I subtracted the area of the sections that were covered by another prism.*
- If all three prisms were separate, would the sum of their surface areas be the same as the surface area you determined in this example?
  - *No, if the prisms were separate, there would be more surfaces shown. The three separate prisms would have a greater surface area than this example. The area would be greater by the area of four* $4\text{ in.} \times 4\text{ in.}$ *squares* (64 in^2).

Example 2 (5 minutes)

> *Scaffolding:*
> As in Lesson 23, students can draw nets of the figures to help them visualize the area of the faces. They could determine the area of these without the holes first and subtract the surface area of the holes.

Example 2

a. **Determine the surface area of the cube.**

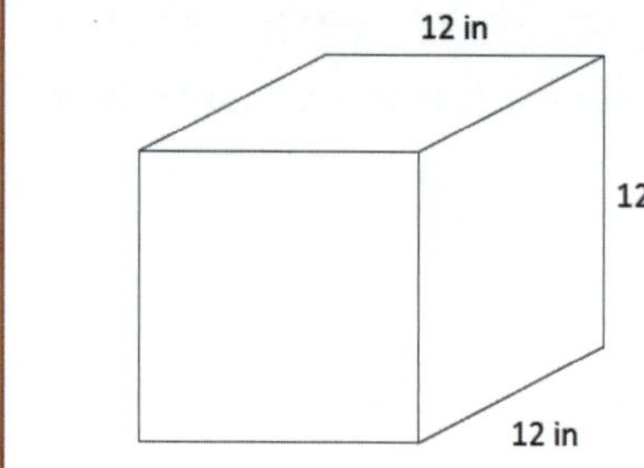

$$\textbf{Surface area} = 6s^2$$
$$SA = 6(12\text{ in.})^2$$
$$SA = 6(144\text{ in}^2)$$
$$SA = 864\text{ in}^2$$

- Explain how $6(12\text{ in.})^2$ represents the surface area of the cube.
  - *The area of one face, one square with a side length of* 12 in.*, is* $(12\text{ in.})^2$*, and so a total area of all six faces is* $6(12\text{ in.})^2$.

b. **A square hole with a side length of 4 inches is cut through the cube. Determine the new surface area.**

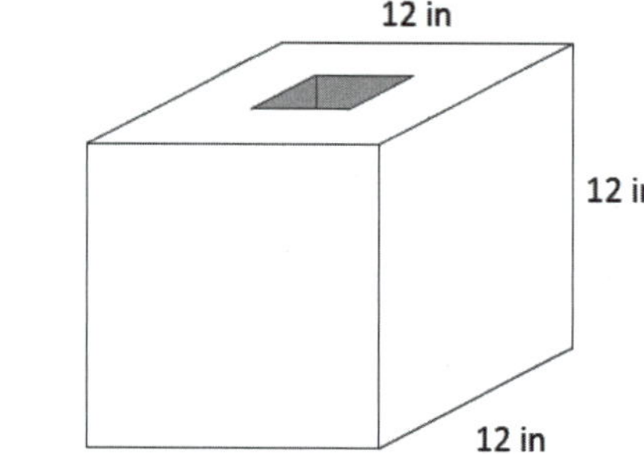

Area of interior lateral sides

$$= 4(12\text{ in.} \times 4\text{ in.})$$
$$= 192\text{ in}^2$$

Surface area of cube with holes

$$= 6\,(12\text{ in.})^2 - 2(4\text{ in.} \times 4\text{ in.}) + 4(12\text{ in.} \times 4\text{ in.})$$
$$= 864\text{ in}^2 - 32\text{ in}^2 + 192\text{ in}^2$$
$$= 1{,}024\text{ in}^2$$

- How does cutting a hole in the cube change the surface area?
  - *We have to subtract the area of the square at the surface from each end.*
  - *We also have to add the area of each of the interior faces to the total surface area.*
- What happens to the surfaces that now show inside the cube?
  - *These are now part of the surface area.*
- What is the shape of the piece that was removed from the cube?
  - *A rectangular prism was cut out of the cube with the following dimensions:* $4\text{ in.} \times 4\text{ in.} \times 12\text{ in.}$
- How can we use this to help us determine the new total surface area?
  - *We can find the surface area of the cube and the surface area of the rectangular prism, but we will have to subtract the area of the square bases from the cube and also exclude these bases in the area of the rectangular prism.*
- Why is the surface area larger when holes have been cut into the cube?
  - *There are more surfaces showing now. All of the surfaces need to be included in the surface area.*

- Explain how the expression $6(12\text{ in.})^2 - 2(4\text{ in.} \times 4\text{ in.}) + 4(12\text{ in.} \times 4\text{ in.})$ represents the surface area of the cube with the hole.
  - *From the total surface area of a whole (uncut) cube,* $6(12\text{ in.})^2$*, the area of the bases (the cuts made to the surface of the cube) are subtracted:* $6(12\text{ in.})^2 - 2(4\text{ in.} \times 4\text{ in.})$*. To this expression we add the area of the four lateral faces of the cutout prism,* $4(12\text{ in.} \times 4\text{ in.})$*. The complete expression then is* $6(12\text{ in.})^2 - 2(4\text{ in.} \times 4\text{ in.}) + 4(12\text{ in.} \times 4\text{ in.})$

Example 3 (5 minutes)

Example 3

A right rectangular pyramid has a square base with a side length of 10 inches. The surface area of the pyramid is 260 in^2. Find the height of the four lateral triangular faces.

Area of base $= 10\text{ in.} \times 10\text{ in.} = 100\text{ in}^2$

Area of the four faces $= 260\text{ in}^2 - 100\text{ in}^2 = 160\text{ in}^2$

The total area of the four faces is 160 in^2***. Therefore, the area of each triangular face is*** 40 in^2***.***

Area of lateral side $= \frac{1}{2}bh$

$40\text{ in}^2 = \frac{1}{2}(10\text{ in.})h$

$40\text{ in}^2 = (5\text{ in.})h$

$h = 8\text{ in.}$

The height of each lateral triangular face is 8 ***inches.***

- What strategies could you use to help you solve this problem?
  - *I could draw a picture of the pyramid and label the sides so that I can visualize what the problem is asking me to do.*

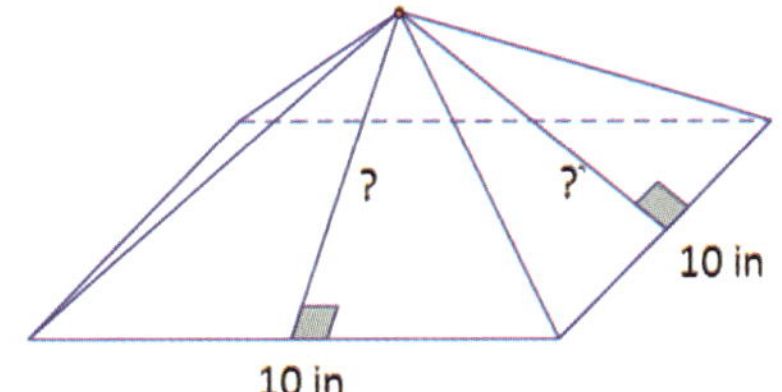

- What information have we been given? How can we use the information?
  - *We know the total surface area, and we know the length of the sides of the square.*
  - *We can use the length of the sides of the square to give us the area of the square base.*
- How does the area of the base help us determine the height of the lateral triangular face?
  - *First, we can subtract the area of the base from the total surface area in order to determine what is left for the lateral sides.*
  - *Next, we can divide the remaining area by* 4 *to get the area of just one triangular face.*
  - *Finally, we can work backward. We have the area of the triangle, and we know the base is* 10 *in., so we can solve for the height.*

Exercises 1–8 (20 minutes)

Students work in pairs to complete the exercises.

Exercises

Determine the surface area of each figure. Assume all faces are rectangles unless it is indicated otherwise.

1.

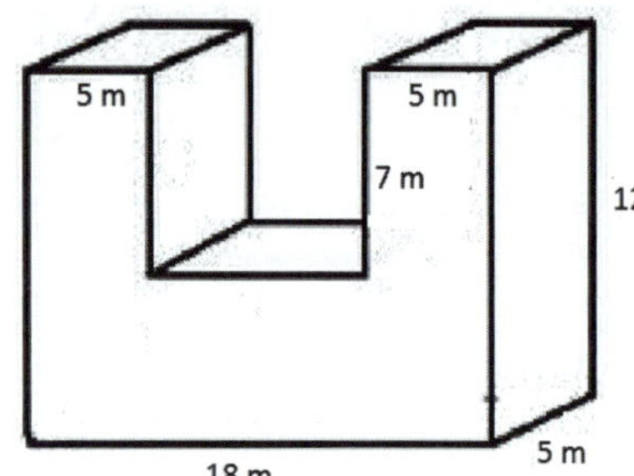

Top and bottom $= 2(18\text{ m} \times 5\text{ m}) = 180\text{ m}^2$

Extra interior sides $= 2(5\text{ m} \times 7\text{ m}) = 70\text{ m}^2$

Left and right sides $= 2(5\text{ m} \times 12\text{ m}) = 120\text{ m}^2$

Front and back sides
$$\begin{aligned} &= 2\big((18\text{ m} \times 12\text{ m}) - (8\text{ m} \times 7\text{ m})\big) \\ &= 2(216\text{ m}^2 - 56\text{ m}^2) \\ &= 2(160\text{ m}^2) \\ &= 320\text{ m}^2 \end{aligned}$$

Surface area
$$\begin{aligned} &= 180\text{ m}^2 + 70\text{ m}^2 + 120\text{ m}^2 + 320\text{ m}^2 \\ &= 690\text{ m}^2 \end{aligned}$$

2. **In addition to your calculation, explain how the surface area of the following figure was determined.**

The surface area of the prism is found by taking the sum of the areas of the trapezoidal front and back and the areas of the back of the four different-sized rectangles that make up the lateral faces.

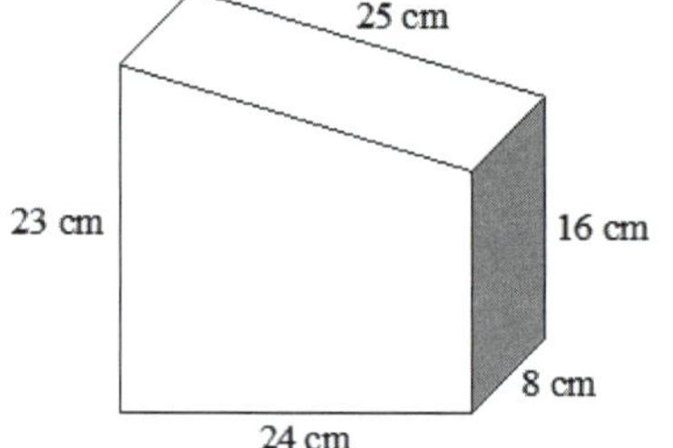

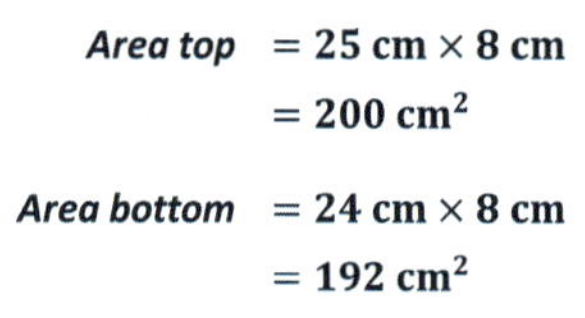

Area top
$$\begin{aligned} &= 25\text{ cm} \times 8\text{ cm} \\ &= 200\text{ cm}^2 \end{aligned}$$

Area bottom
$$\begin{aligned} &= 24\text{ cm} \times 8\text{ cm} \\ &= 192\text{ cm}^2 \end{aligned}$$

Area sides
$$\begin{aligned} &= (23\text{ cm} \times 8\text{ cm}) + (16\text{ cm} \times 8\text{ cm}) \\ &= 312\text{ cm}^2 \end{aligned}$$

Area front and back
$$\begin{aligned} &= 2\left(\frac{1}{2}(16\text{ cm} + 23\text{ cm})(24\text{ cm})\right) \\ &= 2(468\text{ cm}^2) \\ &= 936\text{ cm}^2 \end{aligned}$$

Surface area
$$\begin{aligned} &= 200\text{ cm}^2 + 192\text{ cm}^2 + 312\text{ cm}^2 + 936\text{ cm}^2 \\ &= 1{,}640\text{ cm}^2 \end{aligned}$$

3.

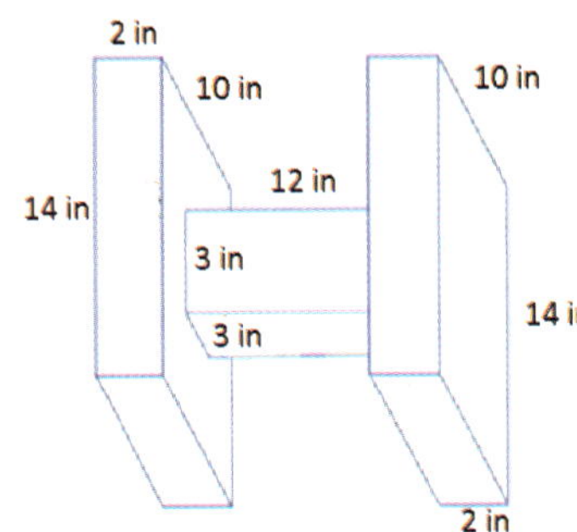

Surface area of one of the prisms on the sides:

Area of front and back $= 2(2 \text{ in.} \times 14 \text{ in.})$
$= 56 \text{ in}^2$

Area of top and bottom $= 2(2 \text{ in.} \times 10 \text{ in.})$
$= 40 \text{ in}^2$

Area of side $= 14 \text{ in.} \times 10 \text{ in.} = 140 \text{ in}^2$

Area of side with hole $= 14 \text{ in.} \times 10 \text{ in.} - 3 \text{ in.} \times 3 \text{ in.}$
$= 131 \text{ in}^2$

There are two such rectangular prisms, so the surface area of both is 734 in^2.

Surface area of middle prism:

Area of front and back $= 2(3 \text{ in.} \times 12 \text{ in.}) = 72 \text{ in}^2$

Area of sides $= 2(3 \text{ in.} \times 12 \text{ in.}) = 72 \text{ in}^2$

Surface area of middle prism $= 72 \text{ in}^2 + 72 \text{ in}^2 = 144 \text{ in}^2$

The total surface area of the figure is $734 \text{ in}^2 + 144 \text{ in}^2 = 878 \text{ in}^2$.

4. In addition to your calculation, explain how the surface area was determined.

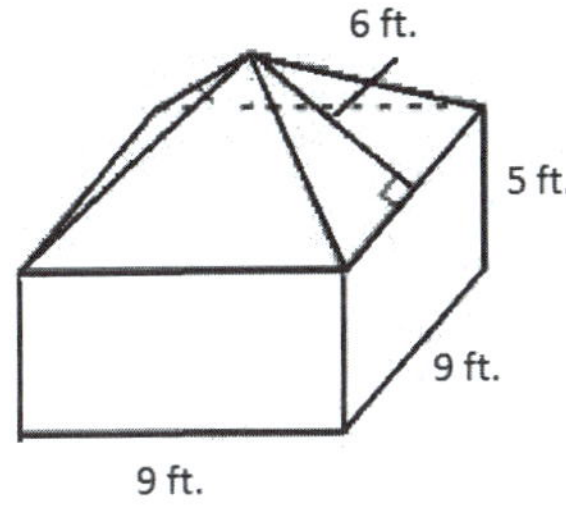

The surface area of the prism is found by taking the area of the base of the rectangular prism and the area of its four lateral faces and adding it to the area of the four lateral faces of the pyramid.

Area of base $= 9 \text{ ft.} \times 9 \text{ ft.}$
$= 81 \text{ ft}^2$

Area of rectangular sides $= 4(9 \text{ ft.} \times 5 \text{ ft.})$
$= 180 \text{ ft}^2$

Area of triangular sides $= 4\left(\frac{1}{2}(9 \text{ ft.})(6 \text{ ft.})\right)$
$= 108 \text{ ft}^2$

Surface area $= 81 \text{ ft}^2 + 180 \text{ ft}^2 + 108 \text{ ft}^2$
$= 369 \text{ ft}^2$

5. A hexagonal prism has the following base and has a height of 8 units. Determine the surface area of the prism.

Area of bases $= 2(48 + 6 + 20 + 10) = 168$

Area of 4 unit sides $= 4(5 \times 8)$
$= 160$

Area of other sides $= (4 \times 8) + (12 \times 8)$
$= 128$

Surface area $= 168 + 160 + 128$
$= 456$

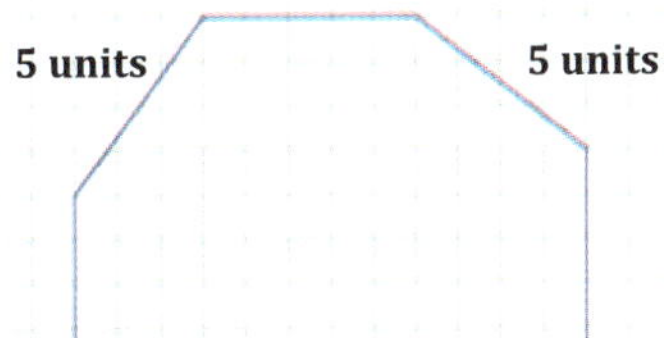

The surface area of the hexagonal prism is 456 units^2.

6. Determine the surface area of each figure.

a.

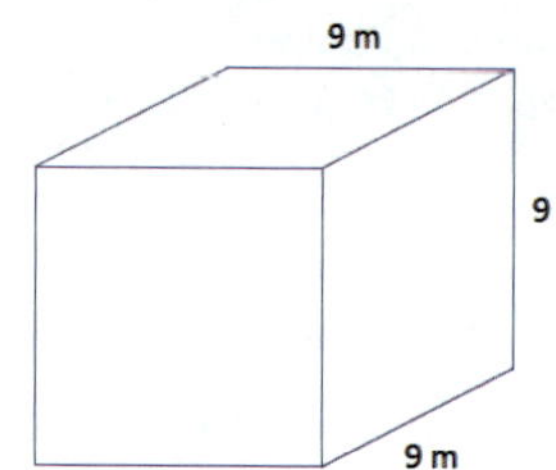

$SA = 6s^2$
$= 6(9\text{ m})^2$
$= 6(81\text{m}^2)$
$= 486\text{ m}^2$

b. A cube with a square hole with 3 m side lengths has been cut through the cube.

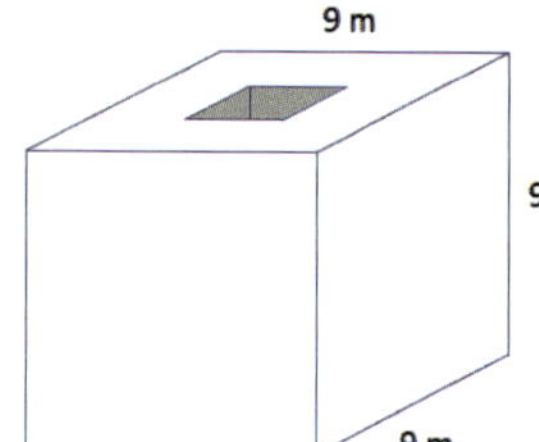

Lateral sides of the hole $= 4(9\text{ m} \times 3\text{ m}) = 108\text{ m}^2$

Surface area of cube with holes $= 486\text{ m}^2 - 2(3\text{ m} \times 3\text{ m}) + 108\text{ m}^2$
$= 576\text{ m}^2$

c. A second square hole with 3 m side lengths has been cut through the cube.

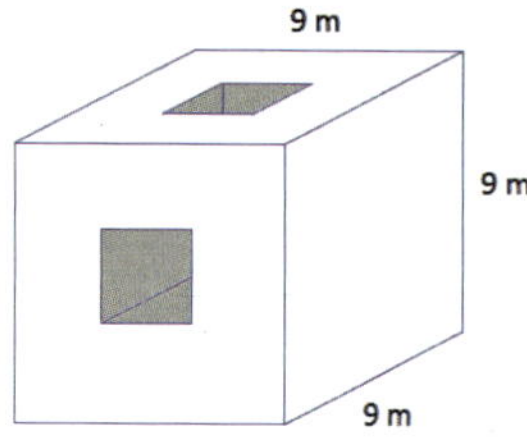

Surface area $= 576\text{ m}^2 - 4(3\text{ m} \times 3\text{ m}) + 2(4(3\text{ m} \times 3\text{ m}))$
$= 612\text{ m}^2$

7. The figure below shows 28 cubes with an edge length of 1 unit. Determine the surface area.

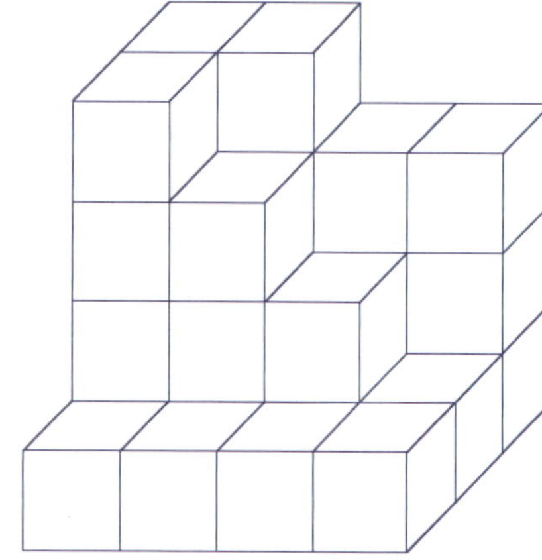

Area top and bottom $= 24$ *units*^{2}

Area sides $= 18$ *units*^{2}

Area front and back $= 28$ *units*^{2}

Surface area
$= 24$ *units*$^2 + 18$ *units*$^2 + 28$ *units*^{2}
$= 70$ *units*^{2}

8. The base rectangle of a right rectangular prism is 4 ft.$\times$ 6 ft. The surface area is 288 ft^2. Find the height.

Let h be the height in feet.

Area of one base: 4 ft.$\times$ 6 ft. $= 24$ ft^2

Area of two bases: $2(24 \text{ ft}^2) = 48 \text{ ft}^2$

Numeric area of four lateral faces: $288 \text{ ft}^2 - 48 \text{ ft}^2 = 240 \text{ ft}^2$

Algebraic area of four lateral faces: $2(6h + 4h) \text{ ft}^2$

Solve for h.

$2(6h + 4h) = 240$

$10h = 120$

$h = 12$

The height is 12 *feet.*

Closing (2 minutes)

- Write down three tips that you would give a friend who is trying to calculate surface area.

Lesson Summary

- **To calculate the surface area of a composite figure, determine the surface area of each prism separately, and add them together. From the sum, subtract the area of the sections that were covered by another prism.**
- **To calculate the surface area with a missing section, find the total surface area of the whole figure. From the total surface area, subtract the area of the missing parts. Then, add the area of the lateral faces of the cutout prism.**

Exit Ticket (5 minutes)

Name ______________________________ Date ______________

Lesson 24: Surface Area

Exit Ticket

Determine the surface area of the right rectangular prism after the two square holes have been cut. Explain how you determined the surface area.

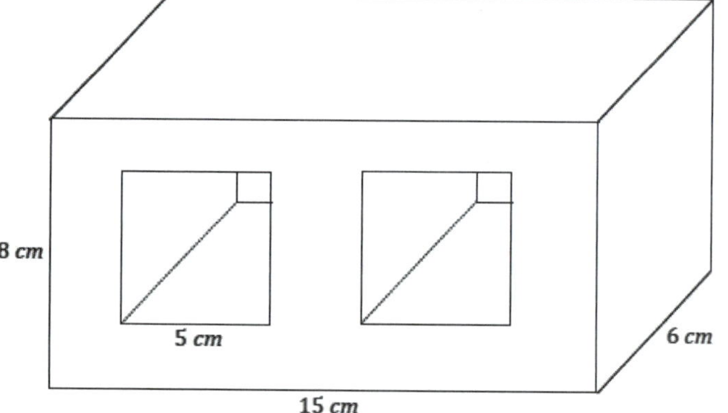

Exit Ticket Sample Solutions

Determine the surface area of the right rectangular prism after the two square holes have been cut. Explain how you determined the surface area.

Area of top and bottom $= 2(15 \text{ cm} \times 6 \text{ cm})$
$= 180 \text{ cm}^2$

Area of sides $= 2(6 \text{ cm} \times 8 \text{ cm})$
$= 96 \text{ cm}^2$

Area of front and back $= 2(15 \text{ cm} \times 8 \text{ cm}) - 4(5 \text{ cm} \times 5 \text{ cm})$
$= 140 \text{ cm}^2$

Area inside $= 8(5 \text{ cm} \times 6 \text{ cm})$
$= 240 \text{ cm}^2$

Surface area $= 180 \text{ cm}^2 + 96 \text{ cm}^2 + 140 \text{ cm}^2 + 240 \text{ cm}^2$
$= 656 \text{ cm}^2$

Take the sum of the areas of the four lateral faces and the two bases of the main rectangular prism, and subtract the areas of the four square cuts from the area of the front and back of the main rectangular prism. Finally, add the lateral faces of the prisms that were cut out of the main prism.

Problem Set Sample Solutions

Determine the surface area of each figure.

1. In addition to the calculation of the surface area, describe how you found the surface area.

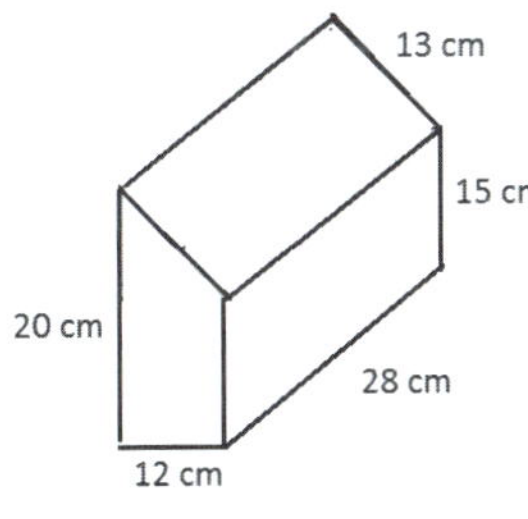

Area of top $= 28 \text{ cm} \times 13 \text{ cm}$
$= 364 \text{ cm}^2$

Area of bottom $= 28 \text{ cm} \times 12 \text{ cm}$
$= 336 \text{ cm}^2$

Area of left and right sides $= 28 \text{ cm} \times 20 \text{ cm} + 15 \text{ cm} \times 28 \text{ cm}$
$= 980 \text{ cm}^2$

Area of front and back sides $= 2\left((12 \text{ cm} \times 15 \text{ cm}) + \frac{1}{2}(5 \text{ cm} \times 12 \text{ cm})\right)$
$= 2(180 \text{ cm}^2 + 30 \text{ cm}^2)$
$= 2(210 \text{ cm}^2)$
$= 420 \text{ cm}^2$

Surface area $= 364 \text{ cm}^2 + 336 \text{ cm}^2 + 980 \text{ cm}^2 + 420 \text{ cm}^2$
$= 2,100 \text{ cm}^2$

Split the area of the two trapezoidal bases into triangles and rectangles, take the sum of the areas, and then add the areas of the four different-sized rectangles that make up the lateral faces.

2.

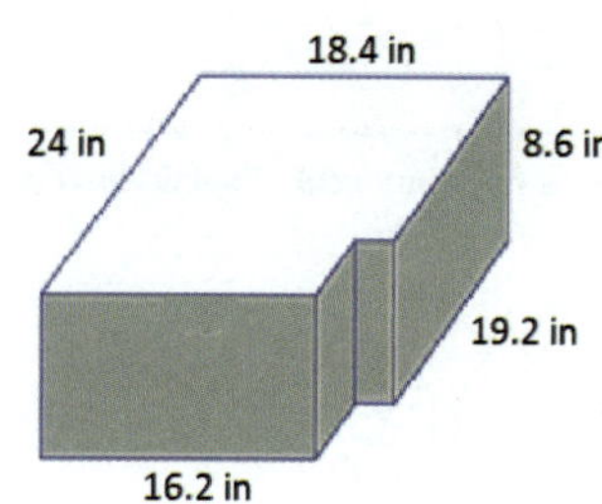

Area of front and back $= 2(18.4 \text{ in.} \times 8.6 \text{ in.}) = 316.48 \text{ in}^2$

Area of sides $= 2(8.6 \text{ in.} \times 24 \text{ in.}) = 412.8 \text{ in}^2$

Area of top and bottom $= 2((18.4 \text{ in.} \times 24 \text{ in.}) - (4.8 \text{ in.} \times 2.2 \text{ in.}))$
$= 2(441.6 \text{ in}^2 - 10.56 \text{ in}^2)$
$= 2(431.04 \text{ in}^2)$
$= 862.08 \text{ in}^2$

Surface area $= 316.48 \text{ in}^2 + 412.8 \text{ in}^2 + 862.08 \text{ in}^2$
$= 1{,}591.36 \text{ in}^2$

3.

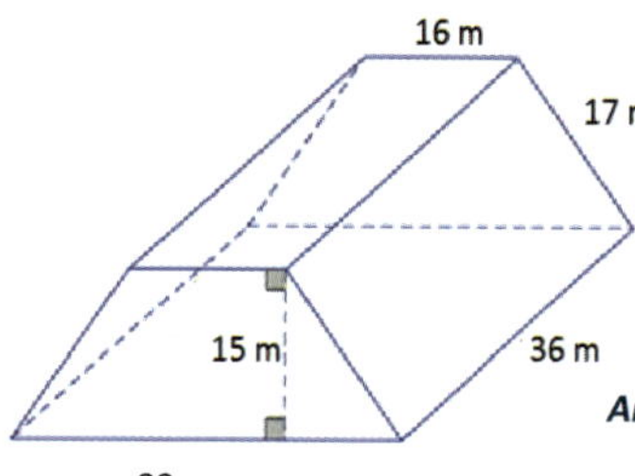

Area of front and back $= 2\left(\frac{1}{2}(32 \text{ m} + 16 \text{ m})15 \text{ m}\right)$
$= 720 \text{ m}^2$

Area of top $= 16 \text{ m} \times 36 \text{ m}$
$= 576 \text{ m}^2$

Area of left and right sides $= 2(17 \text{ m} \times 36 \text{ m})$
$= 2(612 \text{ m}^2)$
$= 1{,}224 \text{ m}^2$

Area of bottom $= 32 \text{ m} \times 36 \text{ m}$
$= 1{,}152 \text{ m}^2$

Surface area $= 720 \text{ m}^2 + 1{,}152 \text{ m}^2 + 1{,}224 \text{ m}^2 + 576 \text{ m}^2$
$= 3{,}672 \text{ m}^2$

4. **Determine the surface area after two square holes with a side length of 2 m are cut through the solid figure composed of two rectangular prisms.**

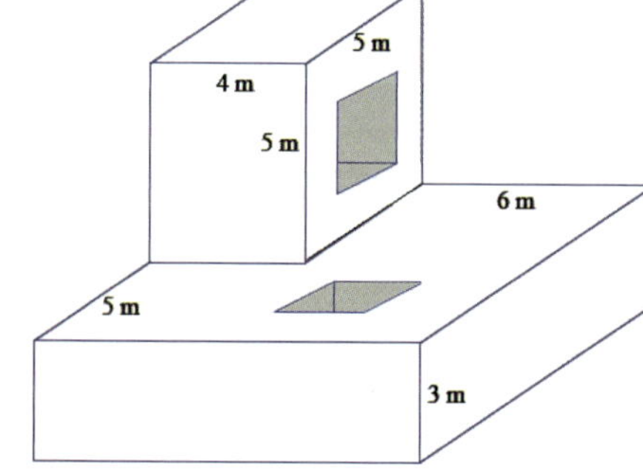

Surface area of top prism before the hole is drilled:

Area of top $= 4 \text{ m} \times 5 \text{ m}$
$= 20 \text{ m}^2$

Area of front and back $= 2(4 \text{ m} \times 5 \text{ m})$
$= 40 \text{ m}^2$

Area of sides $= 2(5 \text{ m} \times 5 \text{ m})$
$= 50 \text{ m}^2$

Surface area of bottom prism before the hole is drilled:

Area of top $= 10 \text{ m} \times 10 \text{ m} - 20 \text{ m}^2$
$= 80 \text{ m}^2$

Area of bottom $= 10 \text{ m} \times 10 \text{ m}$
$= 100 \text{ m}^2$

Area of front and back $= 2(10 \text{ m} \times 3 \text{ m})$
$= 60 \text{ m}^2$

Area of sides $= 2(10 \text{ m} \times 3 \text{ m})$
$= 60 \text{ m}^2$

Surface area of interiors:

Area of interiors $= 4(2 \text{ m} \times 4 \text{ m}) + 4(2 \text{ m} \times 3 \text{ m})$
$= 56 \text{ m}^2$

Surface Area $= 110 \text{ m}^2 + 300 \text{ m}^2 + 56 \text{ m}^2 - 16 \text{ m}^2$
$= 450 \text{ m}^2$

5. The base of a right prism is shown below. Determine the surface area if the height of the prism is 10 cm. Explain how you determined the surface area.

Take the sum of the areas of the two bases made up of two right triangles, and add to it the sum of the areas of the lateral faces made up of rectangles of different sizes.

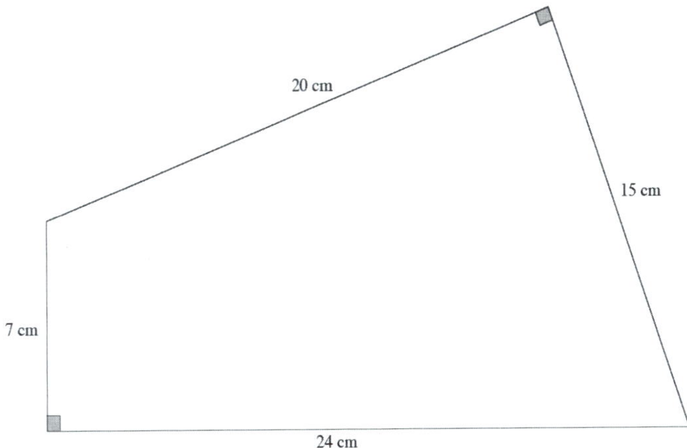

Area of sides $= (20 \text{ cm} \times 10 \text{ cm}) + (15 \text{ cm} \times 10 \text{ cm}) + (24 \text{ cm} \times 10 \text{ cm}) + (7 \text{ cm} \times 10 \text{ cm})$

$= 200 \text{ cm}^2 + 150 \text{ cm}^2 + 240 \text{ cm}^2 + 70 \text{ cm}^2$

$= 660 \text{ cm}^2$

Area of bases $= 2\left(\frac{1}{2}(7 \text{ cm} \times 24 \text{ cm}) + \frac{1}{2}(20 \text{ cm} \times 15 \text{ cm})\right)$

$= (7 \text{ cm} \times 24 \text{ cm}) + (20 \text{ cm} \times 15 \text{ cm})$

$= 168 \text{ cm}^2 + 300 \text{ cm}^2$

$= 468 \text{ cm}^2$

Surface area $= 660 \text{ cm}^2 + 468 \text{ cm}^2$

$= 1,128 \text{ cm}^2$

Topic E

Problems Involving Volume

7.G.B.6

Focus Standard:	7.G.B.6	Solve real-world and mathematical problems involving area, volume, and surface area of two- and three-dimensional objects composed of triangles, quadrilaterals, polygons, cubes, and right prisms.
Instructional Days:	3	
Lesson 25:	Volume of Right Prisms (P)[1]	
Lesson 26:	Volume of Composite Three-Dimensional Objects (P)	
Lesson 27:	Real-World Volume Problems (P)	

Until Grade 6, students have studied volume using right rectangular prisms. In Grade 7 Module 3, students began to explore how to find the volumes of prisms with bases other than rectangles or triangles. In Lesson 25, students complete several context-rich problems involving volume. Some problems require students to use what they know about the additive property of volume (**5.MD.C.5c**) and to use displacement to make indirect measurements (e.g., a typical context might include a stone placed in a container of water). In Lesson 26, students calculate the volume of composite three-dimensional figures, some of which have missing sections (prism-shaped sections). In Lesson 27, students use a rate of flowing liquid to solve modeling problems such as how long it takes to fill a pool or how much water is used to take a shower.

[1]Lesson Structure Key: **P**-Problem Set Lesson, **M**-Modeling Cycle Lesson, **E**-Exploration Lesson, **S**-Socratic Lesson

Lesson 25: Volume of Right Prisms

Student Outcomes

- Students use the formula $V = Bh$ to determine the volume of a right prism. Students identify the base and compute the area of the base by decomposing it into pieces.

Lesson Notes

In Module 3 (Lesson 23), students learned that a polygonal base of a given right prism can be decomposed into triangular and rectangular regions that are bases of a set of right prisms that have heights equal to the height of the given right prism. By extension, the volume of any right prism can be found by multiplying the area of its base times the height of the prism. Students revisit this concept in the Opening Exercise before moving on to examples.

Classwork

Opening Exercise (3 minutes)

MP.3

Opening Exercise

Take your copy of the following figure, and cut it into four pieces along the dotted lines. (The vertical line is the altitude, and the horizontal line joins the midpoints of the two sides of the triangle.)

Arrange the four pieces so that they fit together to form a rectangle.

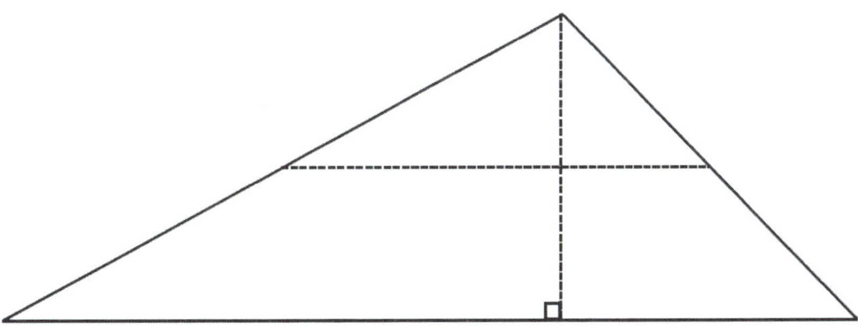

If a prism were formed out of each shape, the original triangle, and your newly rearranged rectangle, and both prisms had the same height, would they have the same volume? Discuss with a partner.

Discussion (5 minutes)

Any triangle can be cut and rearranged to form a rectangle by using the altitude from a vertex to the longest side and a line segment parallel to the longest side halfway between the vertex and the longest side.

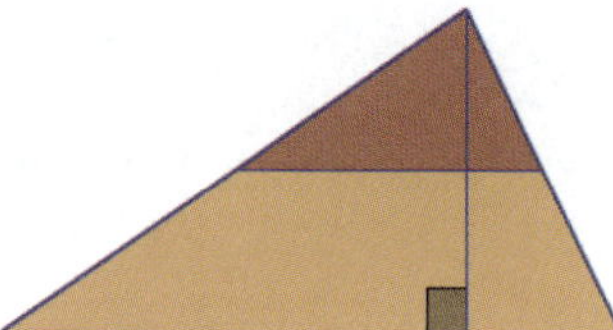
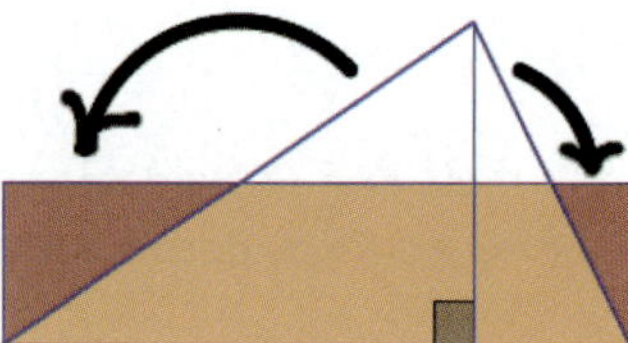

Vertical slices through a right triangular prism show that it can be cut and rearranged as a right rectangular prism.

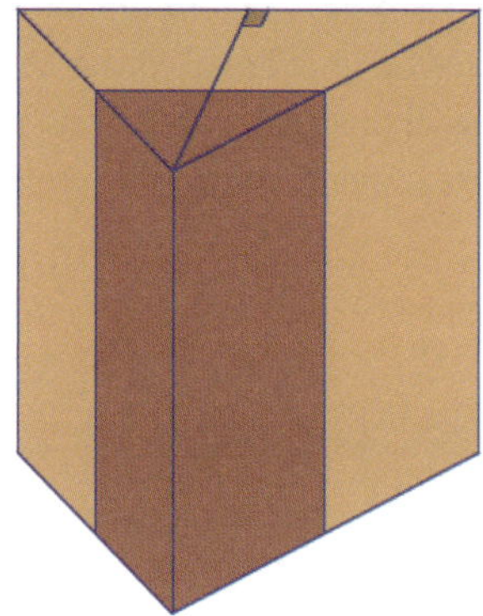
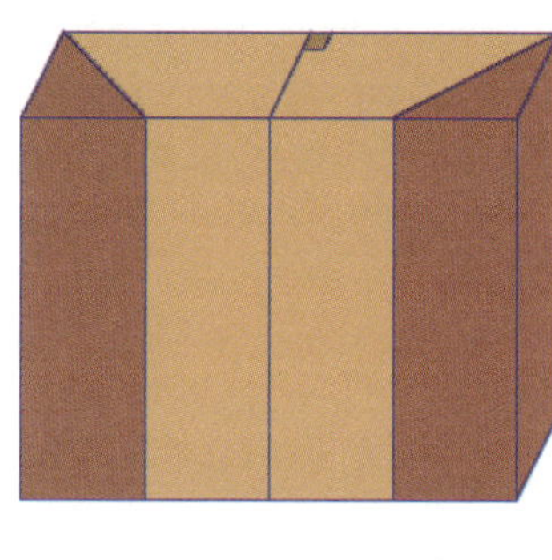

The triangular prism and the right rectangular prism have the same base area, height, and volume:

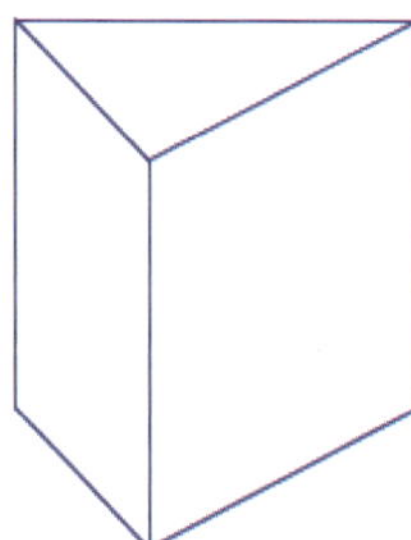
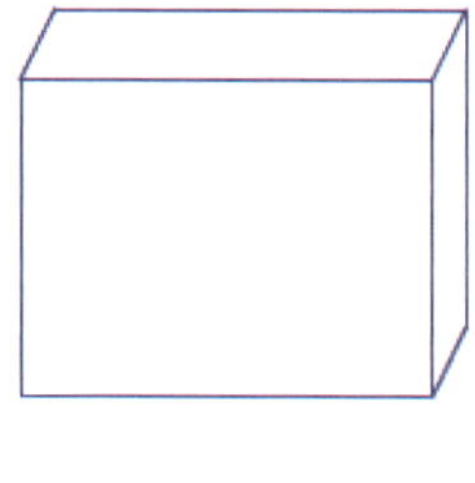

$$\text{Volume of right triangular prism} = \text{volume of right rectangular prism} = \text{area of base} \times \text{height}.$$

The union of right triangular prisms with the same height:

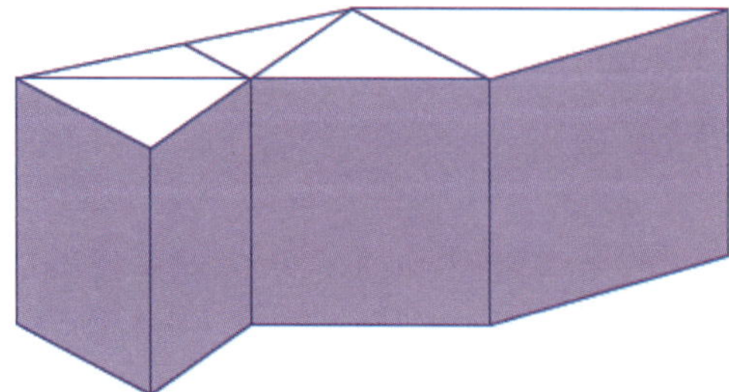

$$\text{Volume} = \text{sum of volumes} = \text{sum of area of base} \times \text{height} = \text{total area of base} \times \text{height}.$$

Any right polygonal prism can be packed without gaps or overlaps by right triangular prisms of the same height:

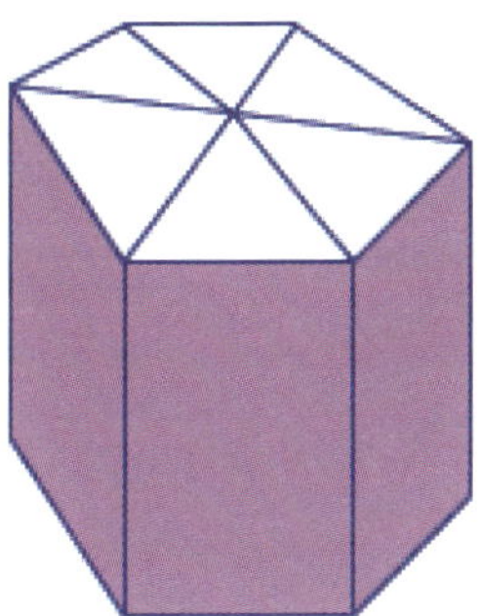

For any right polygonal prism: $\text{Volume} = \text{area of base} \times \text{height}$

Exercise 1 (3 minutes)

Exercise 1

a. **Show that the following figures have equal volumes.**

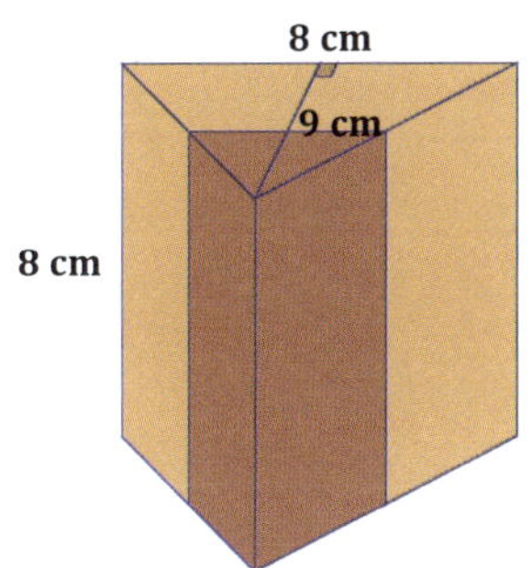

8 cm

8 cm

4.5 cm

Volume of triangular prism:

$$\frac{1}{2}(8\text{ cm} \times 9\text{ cm}) \times 8\text{ cm} = 288\text{ cm}^3$$

Volume of rectangular prism:

$$8\text{ cm} \times 8\text{ cm} \times 4.5\text{ cm} = 288\text{ cm}^3$$

b. **How can it be shown that the prisms will have equal volumes without completing the entire calculation?**

If one base can be cut up and rearranged to form the other base, then the bases have the same area. If the prisms have the same height, then the cutting and rearranging of the bases can show how to slice and rearrange one prism so that it looks like the other prism. Since slicing and rearranging does not change volume, the volumes are the same.

Example 1 (5 minutes)

Example 1

Calculate the volume of the following prism.

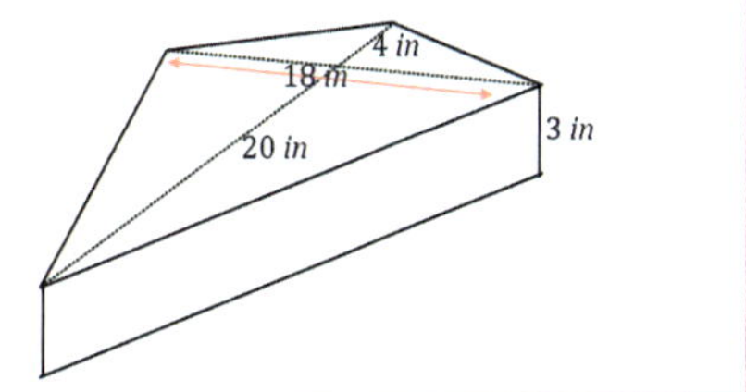

- What is the initial difficulty of determining the volume of this prism?
  - *The base is not in the shape of a rectangle or triangle.*
- Do we have a way of finding the area of a kite in one step? If not, how can we find the area of a kite?
  - *We do not have a way to find the area of a kite in one step. We can break the kite up into smaller shapes. One way would be to break it into two triangles, but it can also be broken into four triangles.*

Once students understand that the base must be decomposed into triangles, allow them time to solve the problem.

- Provide a numeric expression that determines the area of the kite-shaped base.
  - *Area of the base:* $\frac{1}{2}(20 \text{ in.} \cdot 18 \text{ in.}) + \frac{1}{2}(4 \text{ in.} \cdot 18 \text{ in.})$
- Find the volume of the prism.
  - *Volume of the prism:* $\left(\frac{1}{2}(20 \text{ in.} \cdot 18 \text{ in.}) + \frac{1}{2}(4 \text{ in.} \cdot 18 \text{ in.})\right)(3 \text{ in.}) = 648 \text{ in}^3$

Example 2 (7 minutes)

Example 2

A container is shaped like a right pentagonal prism with an open top. When a cubic foot of water is dumped into the container, the depth of the water is 8 inches. Find the area of the pentagonal base.

- How can we use volume to solve this problem?
  - *We know that the formula for the volume of a prism is $V = Bh$, where B represents the area of the pentagonal base and h is the height of the prism.*
- What information do we know from reading the problem?
  - *The volume is 1 cubic foot, and the height of the water is 8 inches.*
- Knowing the volume formula, can we use this information to solve the problem?
  - *We can use this information to solve the problem, but the information is given using two different dimensions.*
- How can we fix this problem?
  - *We have to change inches into feet so that we have the same units.*

Scaffolding:

Provide an image of a pentagonal prism as needed:

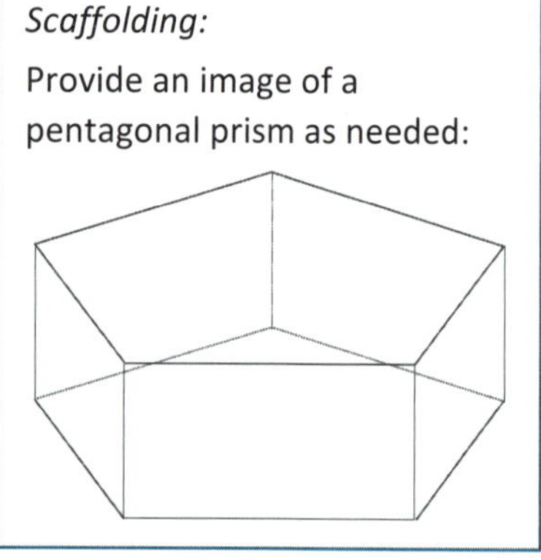

- How can we change inches into feet?
  - *Because 12 inches make 1 foot, we have to divide the 8 inches by 12 to get the height in feet.*
- Convert 8 inches into feet.
  - $\frac{8}{12} \text{ ft.} = \frac{2}{3} \text{ ft.}$
- Now that we know the volume of the water is 1 cubic foot and the height of the water is $\frac{2}{3}$ feet, how can we determine the area of the pentagonal base?
  - *Use the volume formula.*

- Use the information we know to find the area of the base.
  - $1\text{ ft}^3 = B\left(\frac{2}{3}\text{ ft.}\right)$ *Therefore, the area of the pentagonal base is* $\frac{3}{2}\text{ ft}^2$, *or* $1\frac{1}{2}\text{ ft}^2$.

Example 3 (10 minutes)

> **Example 3**
>
> **Two containers are shaped like right triangular prisms, each with the same height. The base area of the larger container is 200% more than the base area of the smaller container. How many times must the smaller container be filled with water and poured into the larger container in order to fill the larger container?**

Solution by manipulating the equation of the volume of the smaller prism:

MP.4

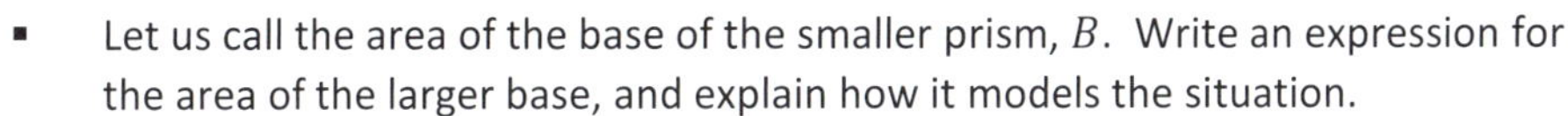

- Let us call the area of the base of the smaller prism, B. Write an expression for the area of the larger base, and explain how it models the situation.
  - *The base area of the larger prism is* $3B$ *because* 200% *more means that its area is* 300% *of* B. *Both prisms have the same height,* h.
- Compute the volume of the smaller prism.
  - *The volume of the smaller prism is* $V_S = Bh$.
- What is the volume of the larger prism?
  - *The volume of the larger prism is* $V_L = 3Bh$.
- How many times greater is the volume of the larger prism relative to the smaller prism?
  - $\frac{3Bh}{Bh} = 3$ *The smaller container must be filled three times in order to fill the larger container.*

> *Scaffolding:*
>
> Students may feel more comfortable manipulating actual values for the dimensions of the prisms as done in the second solution to Example 2. Using sample values may be a necessary stepping stone to understanding how to manipulate the base area of one prism in terms of the other.

Solution by substituting values for the smaller prism's dimensions:

- To solve this problem, create two right triangular prisms. What dimensions should we use for the smaller container?
  - *Answers will vary, but for this example we will use a triangle that has a base of* 10 *inches and a height of* 5 *inches. The prism will have a height of* 2 *inches.*
- What is the area of the base for the smaller container? Explain.
  - *The area of the base of the smaller container is* 25 in^2 *because* $A = \frac{1}{2}(10\text{ in.})(5\text{ in.}) = 25\text{ in}^2$.
- What is the volume of the smaller container? Explain.
  - *The volume of the smaller container is* 50 in^3 *because* $V = 25\text{ in}^2 \times 2\text{ in.} = 50\text{ in}^3$.
- What do we know about the larger container?
  - *The area of the larger container's base is* 200% *more than the area of the smaller container's base.*
  - *The height of the larger container is the same as the height of the smaller container.*

- If the area of the larger container's base is 200% more than the area of the smaller container's base, what is the area of the larger container's base?
  - *The area of the larger container's base would be* $25 \text{ in}^2 + 2(25 \text{ in}^2) = 75 \text{ in}^2$.
- What is the volume of the larger container? Explain.
  - *The volume of the larger container is* 150 in^3 *because* $V = 75 \text{ in}^2 \times 2 \text{ in.} = 150 \text{ in}^3$.
- How many times must the smaller container be filled with water and poured into the larger container in order to fill the larger container? Explain.
  - *The smaller container would have to be filled three times in order to fill the larger container. Each time you fill the smaller container, you will have* 50 in^3*; therefore, you will need to fill the smaller container three times to get a volume of* 150 in^3.
- Would your answer be different if we used different dimensions for the containers? Why or why not?
  - *Our answer would not change if we used different dimensions. Because the area of the base of the larger container is triple the area of the base of the smaller container—and because the two heights are the same—the volume of the larger container is triple that of the smaller container.*

Exercise 2 (6 minutes)

MP.3

Exercise 2

Two aquariums are shaped like right rectangular prisms. The ratio of the dimensions of the larger aquarium to the dimensions of the smaller aquarium is $3:2$.

Addie says the larger aquarium holds 50% more water than the smaller aquarium.

Berry says that the larger aquarium holds 150% more water.

Cathy says that the larger aquarium holds over 200% more water.

Are any of the girls correct? Explain your reasoning.

Cathy is correct. If the ratio of the dimensions of the larger aquarium to the dimensions of the smaller aquarium is $3:2$, then the volume must be 1.5^3, or 3.375 times greater than the smaller aquarium. Therefore, the larger aquarium's capacity is 337.5% times as much water, which is 237.5% more than the smaller aquarium. Cathy said that the larger aquarium holds over 200% more water than the smaller aquarium, so she is correct.

Closing (1 minute)

Lesson Summary

- **The formula for the volume of a prism is $V = Bh$, where B is the area of the base of the prism and h is the height of the prism.**
- **A base that is neither a rectangle nor a triangle must be decomposed into rectangles and triangles in order to find the area of the base.**

Exit Ticket (5 minutes)

Name ______________________________ Date ________________

Lesson 25: Volume of Right Prisms

Exit Ticket

Determine the volume of the following prism. Explain how you found the volume.

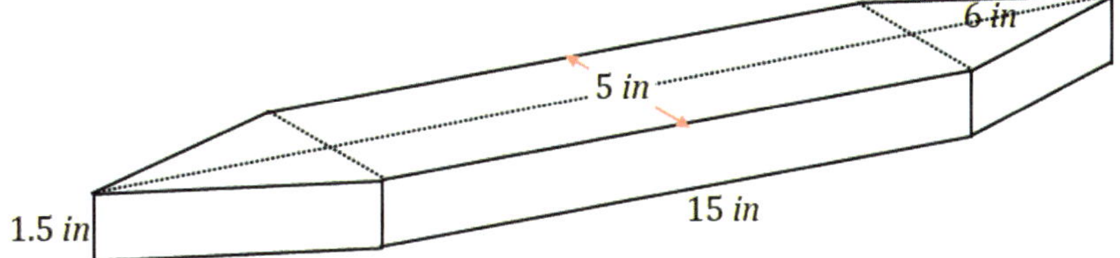

Exit Ticket Sample Solutions

Determine the volume of the following prism. Explain how you found the volume.

To find the volume of the prism, the base must be decomposed into triangles and rectangles since there is no way to find the area of the base as is. The base can be decomposed into two triangles and a rectangle, and their areas must be summed to find the area of the base. Once the area of the base is determined, it should be multiplied by the height to find the volume of the entire prism.

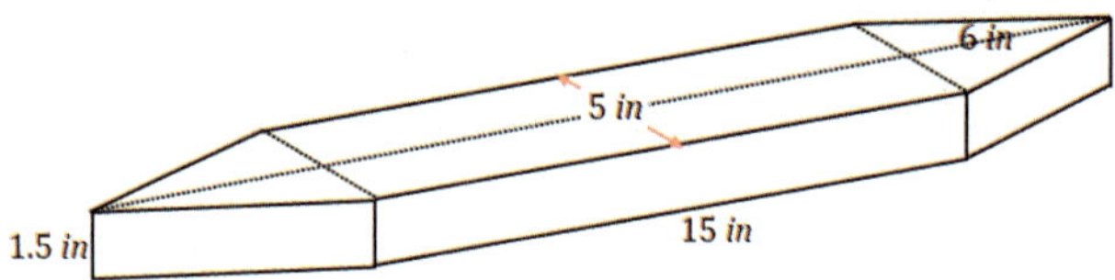

Area of both triangles: $2\left(\frac{1}{2}(5\text{ in.}\times 6\text{ in.})\right) = 30\text{ in}^2$

Area of the rectangle: $15\text{ in.}\times 5\text{ in.} = 75\text{ in}^2$

Total area of the base: $(30+75)\text{ in}^2 = 105\text{ in}^2$

Volume of the prism: $(105\text{ in}^2)(1.5\text{ in.}) = 157.5\text{ in}^3$

Problem Set Sample Solutions

1. The pieces in Figure 1 are rearranged and put together to form Figure 2.

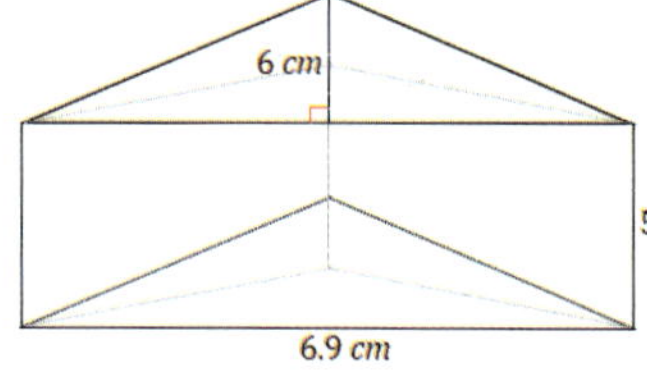

Figure 1

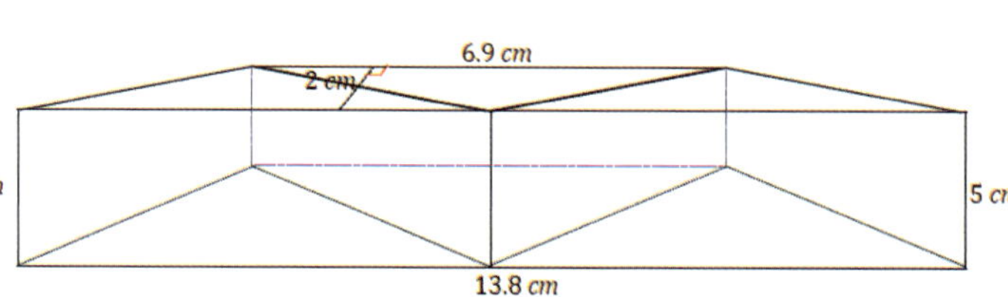

Figure 2

a. Use the information in Figure 1 to determine the volume of the prism.

Volume: $\frac{1}{2}(6.9\text{ cm}\cdot 6\text{ cm})(5\text{ cm}) = 103.5\text{ cm}^3$

b. Use the information in Figure 2 to determine the volume of the prism.

Volume: $\left(\frac{1}{2}(6.9\text{ cm}+13.8\text{ cm})\cdot 2\text{ cm}\right)(5\text{ cm}) = 103.5\text{ cm}^3$

c. If we were not told that the pieces of Figure 1 were rearranged to create Figure 2, would it be possible to determine whether the volumes of the prisms were equal without completing the entire calculation for each?

Both prisms have the same height, so as long as it can be shown that both bases have the same area, both prisms must have equal volumes. We could calculate the area of the triangle base and the trapezoid base and find that they are equal in area and be sure that both volumes are equal.

2. Two right prism containers each hold 75 gallons of water. The height of the first container is 20 inches. The height of the second container is 30 inches. If the area of the base in the first container is 6 ft^2, find the area of the base in the second container. Explain your reasoning.

We know that the volume of each of the two containers is 75 gallons; therefore, the volumes must be equal. In order to find the volume of the first container, we could multiply the area of the base (6 ft^2) by the height (20 inches). To find the volume of the second container, we would also multiply the area of its base, which we will call A (area in square feet), and the height (30 inches). These two expressions must equal each other since both containers have the same volume.

$$6 \times 20 = A \times 30$$
$$120 = 30A$$
$$120 \div 30 = 30A \div 30$$
$$4 = A$$

Therefore, the area of the base in the second container is 4 ft^2. Note: The units for the volume are $1 \text{ ft.} \times 1 \text{ ft.} \times 1 \text{ in.}$ in this computation. Converting the inches to feet would make the computation in cubic feet, but it would not change the answer for A.

3. Two containers are shaped like right rectangular prisms. Each has the same height, but the base of the larger container is 50% more in each direction. If the smaller container holds 8 gallons when full, how many gallons does the larger container hold? Explain your reasoning.

The larger container holds 18 gallons because each side length of the base is 1.5 times larger than the smaller container's dimensions. Therefore, the area of the larger container's base is 1.5^2, or 2.25 times larger than the smaller container. Because the height is the same in both containers, the volume of the larger container must be 2.25 times larger than the smaller container. $8 \text{ gal.} \times 2.25 = 18 \text{ gal.}$

4. A right prism container with the base area of 4 ft^2 and height of 5 ft. is filled with water until it is 3 ft. deep. If a solid cube with edge length 1 ft. is dropped to the bottom of the container, how much will the water rise?

The volume of the cube is 1 ft^3. Let the number of feet the water will rise be x. Then, the volume of the water over the 3 ft. mark is $4x \text{ ft}^3$ because this represents the area of the base (4 ft^2) times the height (x). Because the volume of the cube is 1 ft^3, $4x \text{ ft}^3$ must equal 1 ft^3.

$$4x = 1$$
$$4x \div 4 = 1 \div 4$$
$$x = \frac{1}{4}$$

Therefore, the water will rise $\frac{1}{4} \text{ ft.}$, or 3 inches.

5. A right prism container with a base area of 10 ft^2 and height 9 ft. is filled with water until it is 6 ft. deep. A large boulder is dropped to the bottom of the container, and the water rises to the top, completely submerging the boulder without causing overflow. Find the volume of the boulder.

The increase in volume is the same as the volume of the boulder. The height of the water increases 3 ft. Therefore, the increase in volume is 10 ft^2 (area of the base) multiplied by 3 ft. (i.e., the change in height).

$$V = 10 \text{ ft}^2 \times 3 \text{ ft.} = 30 \text{ ft}^3$$

Because the increase in volume is 30 ft^3, the volume of the boulder is 30 ft^3.

6. A right prism container with a base area of 8 ft^2 and height 6 ft. is filled with water until it is 5 ft. deep. A solid cube is dropped to the bottom of the container, and the water rises to the top. Find the length of the cube.

When the cube is dropped into the container, the water rises 1 foot, which means the volume increases 8 cubic feet. Therefore, the volume of the cube must be 8 cubic feet. We know that the length, width, and height of a cube are equal, so the length of the cube is 2 feet because $2 \text{ ft.} \times 2 \text{ ft.} \times 2 \text{ ft.} = 8 \text{ ft}^3$, which is the volume of the cube.

7. A rectangular swimming pool is 30 feet wide and 50 feet long. The pool is 3 feet deep at one end, and 10 feet deep at the other.

   a. Sketch the swimming pool as a right prism.

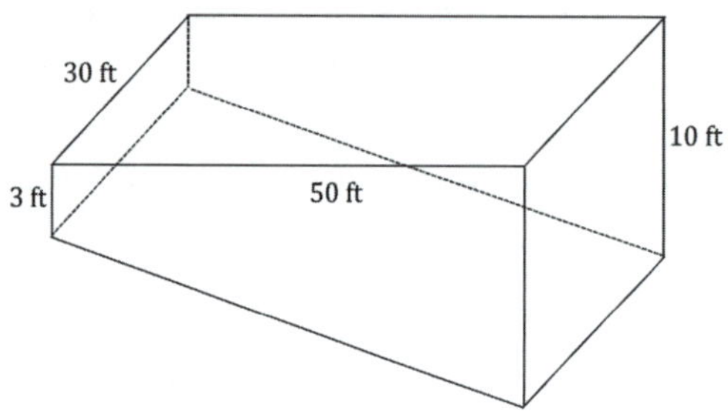

   b. What kind of right prism is the swimming pool?

   The swimming pool is a right trapezoidal prism.

   c. What is the volume of the swimming pool in cubic feet?

   $$\text{Area of base} = \frac{50 \text{ ft.}(10 \text{ ft.} + 3 \text{ ft.})}{2} = 325 \text{ ft}^2$$

   $$\text{Volume of pool} = 325 \text{ ft}^2 \times 30 \text{ ft.} = 9{,}750 \text{ ft}^3$$

   d. How many gallons will the swimming pool hold if each cubic feet of water is about 7.5 gallons?

   $(9{,}750 \text{ ft}^3)\dfrac{(7.5 \text{ gal.})}{1 \text{ ft}^3} = 73{,}125 \text{ gal.}$ *The pool will hold $73{,}125$ gal.*

8. A milliliter (mL) has a volume of 1 cm^3. A 250 mL measuring cup is filled to 200 mL. A small stone is placed in the measuring cup. The stone is completely submerged, and the water level rises to 250 mL.

   a. What is the volume of the stone in cm^3?

   When the stone is dropped into the measuring cup, the increase in volume is $250 \text{ mL} - 200 \text{ mL} = 50 \text{ mL}$. We know that 1 mL has a volume of 1 cm^3; therefore, the stone has a volume of 50 cm^3.

   b. Describe a right rectangular prism that has the same volume as the stone.

   Answers will vary. Possible answers are listed below.

   $$1 \text{ cm} \times 1 \text{ cm} \times 50 \text{ cm}$$
   $$1 \text{ cm} \times 2 \text{ cm} \times 25 \text{ cm}$$
   $$1 \text{ cm} \times 5 \text{ cm} \times 10 \text{ cm}$$
   $$2 \text{ cm} \times 5 \text{ cm} \times 5 \text{ cm}$$

Opening Exercise

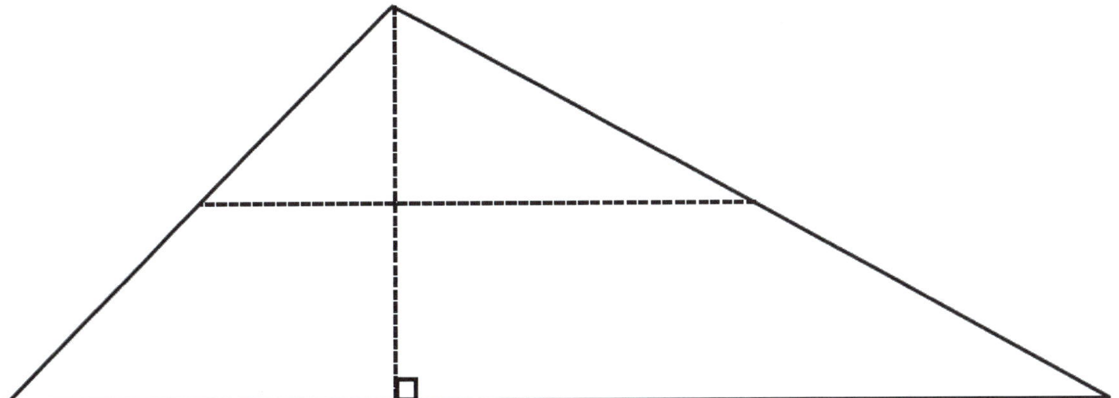

Lesson 26: Volume of Composite Three-Dimensional Objects

Student Outcomes

- Students compute volumes of three-dimensional objects composed of right prisms by using the fact that volume is additive.

Lesson Notes

Lesson 26 is an extension of work done in the prior lessons on volume as well as an extension of work started in the final lesson of Module 3 (Lesson 26). Students have more exposure to composite figures such as prisms with prism-shaped holes or prisms that have smaller prisms removed from their volumes. Furthermore, in applicable situations, students compare different methods to determine composite volume. This is necessary when the entire prism can be decomposed into multiple prisms or when the prism hole shares the height of the main prism.

Classwork

Example 1 (4 minutes)

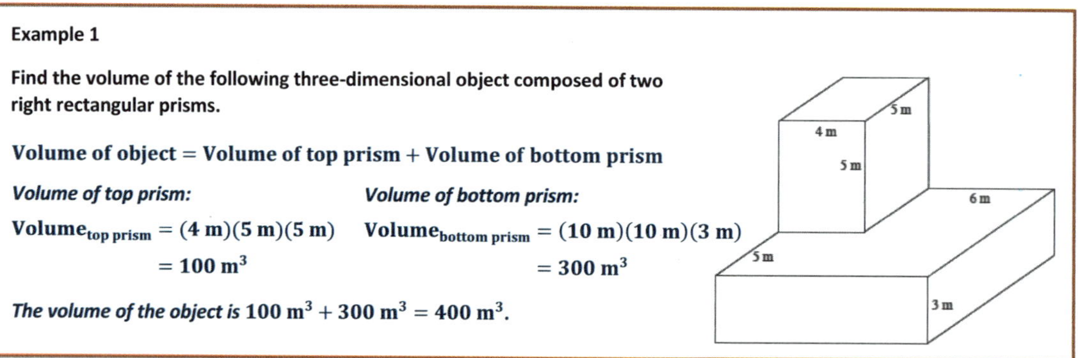

Example 1

Find the volume of the following three-dimensional object composed of two right rectangular prisms.

$\text{Volume of object} = \text{Volume of top prism} + \text{Volume of bottom prism}$

Volume of top prism:

$\text{Volume}_{\text{top prism}} = (4\text{ m})(5\text{ m})(5\text{ m}) = 100\text{ m}^3$

Volume of bottom prism:

$\text{Volume}_{\text{bottom prism}} = (10\text{ m})(10\text{ m})(3\text{ m}) = 300\text{ m}^3$

The volume of the object is $100\text{ m}^3 + 300\text{ m}^3 = 400\text{ m}^3$.

There are different ways the volume of a composite figure may be calculated. If the figure is like the one shown in Example 1, where the figure can be decomposed into separate prisms and it would be impossible for the prisms to share any one dimension, the individual volumes of the decomposed prisms can be determined and then summed. If, however, the figure is similar to the figure in Exercise 1, there are two possible strategies. In Exercise 1, the figure can be decomposed into two individual prisms, but a dimension is shared between the two prisms—in this case, the height. Instead of calculating the volume of each prism and then taking the sum, we can calculate the area of the entire base by decomposing it into shapes we know and then multiplying the area of the base by the height.

Exercise 1 (4 minutes)

Exercise 1

Find the volume of the following three-dimensional figure composed of two right rectangular prisms.

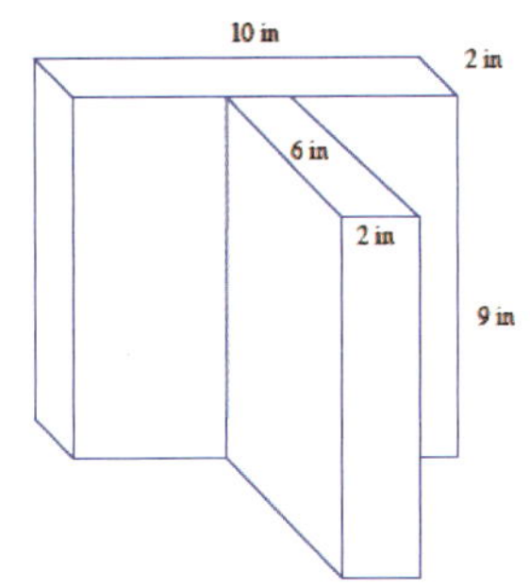

$\text{Area of Base}_{\text{back prism}} = (2 \text{ in.})(10 \text{ in.})$
$= 20 \text{ in.}^2$

$\text{Area of Base}_{\text{front prism}} = (6 \text{ in.})(2 \text{ in.})$
$= 12 \text{ in}^2$

$\text{Area of Base} = 20 \text{ in}^2 + 12 \text{ in}^2 = 32 \text{ in}^2$

The volume of the object is $(9 \text{ in.})(32 \text{ in}^2) = 288 \text{ in}^3$.

Exercise 2 (10 minutes)

MP.1

Exercise 2

The right trapezoidal prism is composed of a right rectangular prism joined with a right triangular prism. Find the volume of the right trapezoidal prism shown in the diagram using two different strategies.

Strategy 1

The volume of the trapezoidal prism is equal to the sum of the volumes of the rectangular and triangular prisms.

$\text{Volume of object} = \text{Volume of rectangular prism} + \text{Volume of triangular prism}$

Volume of rectangular prism:

$$\begin{aligned}\text{Volume}_{\text{rectangular prism}} &= Bh \\ &= (lw)h \\ &= (3 \text{ cm} \cdot 2 \text{ cm}) \cdot 1\frac{1}{2} \text{ cm} \\ &= 9 \text{ cm}^3\end{aligned}$$

Volume of triangular prism:

$$\begin{aligned}\text{Volume}_{\text{triangular prism}} &= Bh = \left(\frac{1}{2}lw\right)h \\ &= \left(\frac{1}{2} \cdot 3 \text{ cm} \cdot 2\frac{1}{4} \text{ cm}\right) \cdot 1\frac{1}{2} \text{ cm} \\ &= 5\frac{1}{16} \text{ cm}^3\end{aligned}$$

The volume of the object is $9 \text{ cm}^3 + 5\frac{1}{16} \text{ cm}^3 = 14\frac{1}{16} \text{ cm}^3$.

Strategy 2

The volume of a right prism is equal to the area of its base times its height. The base consists of a rectangle and a triangle.

$\text{Volume of object} = Bh$

$B = \text{Area}_{\text{rectangle}} + \text{Area}_{\text{triangle}}$

$\text{Area}_{\text{rectangle}} = 3 \text{ cm} \cdot 2 \text{ cm} = 6 \text{ cm}^2$

$\text{Area}_{\text{triangle}} = \frac{1}{2} \cdot 3 \text{ cm} \cdot 2\frac{1}{4} \text{ cm} = 3\frac{3}{8} \text{ cm}^2$

$B = 6 \text{ cm}^2 + 3\frac{3}{8} \text{ cm}^2 = 9\frac{3}{8} \text{ cm}^2$

Volume of object:

$$\begin{aligned}\text{Volume}_{\text{object}} &= Bh \\ &= \left(9\frac{3}{8} \text{ cm}^2\right)\left(1\frac{1}{2} \text{ cm}\right) \\ &= 14\frac{1}{16} \text{ cm}^3\end{aligned}$$

The volume of the object is $14\frac{1}{16} \text{ cm}^3$.

- Write a numeric expression to represent the volume of the figure in Strategy 1.
  - $(3 \text{ cm} \cdot 2 \text{ cm}) \cdot 1\frac{1}{2} \text{cm} + \left(\frac{1}{2} \cdot 3 \text{ cm} \cdot 2\frac{1}{4} \text{cm}\right) \cdot 1\frac{1}{2} \text{cm}$
- Write a numeric expression to represent the volume of the figure in Strategy 2.
  - $\left(3 \text{ cm} \cdot 2 \text{ cm} + \frac{1}{2} \cdot 3 \text{ cm} \cdot 2\frac{1}{4} \text{cm}\right)\left(1\frac{1}{2} \text{cm}\right)$
- How do the numeric expressions represent the problem differently?
  - *The first expression is appropriate to use when individual volumes of the decomposed figure are being added together, whereas the second expression is used when the area of the base of the composite figure is found and then multiplied by the height to determine the volume.*
- What property allows us to show that these representations are equivalent?
  - *The distributive property.*

Example 2 (10 minutes)

Example 2

Find the volume of the right prism shown in the diagram whose base is the region between two right triangles. Use two different strategies.

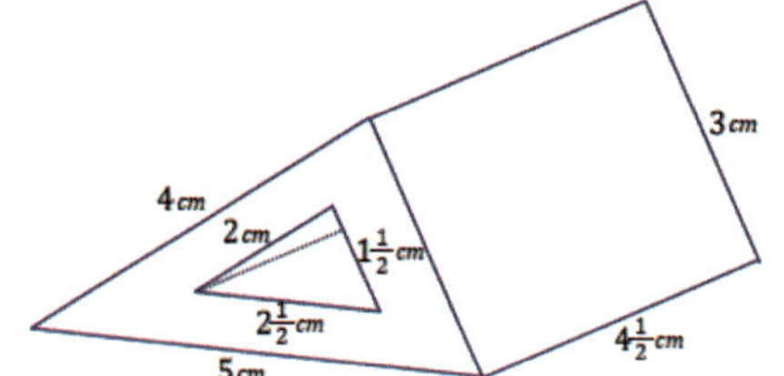

Strategy 1

The volume of the right prism is equal to the difference of the volumes of the two triangular prisms.

$\textbf{Volume of object} = \textbf{Volume}_{\textbf{large prism}} - \textbf{Volume}_{\textbf{small prism}}$

Volume of large prism:

$$\textbf{Volume}_{\textbf{large prism}} = \left(\frac{1}{2} \cdot 3 \text{ cm} \cdot 4 \text{ cm}\right) 4\frac{1}{2} \text{cm} = 27 \text{ cm}^3$$

Volume of small prism:

$$\textbf{Volume}_{\textbf{small prism}} = \left(\frac{1}{2} \cdot 1\frac{1}{2} \text{cm} \cdot 2 \text{ cm}\right) 4\frac{1}{2} \text{cm} = 6\frac{3}{4} \text{cm}^3$$

The volume of the object is $20\frac{1}{4} \text{cm}^3$.

Strategy 2

The volume of a right prism is equal to the area of its base times its height. The base is the region between two right triangles.

$\textbf{Volume of object} = Bh$

$B = \textbf{Area}_{\textbf{large triangle}} - \textbf{Area}_{\textbf{small triangle}}$

$$\textbf{Area}_{\textbf{large triangle}} = \frac{1}{2} \cdot 3 \text{ cm} \cdot 4 \text{ cm} = 6 \text{ cm}^2$$

$$\textbf{Area}_{\textbf{small triangle}} = \frac{1}{2} \cdot 1\frac{1}{2} \text{ cm} \cdot 2 \text{ cm} = 1\frac{1}{2} \text{cm}^2$$

$$B = 6 \text{ cm}^2 - 1\frac{1}{2} \text{cm}^2 = 4\frac{1}{2} \text{cm}^2$$

Volume of object:

$$\textbf{Volume}_{\textbf{object}} = Bh = \left(4\frac{1}{2} \text{cm}^2 \cdot 4\frac{1}{2} \text{cm}\right) = 20\frac{1}{4} \text{cm}^3$$

The volume of the object is $20\frac{1}{4} \text{cm}^3$.

- Write a numeric expression to represent the volume of the figure in Strategy 1.
  - $\left(\frac{1}{2} \cdot 3 \text{ cm} \cdot 4 \text{ cm}\right) 4\frac{1}{2} \text{cm} - \left(\frac{1}{2} \cdot 1\frac{1}{2} \text{cm} \cdot 2 \text{ cm}\right) 4\frac{1}{2} \text{cm}$
- Write a numeric expression to represent the volume of the figure in Strategy 2.
  - $4\frac{1}{2} \text{cm}$
- How do the numeric expressions represent the problem differently?
  - *The first expression is appropriate to use when the volume of the smaller prism is being subtracted away from the volume of the larger prism, whereas the second expression is used when the area of the base of the composite figure is found and then multiplied by the height to determine the volume.*
- What property allows us to show that these representations are equivalent?
  - *The distributive property*

Example 3 (10 minutes)

Example 3

A box with a length of 2 ft., a width of 1.5 ft., and a height of 1.25 ft. contains fragile electronic equipment that is packed inside a larger box with three inches of styrofoam cushioning material on each side (above, below, left side, right side, front, and back).

a. **Give the dimensions of the larger box.**

Length 2.5 ft.*, width* 2 ft.*, and height* 1.75 ft.

b. **Design styrofoam right rectangular prisms that could be placed around the box to provide the cushioning (i.e., give the dimensions and how many of each size are needed).**

Possible answer: Two pieces with dimensions $2.5 \text{ ft.} \cdot 2 \text{ ft.} \cdot 3 \text{ in.}$ *and four pieces with dimensions* $2 \text{ ft.} \cdot 1.25 \text{ ft.} \cdot 3 \text{ in.}$

c. **Find the volume of the styrofoam cushioning material by adding the volumes of the right rectangular prisms in the previous question.**

$V_1 = 2(2.5 \text{ ft.} \cdot 2 \text{ ft.} \cdot 0.25 \text{ ft.}) = 2.5 \text{ ft}^3$ $\quad V_2 = 4(2 \text{ ft.} \cdot 1.25 \text{ ft.} \cdot 0.25 \text{ ft.}) = 2.5 \text{ ft}^3$

$$V_1 + V_2 = 2.5 \text{ ft}^3 + 2.5 \text{ ft}^3 = 5 \text{ ft}^3$$

d. **Find the volume of the styrofoam cushioning material by computing the difference between the volume of the larger box and the volume of the smaller box.**

$(2.5 \text{ ft.} \cdot 2 \text{ ft.} \cdot 1.75 \text{ ft.}) - (2 \text{ ft.} \cdot 1.5 \text{ ft.} \cdot 1.25 \text{ ft.}) = 8.75 \text{ ft}^3 - 3.75 \text{ ft}^3 = 5 \text{ ft}^3$

Closing (2 minutes)

> **Lesson Summary**
>
> To find the volume of a three-dimensional composite object, two or more distinct volumes must be added together (if they are joined together) or subtracted from each other (if one is a missing section of the other). There are two strategies to find the volume of a prism:
>
> - Find the area of the base and then multiply times the prism's height.
> - Decompose the prism into two or more smaller prisms of the same height and add the volumes of those smaller prisms.

Exit Ticket (5 minutes)

Name ________________________________ Date ________________

Lesson 26: Volume of Composite Three-Dimensional Objects

Exit Ticket

A triangular prism has a rectangular prism cut out of it from one base to the opposite base, as shown in the figure. Determine the volume of the figure, provided all dimensions are in millimeters.

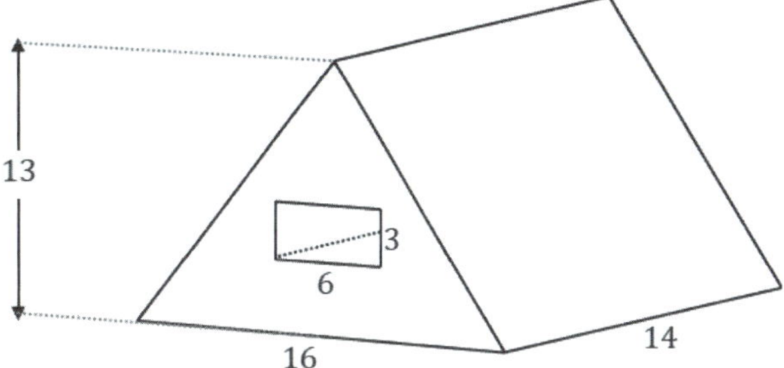

Is there any other way to determine the volume of the figure? If so, please explain.

Exit Ticket Sample Solutions

A triangular prism has a rectangular prism cut out of it from one base to the opposite base, as shown in the figure. Determine the volume of the figure, provided all dimensions are in millimeters.

Is there any other way to determine the volume of the figure? If so, please explain.

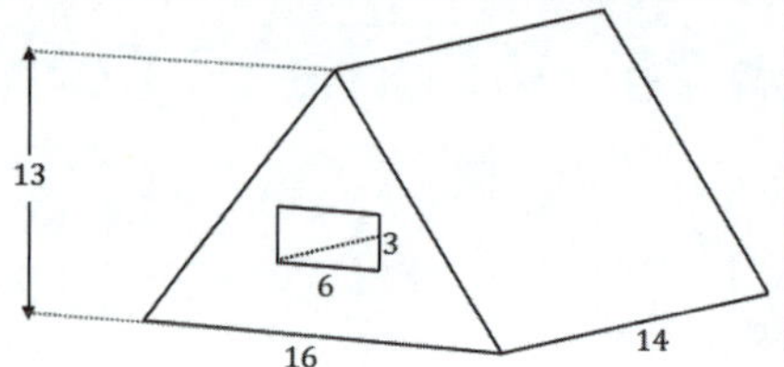

Possible response:

Volume of the triangular prism: $\left(\frac{1}{2} \cdot 16 \text{ mm} \cdot 13 \text{ mm}\right)(14 \text{ mm}) = 1,456 \text{ mm}^3$

Volume of the rectangular prism: $(6 \text{ mm} \cdot 3 \text{ mm} \cdot 14 \text{ mm}) = 252 \text{ mm}^3$

Volume of the composite prism: $1,456 \text{ mm}^3 - 252 \text{ mm}^3 = 1,204 \text{ mm}^3$

The calculations above subtract the volume of the cutout prism from the volume of the main prism. Another strategy would be to find the area of the base of the figure, which is the area of the triangle less the area of the rectangle, and then multiply by the height to find the volume of the prism.

Problem Set Sample Solutions

1. **Find the volume of the three-dimensional object composed of right rectangular prisms.**

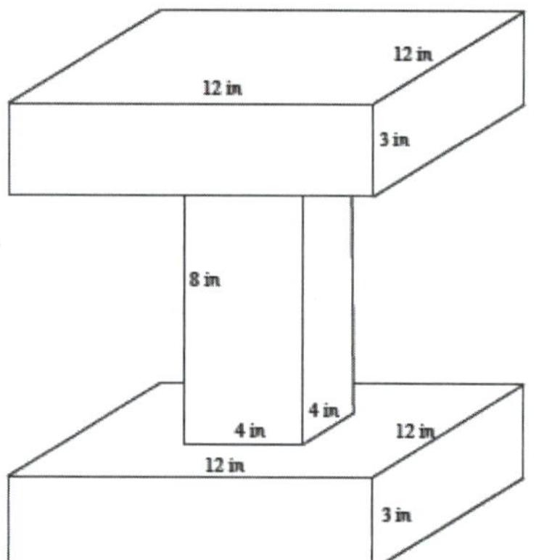

$\text{Volume}_{\text{object}} = \text{Volume}_{\text{top and bottom prisms}} + \text{Volume}_{\text{middle prism}}$

Volume of top and bottom prisms:

$V = 2(12 \text{ in.} \cdot 12 \text{ in.} \cdot 3 \text{ in.})$
$= 864 \text{ in}^3$

Volume of middle prism:

$V = 4 \text{ in.} \cdot 4 \text{ in.} \cdot 8 \text{ in.}$
$= 128 \text{ in}^3$

The volume of the object is $864 \text{ in}^3 + 128 \text{ in}^3 = 992 \text{ in}^3$.

2. **A smaller cube is stacked on top of a larger cube. An edge of the smaller cube measures $\frac{1}{2}$ cm in length, while the larger cube has an edge length three times as long. What is the total volume of the object?**

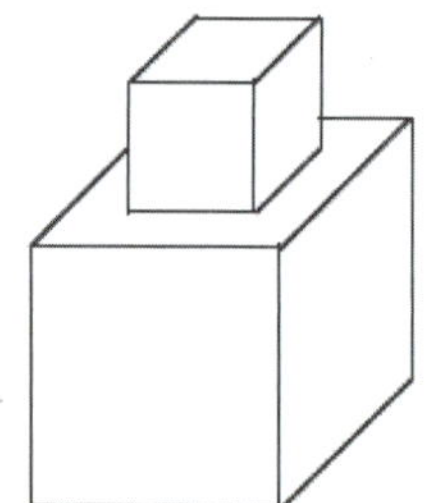

$\text{Volume}_{\text{object}} = \text{Volume}_{\text{small cube}} + \text{Volume}_{\text{large cube}}$

$\text{Volume}_{\text{small cube}} = \left(\frac{1}{2} \text{ cm}\right)^3$
$= \frac{1}{8} \text{ cm}^3$

$\text{Volume}_{\text{large cube}} = \left(\frac{3}{2} \text{ cm}\right)^3$
$= \frac{27}{8} \text{ cm}^3$

$V = \frac{1}{8} \text{ cm}^3 + \frac{27}{8} \text{ cm}^3$
$= 3\frac{1}{2} \text{ cm}^3$

The total volume of the object is $3\frac{1}{2} \text{ cm}^3$.

3. Two students are finding the volume of a prism with a rhombus base but are provided different information regarding the prism. One student receives Figure 1, while the other receives Figure 2.

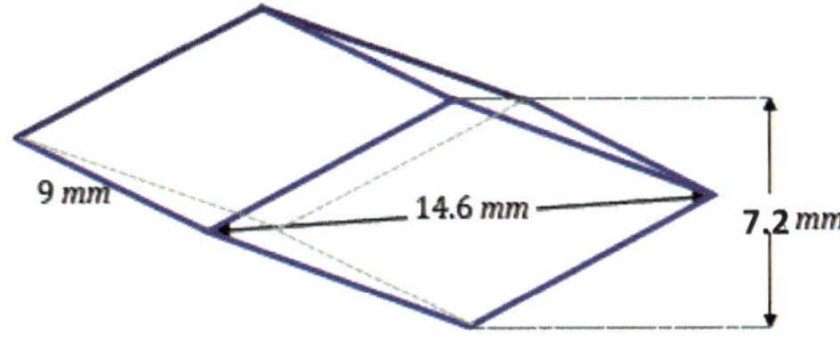

Figure 1

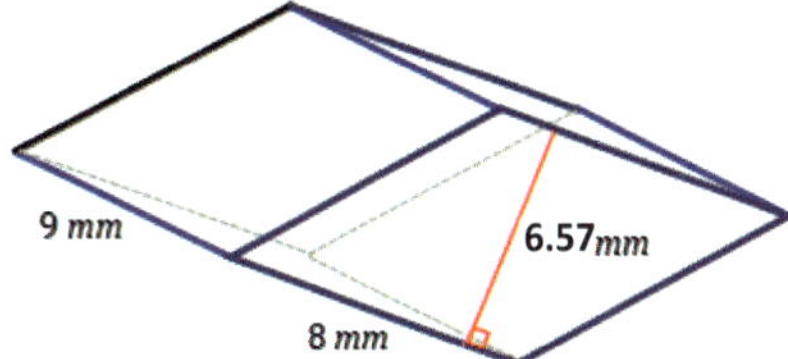

Figure 2

a. Find the expression that represents the volume in each case; show that the volumes are equal.

Figure 1

$2\left(\frac{1}{2}\cdot 14.6\text{ mm}\cdot 3.6\text{ mm}\right)\cdot 9\text{ mm}$

473.04 mm^3

Figure 2

$((8\text{ mm}\cdot 6.57\text{ mm})\cdot 9\text{ mm})$

473.04 mm^3

b. How does each calculation differ in the context of how the prism is viewed?

In Figure 1, the prism is treated as two triangular prisms joined together. The volume of each triangular prism is found and then doubled, whereas in Figure 2, the prism has a base in the shape of a rhombus, and the volume is found by calculating the area of the rhomboid base and then multiplying by the height.

4. Find the volume of wood needed to construct the following side table composed of right rectangular prisms.

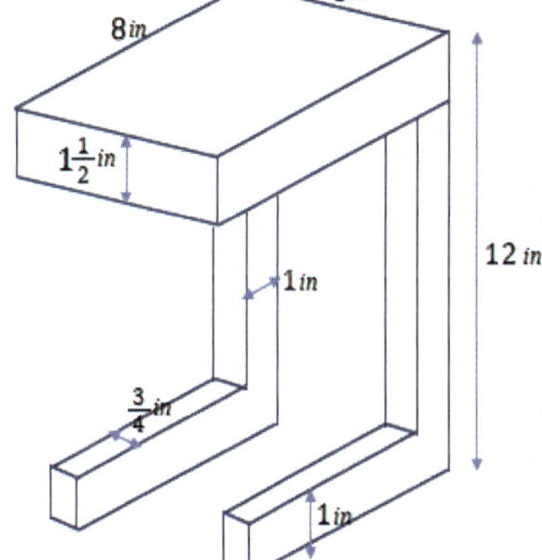

Volume of bottom legs: $V = 2(8\text{ in.}\cdot 1\text{ in.}\cdot 0.75\text{ in.})$
$= 12\text{ in}^3$

Volume of vertical legs: $V = 2(1\text{ in.}\cdot 9.5\text{ in.}\cdot 0.75\text{ in.})$
$= 14.25\text{ in}^3$

Volume of tabletop: $V = 8\text{ in.}\cdot 6\text{ in.}\cdot 1.5\text{ in.}$
$= 72\text{ in}^3$

The volume of the table is

$12\text{ in}^3 + 14.25\text{ in}^3 + 72\text{ in}^3 = 98.25\text{ in}^3$.

5. A plastic die (singular for dice) for a game has an edge length of 1.5 cm. Each face of the cube has the number of cubic cutouts as its marker is supposed to indicate (i.e., the face marked 3 has 3 cutouts). What is the volume of the die?

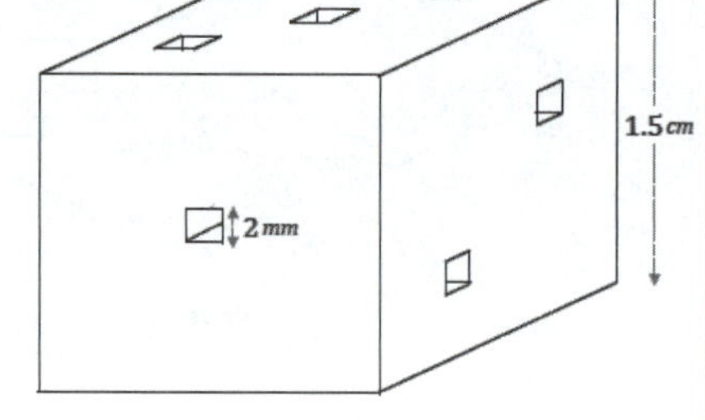

Number of cubic cutouts:

$1 + 2 + 3 + 4 + 5 + 6 = 21$

Volume of cutout cubes:

$V = 21(2 \text{ mm})^3$

$V = 168 \text{ mm}^3 = 0.168 \text{ cm}^3$

Volume of large cube:

$V = (1.5 \text{ cm})^3$

$V = 3.375 \text{ cm}^3$

The total volume of the die is

$3.375 \text{ cm}^3 - 0.168 \text{ cm}^3 = 3.207 \text{ cm}^3$.

6. A wooden cube with an edge length of 6 inches has square holes (holes in the shape of right rectangular prisms) cut through the centers of each of the three sides as shown in the figure. Find the volume of the resulting solid if the square for the holes has an edge length of 1 inch.

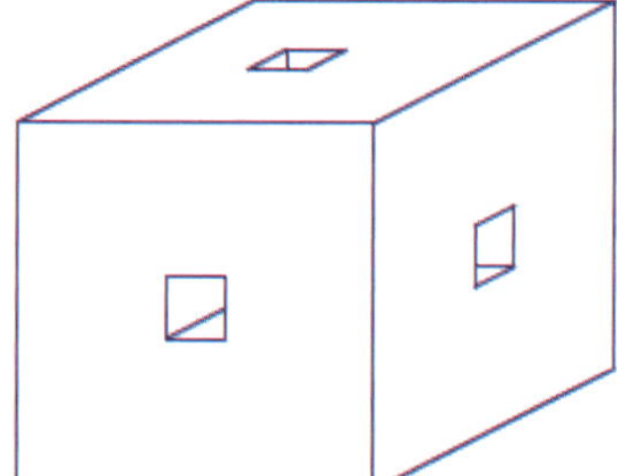

Think of making the square holes between opposite sides by cutting three times: The first cut removes 6 in^3, and the second and third cuts each remove 5 in^3. The resulting solid has a volume of $(6 \text{ in.})^3 - 6 \text{ in}^3 - 5 \text{ in}^3 - 5 \text{ in}^3 = 200 \text{ in}^3$.

7. A right rectangular prism has each of its dimensions (length, width, and height) increased by 50%. By what percent is its volume increased?

$V = l \cdot w \cdot h$

$V' = 1.5l \cdot 1.5w \cdot 1.5h$

$V' = 3.375lwh$

The larger volume is 337.5% of the smaller volume. The volume has increased by 237.5%.

8. A solid is created by putting together right rectangular prisms. If each of the side lengths is increased by 40%, by what percent is the volume increased?

If each of the side lengths is increased by 40%, then the volume of each right rectangular prism is multiplied by $1.4^3 = 2.744$. Since this is true for each right rectangular prism, the volume of the larger solid, V', can be found by multiplying the volume of the smaller solid, V, by $2.744 = 274.4\%$ (i.e., $V' = 2.744V$). This is an increase of 174.4%.

Lesson 27: Real-World Volume Problems

Student Outcomes

- Students use the volume formula for a right prism ($V = Bh$) to solve volume problems involving rate of flow.

Lesson Notes

Students apply their knowledge of volume to real-world contexts, specifically problems involving rate of flow of liquid. These problems are similar to problems involving distance, speed, and time; instead of manipulating the formula $d = rt$, students work with the formula $V = rt$, where r is the volumetric flow rate, and develop an understanding of its relationship to $V = Bh$. Specifically,

$$V = Bh = rt \text{ and } \frac{V}{t} = \frac{Bh}{t} = r,$$

where r is the volumetric flow rate.

Classwork

Opening (6 minutes)

- Imagine a car is traveling at 50 mph. How far does it go in 30 minutes?
  - *It travels 50 miles in one hour; therefore, it travels 25 miles in 30 minutes.*
- You just made use of the formula $d = rt$ to solve that problem. Today, we will use a similar formula.
- Here is a sample of the real-world context that we are studying today. Imagine a faucet turned on to the maximum level flows at a rate of 1 gallon in 25 seconds.
- Let's find out how long it would take to fill a 10-gallon tank at this rate.

Scaffolding:

Hours	Miles
0.5	25
1	50
1.5	75
2	100
2.5	125

Remind students how to use a ratio table by posing questions such as the following: How far does the car travel in half an hour? In two hours?

- First, what are the different quantities in this question?
  - *Rate, time, and volume*
- Create a ratio table for this situation. What is the constant of proportionality?
  - *The constant of proportionality is $\frac{1}{25}$.*
- What is the rate?
  - $\frac{1 \text{ gal}}{25 \text{ s}}$
- When we think about the volume of a liquid that passes in one unit of time, this rate is called the *flow rate* (or *volumetric flow rate*).

Seconds	*Gallons*
25	1
50	2
75	3
100	4
125	5
150	6
175	7
200	8
225	9
250	10

- What is the relationship between the quantities?
  - *Water that flows at a rate (volume per unit of time) for a given amount of time and yields a volume*
  - $V = rt$
- How can we tackle this problem?
  - $\frac{V}{r} = t$

    $(10 \text{ gal.}) \div \frac{1 \text{ gal}}{25 \text{ s}} = 250 \text{ sec.}$, *or* 4 min. *and* 10 sec.
- We will use the formula $V = rt$, where r is flow rate, in other contexts that involve a rate of flow.

Example 1 (8 minutes)

> **Example 1**
>
> **A swimming pool holds $10,000 \text{ ft}^3$ of water when filled. Jon and Anne want to fill the pool with a garden hose. The garden hose can fill a five-gallon bucket in 30 seconds. If each cubic foot is about 7.5 gallons, find the flow rate of the garden hose in gallons per minute and in cubic feet per minute. About how long will it take to fill the pool with a garden hose? If the hose is turned on Monday at 8:00 a.m., approximately when will the pool be filled?**

Scaffolding:

To complete the problem without the step involving conversion of units, use the following problem:

- A swimming pool holds $10{,}000 \text{ ft}^3$ of water when filled. Jon and Anne want to fill the pool with a garden hose. The flow rate of the garden hose is $1\frac{1}{3}\frac{\text{ft}^3}{\text{min}}$. About how long will it take to fill the pool with a garden hose? If the hose is turned on Monday at 8:00 a.m., approximately when will the pool be filled?

- If the hose fills a 5-gallon bucket in 30 seconds, how much would it fill in 1 minute? Find the flow rate in gallons per minute.
- *It would fill* 10 *gallons in* 1 *minute; therefore, the flow rate is* $10\frac{\text{gal}}{\text{min}}$.
- Find the flow rate in cubic feet per minute.
  - *Convert gallons to cubic feet:* $(10 \text{ gal.})\frac{1 \text{ ft}^3}{7.5 \text{ gal}} = 1\frac{1}{3}\text{ft}^3$

    Therefore, the flow rate of the garden hose in cubic feet per minute is

    $1\frac{1}{3}\frac{\text{ft}^3}{\text{min}}$
- How many minutes would it take to fill the $10{,}000 \text{ ft}^3$ pool?
  - $\frac{10{,}000 \text{ ft}^3}{1\frac{1}{3}\left(\frac{\text{ft}^3}{1 \text{ min}}\right)} = 7{,}500 \text{ min.}$
- How many days and hours is 7,500 minutes?
  - $(7{,}500 \text{ min.})\frac{1 \text{ h}}{60 \text{ min}} = 125 \text{ h}$, *or* 5 *days and* 5 *hours*
- At what time will the pool be filled?
  - *The pool begins to fill at 8:00 a.m. on Monday, so* 5 *days and* 5 *hours later on Saturday at 1:00 p.m., the pool will be filled.*

Example 2 (8 minutes)

Example 2

A square pipe (a rectangular prism-shaped pipe) with inside dimensions of 2 in. $\times$ 2 in. has water flowing through it at a flow speed of $3\frac{\text{ft}}{\text{s}}$. The water flows into a pool in the shape of a right triangular prism, with a base in the shape of a right isosceles triangle and with legs that are each 5 feet in length. How long will it take for the water to reach a depth of 4 feet?

- This problem is slightly different than the previous example. In this example, we are given a *flow speed* (also known as *linear flow speed*) instead of a *flow rate.*
- What do you think the term *flow speed* means based on what you know about *flow rate* and what you know about the units of each?
  - *Flow speed in this problem is measured in feet per second. Flow rate in Example 1 is measured in gallons per minute (or cubic feet per minute).*
  - *Flow speed is the distance that the liquid moves in one unit of time.*
- Now, let's go back to our example. The water is traveling at a flow speed of $3\frac{\text{ft}}{\text{s}}$. This means that for each second that the water is flowing out of the pipe, the water travels a distance of 3 ft. Now, we need to determine the volume of water that passes per second; in other words, we need to find the flow rate.
- Each second, the water in a cross-section of the pipe will travel 3 ft. This is the same as the volume of a right rectangular prism with dimensions 2 in. $\times$ 2 in. $\times$ 3 ft.

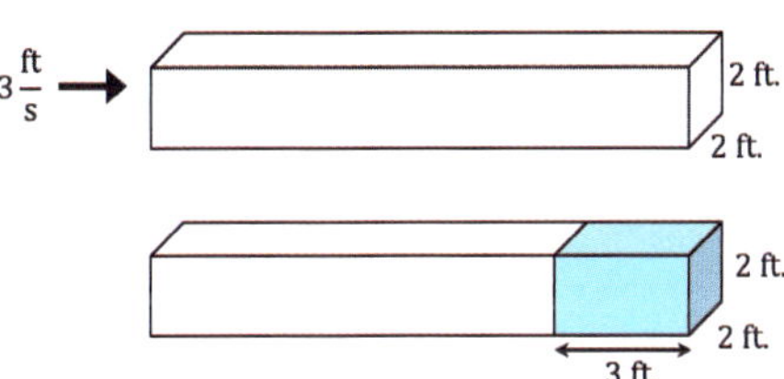

- The volume of this prism in cubic feet is $\frac{1}{6}$ ft. $\times$ $\frac{1}{6}$ ft. $\times$ 3 ft. $=$ $\frac{1}{12}$ ft^3, and the volume of water flowing out of the pipe every second is $\frac{1}{12}$ ft^3. So, the flow rate is $\frac{1}{12}\frac{\text{ft}^3}{\text{s}}$.
- Seconds is a very small unit of time when we think about filling up a pool. What is the flow rate in cubic feet per minute?
  - $\dfrac{\frac{1}{12}\text{ ft}^3}{1\text{ s}} \cdot \dfrac{60\text{ s}}{1\text{ min}} = 5\dfrac{\text{ft}^3}{\text{min}}$
- What is the volume of water that will be in the pool once the water reaches a depth of 4 ft.?
  - *The volume of water in the pool will be* $\frac{1}{2}(5\text{ ft.})(5\text{ ft.})(4\text{ ft.}) = 50\text{ ft}^3$.

- Now that we know our flow rate and the total volume of water in the pool, how long will it take for the pool to fill to a depth of 4 ft.?
  - $\dfrac{50 \text{ ft}^3}{5\frac{\text{ft}^3}{\text{min}}} = 10 \text{ min.}$

    It will take 10 *minutes to fill the pool to a depth of* 4 ft.

Exercise 1 (8 minutes)

Students have to find volumes of two composite right rectangular prisms in this exercise. As they work on finding the volume of the lower level of the fountain, remind students that the volume of the whole top level must be subtracted from the inner volume of the lower level. This does not, however, require the whole height of the top level; the relevant height for the volume that must be subtracted is 2 ft. (see calculation in solution).

Exercise 1

A park fountain is about to be turned on in the spring after having been off all winter long. The fountain flows out of the top level and into the bottom level until both are full, at which point the water is just recycled from top to bottom through an internal pipe. The outer wall of the top level, a right square prism, is five feet in length; the thickness of the stone between the outer and inner wall is 1 ft.; and the depth is 1 ft. The bottom level, also a right square prism, has an outer wall that is 11 ft. long with a 2 ft. thickness between the outer and inner wall and a depth of 2 ft. Water flows through a 3 in. × 3 in. square pipe into the top level of the fountain at a flow speed of $4\frac{\text{ft}}{\text{s}}$. Approximately how long will it take for both levels of the fountain to fill completely?

Volume of top:
$3 \text{ ft.} \times 3 \text{ ft.} \times 1 \text{ ft.} = 9 \text{ ft}^3$

Volume of bottom:
$(7 \text{ ft.} \times 7 \text{ ft.} \times 2 \text{ ft.}) - (5 \text{ ft.} \times 5 \text{ ft.} \times 2 \text{ ft.}) = 48 \text{ ft}^3$

Combined volume of both levels:
$9 \text{ ft}^3 + 48 \text{ ft}^3 = 57 \text{ ft}^3$

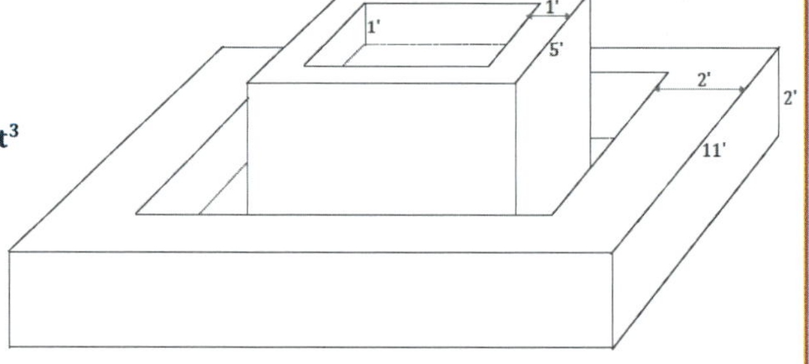

With a flow speed of $4\frac{\text{ft}}{\text{s}}$ through a 3 in. × 3 in. square pipe, the volume of water moving through the pipe in one second is equivalent to the volume of a right rectangular prism with dimensions 3 in. × 3 in. × 4 ft. The volume in feet is $\frac{1}{4} \text{ ft.} \times \frac{1}{4} \text{ ft.} \times 4 \text{ ft.} = \frac{1}{4} \text{ ft}^3$. Therefore, the flow rate is $\frac{1}{4}\frac{\text{ft}^3}{\text{s}}$ because $\frac{1}{4}$ ft^3 of water flows every second.

Volume of water that will flow in one minute:

$$\frac{\frac{1}{4}\text{ft}^3}{1 \text{ s}} \cdot \frac{60 \text{ s}}{1 \text{ min}} = 15\frac{\text{ft}^3}{\text{min}}$$

Time needed to fill both fountain levels: $\dfrac{57 \text{ ft}^3}{15\frac{\text{ft}^3}{\text{min}}} = 3.8 \text{ min.}$; ***it will take* 3.8 *minutes to fill both fountain levels.***

Exercise 2 (7 minutes)

Exercise 2

A decorative bathroom faucet has a 3 in. $\times$ 3 in. square pipe that flows into a basin in the shape of an isosceles trapezoid prism like the one shown in the diagram. If it takes one minute and twenty seconds to fill the basin completely, what is the approximate speed of water flowing from the faucet in feet per second?

Volume of the basin in cubic inches:

$$\frac{1}{2}(3 \text{ in.} + 15 \text{ in.})(10 \text{ in.}) \times 4.5 \text{ in.} = 405 \text{ in}^3$$

Approximate volume of the basin in cubic feet:

$$(405 \text{ in}^3)\left(\frac{1 \text{ ft}^3}{1,728 \text{ in}^3}\right) = 0.234375 \text{ ft}^3$$

Based on the rate of water flowing out the faucet, the volume of water can also be calculated as follows:

Let s represent the distance that the water is traveling in 1 second.

$$3 \text{ in.} \cdot 3 \text{ in.} \cdot s = \frac{1}{4} \text{ft.} \cdot \frac{1}{4} \text{ft.} \cdot s \text{ ft.} = 0.234375 \text{ ft}^3$$

Therefore, the speed of the water flowing from the faucet is $3.75 \frac{\text{ft}}{\text{s}}$.

Closing (1 minute)

- What does it mean for water to flow through a square pipe?
  - *The pipe can be visualized as a right rectangular prism.*
- If water is flowing through a 2 in. $\times$ 2 in. square pipe at a speed of $4 \frac{\text{ft.}}{\text{s}}$, what is the volume of water that flows from the pipe every second? What is the flow rate?
  - $\frac{1}{6} \text{ft.} \cdot \frac{1}{6} \text{ft.} \cdot 4 \text{ ft.} = \frac{1}{9} \text{ft}^3$
  - *The flow rate is* $\frac{1}{9} \frac{\text{ft}^3}{\text{s}}$.

Lesson Summary

The formulas $V = Bh$ and $V = rt$, where r is flow rate, can be used to solve real-world volume problems involving flow speed and flow rate. For example, water flowing through a square pipe can be visualized as a right rectangular prism. If water is flowing through a 2 in. $\times$ 2 in. square pipe at a flow speed of $4 \frac{\text{ft}}{\text{s}}$, then for every second the water flows through the pipe, the water travels a distance of 4 ft. The volume of water traveling each second can be thought of as a prism with a 2 in. $\times$ 2 in. base and a height of 4 ft. The volume of this prism is:

$$\begin{aligned} V &= Bh \\ &= \frac{1}{6} \text{ft.} \times \frac{1}{6} \text{ft.} \times 4 \text{ ft.} \\ &= \frac{1}{9} \text{ ft}^3 \end{aligned}$$

Therefore, $\frac{1}{9}$ ft^3 of water flows every second, and the flow rate is $\frac{1}{9} \frac{\text{ft}^3}{\text{s}}$.

Exit Ticket (7 minutes)

Name ______________________________ Date ______________

Lesson 27: Real-World Volume Problems

Exit Ticket

Jim wants to know how much his family spends on water for showers. Water costs $\$1.50$ for $1{,}000$ gallons. His family averages 4 showers per day. The average length of a shower is 10 minutes. He places a bucket in his shower and turns on the water. After one minute, the bucket has 2.5 gallons of water. About how much money does his family spend on water for showers in a 30-day month?

Exit Ticket Sample Solutions

Jim wants to know how much his family spends on water for showers. Water costs $\$1.50$ for $1,000$ gallons. His family averages 4 showers per day. The average length of a shower is 10 minutes. He places a bucket in his shower and turns on the water. After one minute, the bucket has 2.5 gallons of water. About how much money does his family spend on water for showers in a 30-day month?

Number of gallons of water in one day of showering (four ten-minute showers): $4(10 \text{ min.})\left(\frac{2.5 \text{ gal}}{1 \text{ min}}\right) = 100 \text{ gal.}$

Number of gallons of water in 30 days: $(30 \text{ days})\left(\frac{100 \text{ gal}}{1 \text{ day}}\right) = 3,000 \text{ gal.}$

Cost of showering for 30 days: $(3,000 \text{ gal.})\left(\frac{\$1.50}{1,000 \text{ gal}}\right) = \$4.50$

The family spends $\$4.50$ in a 30-day month on water for showers.

Problem Set Sample Solutions

1. Harvey puts a container in the shape of a right rectangular prism under a spot in the roof that is leaking. Rainwater is dripping into the container at an average rate of 12 drops a minute. The container Harvey places under the leak has a length and width of 5 cm and a height of 10 cm. Assuming each raindrop is roughly 1 cm^3, approximately how long does Harvey have before the container overflows?

   Volume of the container in cubic centimeters:

   $5 \text{ cm} \times 5 \text{ cm} \times 10 \text{ cm} = 250 \text{ cm}^3$

   Number of minutes until the container is filled with rainwater:

   $(250 \text{ cm}^3)\left(\frac{1 \text{ min}}{12 \text{ cm}^3}\right) \approx 20.8 \text{ min.}$

2. A large square pipe has inside dimensions $3 \text{ in.} \times 3 \text{ in.}$, and a small square pipe has inside dimensions $1 \text{ in.} \times 1 \text{ in.}$ Water travels through each of the pipes at the same constant flow speed. If the large pipe can fill a pool in 2 hours, how long will it take the small pipe to fill the same pool?

   If s is the length that the water travels in one minute, then in one minute the large pipe provides $\frac{1}{4} \text{ ft.} \cdot \frac{1}{4} \text{ ft.} \cdot s \text{ ft.}$ of water. In one minute, the small pipe provides one-ninth as much, $\frac{1}{12} \text{ ft.} \cdot \frac{1}{12} \text{ ft.} \cdot s \text{ ft.}$ of water. Therefore, it will take the small pipe nine times as long. It will take the small pipe 18 hours to fill the pool.

3. A pool contains $12,000 \text{ ft}^3$ of water and needs to be drained. At 8:00 a.m., a pump is turned on that drains water at a flow rate of 10 ft^3 per minute. Two hours later, at 10:00 a.m., a second pump is activated that drains water at a flow rate of 8 ft^3 per minute. At what time will the pool be empty?

   Water drained in the first two hours: $\frac{10 \text{ ft}^3}{1 \text{ min}} \cdot 120 \text{ min.} = 1,200 \text{ ft}^3$

   Volume of water that still needs to be drained: $12,000 \text{ ft}^3 - 1,200 \text{ ft}^3 = 10,800 \text{ ft}^3$

   Amount of time needed to drain remaining water with both pumps working:

   $10,800 \text{ ft}^3 \left(\frac{1 \text{ min}}{10 \text{ ft}^3} + \frac{1 \text{ min}}{8 \text{ ft}^3}\right) = 600 \text{ min.}$, *or* 10 h.

   The total time needed to drain the pool is 12 hours, so the pool will drain completely at 8:00 p.m.

4. **In the previous problem, if water starts flowing into the pool at noon at a flow rate of 3 ft^3 per minute, how much longer will it take to drain the pool?**

At noon, the first pump will have been on for four hours, and the second pump will have been on for two hours. The cubic feet of water drained by the two pumps together at noon is

$$240\text{ min.}\left(\frac{10\text{ ft}^3}{1\text{ min}}\right) + 120\text{ min.}\left(\frac{8\text{ ft}^3}{1\text{ min}}\right) = 3{,}360\text{ ft}^3$$

Volume of water that still needs to be drained:

$$12{,}000\text{ ft}^3 - 3{,}360\text{ ft}^3 = 8{,}640\text{ ft}^3$$

If water is entering the pool at $3\frac{\text{ft}^3}{\text{min}}$ *but leaving it at* $18\frac{\text{ft}^3}{\text{min}}$, *the net effect is that water is leaving the pool at* $15\frac{\text{ft}^3}{\text{min}}$.

The amount of time needed to drain the remaining water with both pumps working and water flowing in:

$$8{,}640\text{ ft}^3\left(\frac{1\text{ min}}{15\text{ ft}^3}\right) = 576\text{ min., or } 9\text{ h. and } 36\text{ min.}$$

The pool will finish draining at 9:36 p.m. the same day. It will take an additional 1 *hour and* 36 *minutes to drain the pool.*

5. **A pool contains $6{,}000\text{ ft}^3$ of water. Pump A can drain the pool in 15 hours, Pump B can drain it in 12 hours, and Pump C can drain it in 10 hours. How long will it take all three pumps working together to drain the pool?**

Rate at which Pump A drains the pool: $\frac{1}{15}$ *pool per hour*

Rate at which Pump B drains the pool: $\frac{1}{12}$ *pool per hour*

Rate at which Pump C drains the pool: $\frac{1}{10}$ *pool per hour*

Together, the pumps drain the pool at $\left(\frac{1}{15}+\frac{1}{12}+\frac{1}{10}\right)$ *pool per hour, or* $\frac{1}{4}$ *pool per hour. Therefore, it will take* 4 *hours to drain the pool when all three pumps are working together.*

6. **A $2{,}000$-gallon fish aquarium can be filled by water flowing at a constant rate in 10 hours. When a decorative rock is placed in the aquarium, it can be filled in 9.5 hours. Find the volume of the rock in cubic feet ($1\text{ ft}^3 = 7.5\text{ gal.}$)**

Rate of water flow into aquarium:

$$\frac{2{,}000\text{ gal}}{10\text{ h}} = \frac{200\text{ gal}}{1\text{ h}}$$

Since it takes half an hour less time to fill the aquarium with the rock inside, the volume of the rock is

$$\frac{200\text{ gal}}{1\text{ h}}(0.5\text{ h.}) = 100\text{ gal.}$$

Volume of the rock:

$$100\text{ gal.}\left(\frac{1\text{ ft}^3}{7.5\text{ gal}}\right) \approx 13.3\text{ ft}^3$$; *the volume of the rock is approximately* 13.3 ft^3.

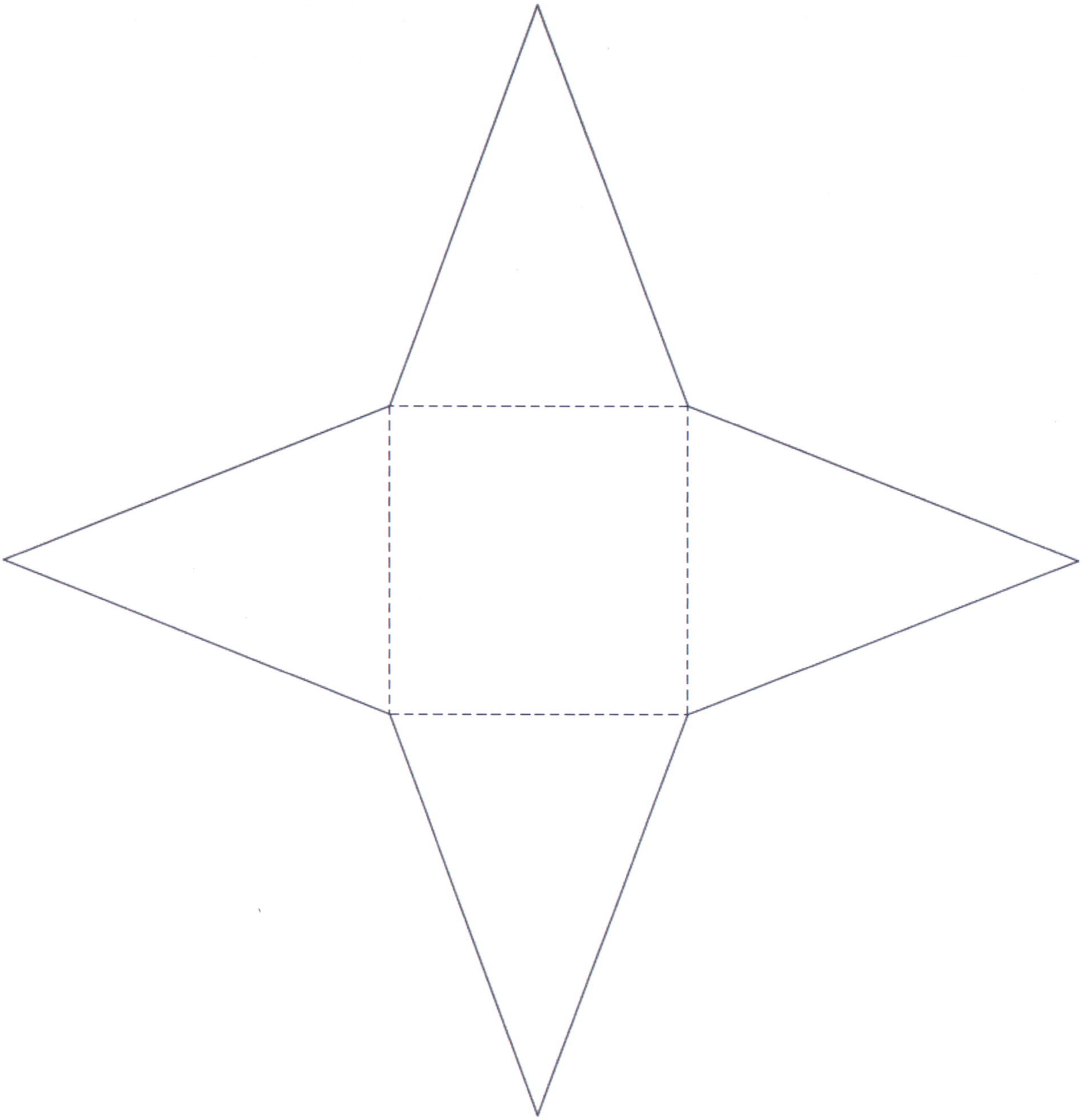

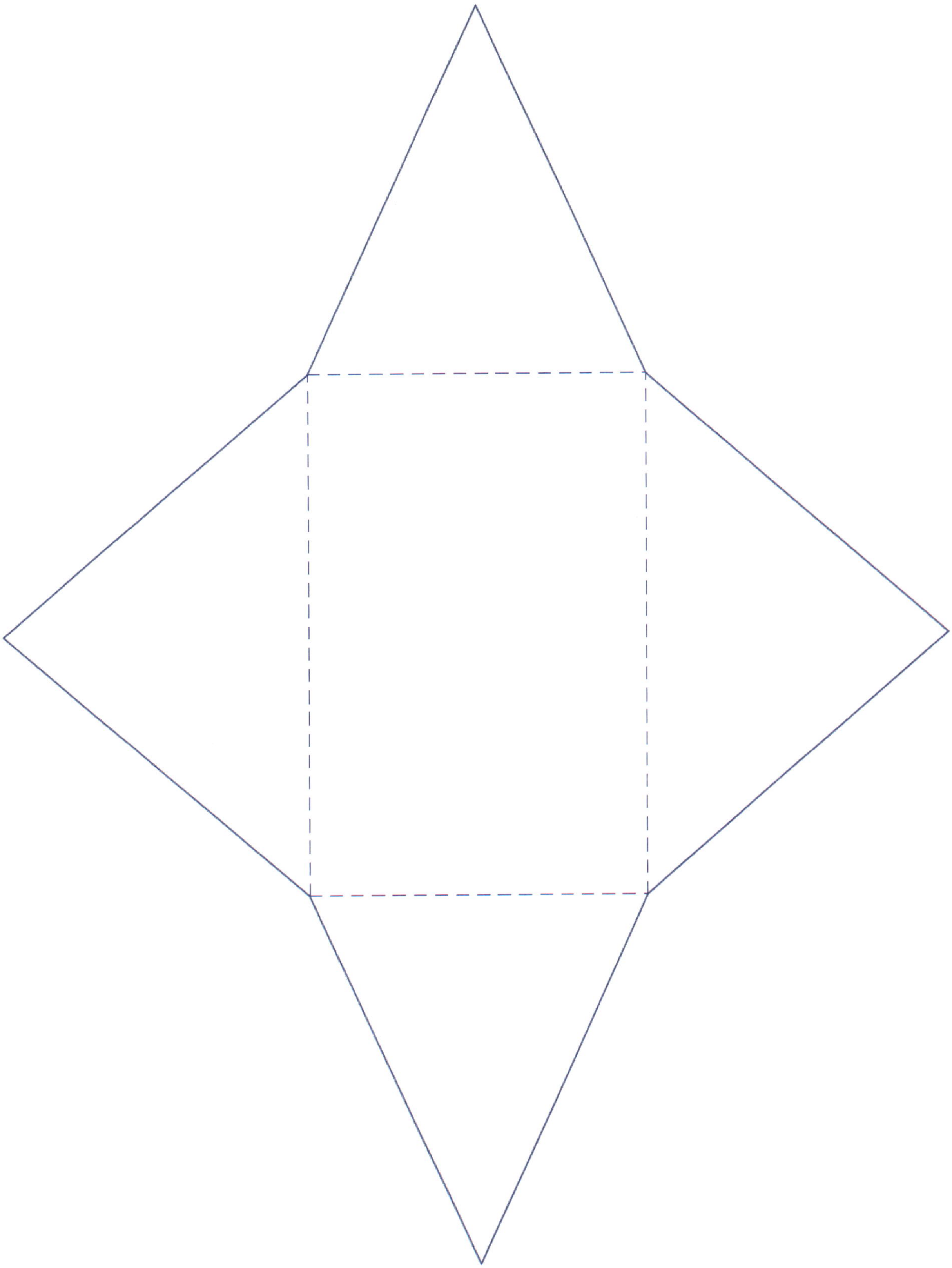

Name ______________________________ Date ______________

1. In the following two questions, lines AB and CD intersect at point O. When necessary, assume that seemingly straight lines are indeed straight lines. Determine the measures of the indicated angles.

   a. Find the measure of $\angle XOC$.

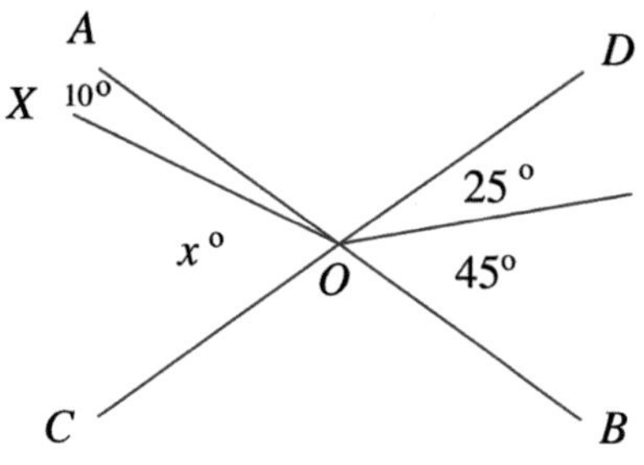

   b. Find the measures of $\angle AOX$, $\angle YOD$, and $\angle DOB$.

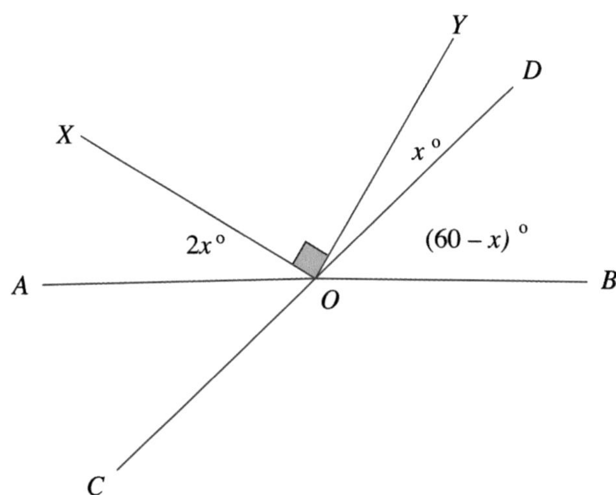

2. Is it possible to draw two different triangles that both have angle measurements of $40°$ and $50°$ and a side length of 5 cm? If it is possible, draw examples of these conditions, and label all vertices and angle and side measurements. If it is not possible, explain why.

3. In each of the following problems, two triangles are given. For each: (1) State if there are sufficient or insufficient conditions to show the triangles are identical, and (2) explain your reasoning.

   a.

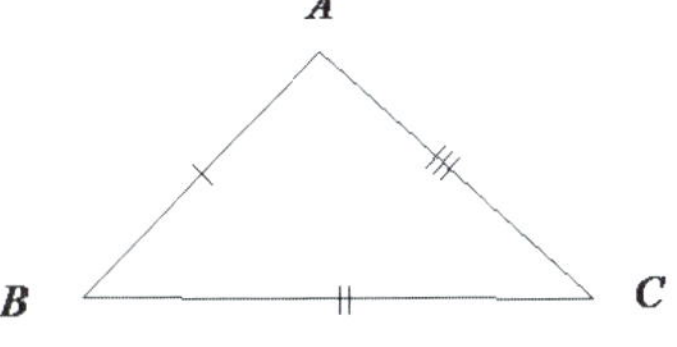

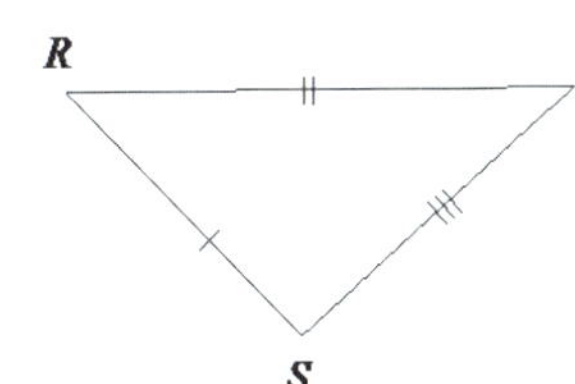

   b.

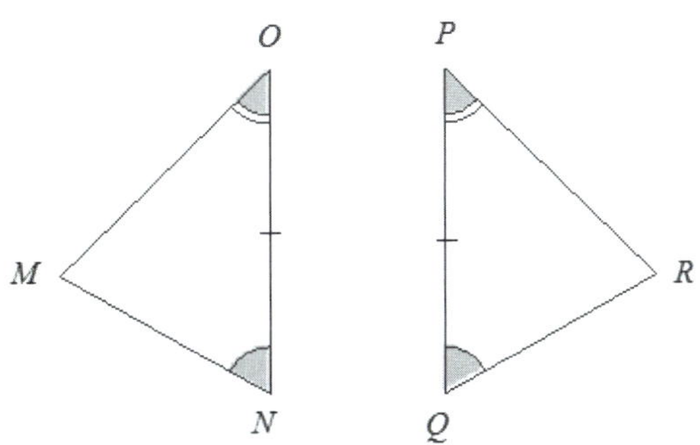

4. In the following diagram, the length of one side of the smaller shaded square is $\frac{1}{3}$ the length of square $ABCD$. What percent of square $ABCD$ is shaded? Provide all evidence of your calculations.

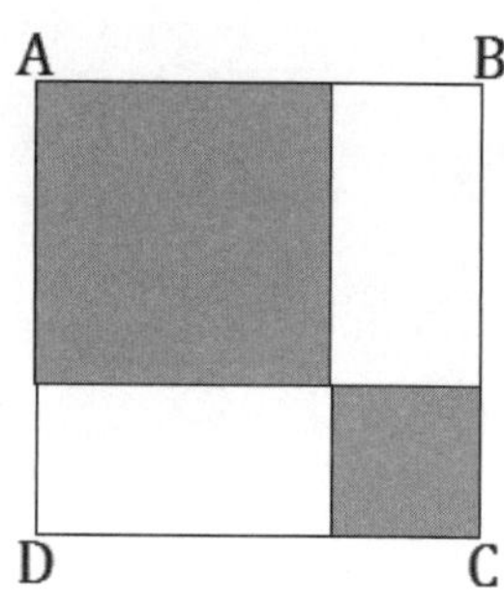

5. Side $\overline{EF}$ of square $DEFG$ has a length of 2 cm and is also the radius of circle F. What is the area of the entire shaded region? Provide all evidence of your calculations.

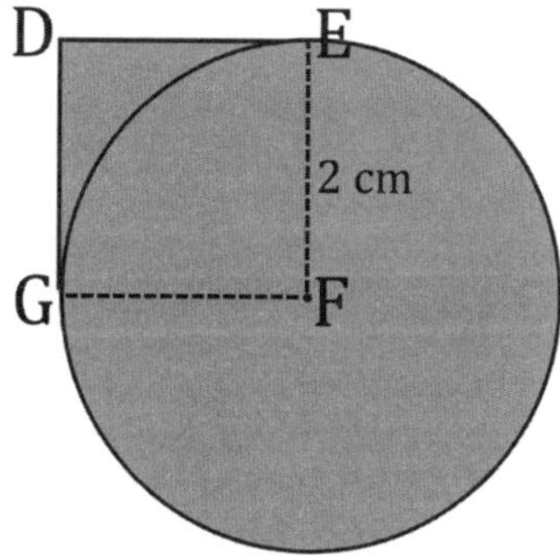

6. For his latest design, a jeweler hollows out crystal cube beads (like the one in the diagram) through which the chain of a necklace is threaded. If the edge of the crystal cube is 10 mm, and the edge of the square cut is 6 mm, what is the volume of one bead? Provide all evidence of your calculations.

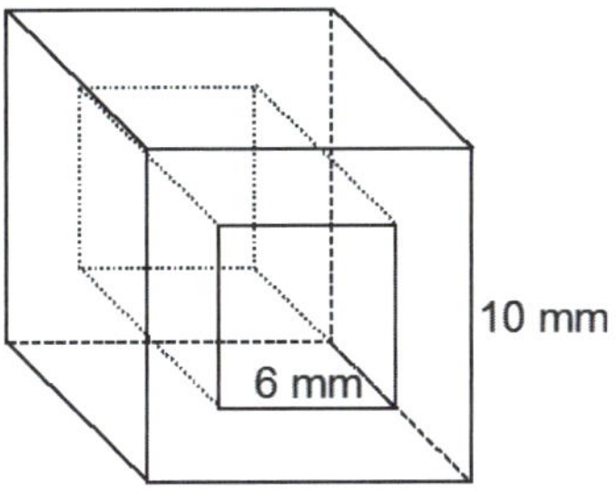

7. John and Joyce are sharing a piece of cake with the dimensions shown in the diagram. John is about to cut the cake at the mark indicated by the dotted lines. Joyce says this cut will make one of the pieces three times as big as the other. Is she right? Justify your response.

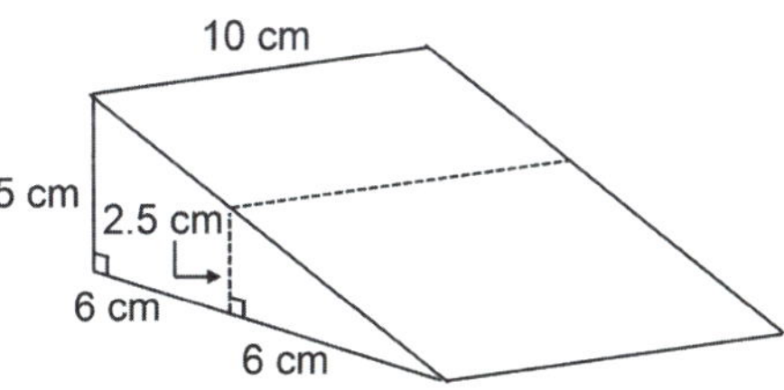

8. A tank measures 4 ft. in length, 3 ft. in width, and 2 ft. in height. It is filled with water to a height of 1.5 ft. A typical brick measures a length of 9 in., a width of 4.5 in., and a height of 3 in. How many whole bricks can be added before the tank overflows? Provide all evidence of your calculations.

9. Three vertical slices perpendicular to the base of the right rectangular pyramid are to be made at the marked locations: (1) through $\overline{AB}$, (2) through $\overline{CD}$, and (3) through vertex E. Based on the relative locations of the slices on the pyramid, make a reasonable sketch of each slice. Include the appropriate notation to indicate measures of equal length.

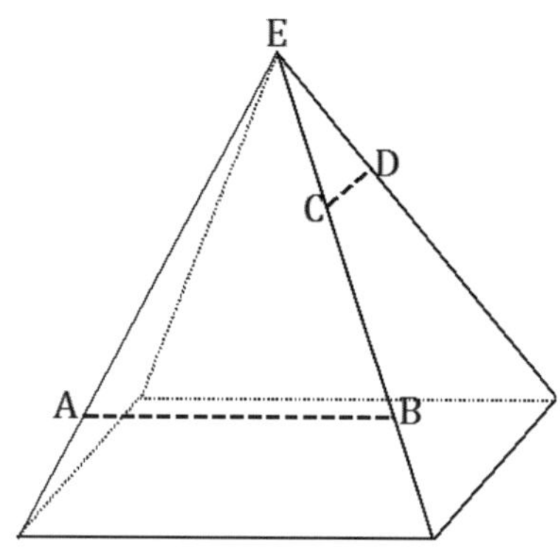

(1) Slice through $\overline{AB}$	(2) Slice through $\overline{CD}$	(3) Slice through vertex E

G7-M6-TE-B6-1.3.0-10.2015

10. Five three-inch cubes and two triangular prisms have been glued together to form the composite three-dimensional figure shown in the diagram. Find the surface area of the figure, including the base. Provide all evidence of your calculations.

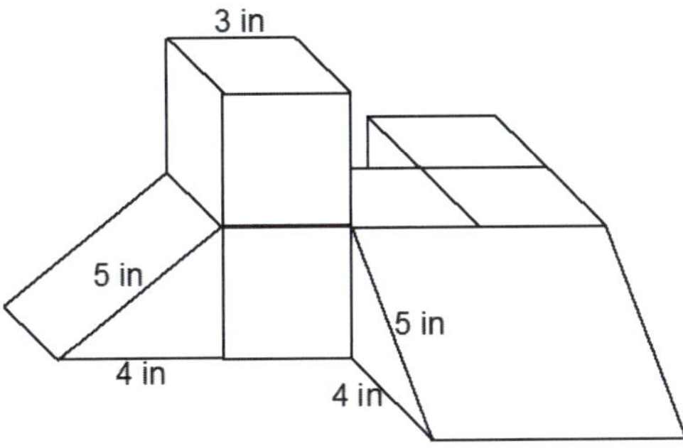

A Progression Toward Mastery					
Assessment Task Item		**STEP 1** Missing or incorrect answer and little evidence of reasoning or application of mathematics to solve the problem.	**STEP 2** Missing or incorrect answer but evidence of some reasoning or application of mathematics to solve the problem.	**STEP 3** A correct answer with some evidence of reasoning or application of mathematics to solve the problem OR an incorrect answer with substantial evidence of solid reasoning or application of mathematics to solve the problem.	**STEP 4** A correct answer supported by substantial evidence of solid reasoning or application of mathematics to solve the problem.
1	a 7.G.B.5	Student correctly sets up the equation to solve the problem, but no further supporting work is shown.	Student finds an incorrect value due to conceptual error (e.g., an equation that does not reflect the angle relationship), but complete supporting work is shown.	Student finds an incorrect value due to a calculation error, but complete supporting work is shown.	Student finds $\angle XOC = 60°$, and complete supporting work is shown.
	b 7.G.B.5	Student correctly sets up the equations to solve for the angles, but no further supporting work is shown.	Student finds the correct value for one angle and shows complete supporting work, but a calculation error leads to one incorrect answer.	Student finds the correct values for two angles and shows complete supporting work, but a calculation error leads to one incorrect answer.	Student finds $\angle AOX = 30°$, $\angle YOD = 15°$, and $\angle DOB = 45°$, and complete supporting work is shown.
2	7.G.A.2	Student states that it is not possible to construct two different triangles under the given conditions, or two identical triangles are constructed.	Student correctly constructs two different triangles according to the given conditions, but they are missing measurement and/or vertex labels.	Student constructs two different triangles that are appropriately labeled, but the corresponding angle measurements are not exactly equal and off within 3° of the given conditions.	Student correctly constructs two different triangles according to the given conditions, and they are appropriately labeled.

3	a 7.G.A.2	Student does not provide a response. OR Student fails to provide evidence of comprehension.	Student correctly identifies triangles as identical or not identical, but no further evidence is provided.	Student correctly identifies triangles as identical or not identical but with the incorrect supporting evidence, such as giving the incorrect condition by which they are identical.	Student correctly identifies triangles as identical or not identical and supports this answer, such as giving the condition by which they are identical or the information that prevents them from being identical.
	b 7.G.A.2	Student does not provide a response. OR Student fails to provide evidence of comprehension.	Student correctly identifies triangles as identical or not identical, but no further evidence is provided.	Student correctly identifies triangles as identical or not identical but with the incorrect supporting evidence, such as giving the incorrect condition by which they are identical.	Student correctly identifies triangles as identical or not identical and supports this answer, such as giving the condition by which they are identical or the information that prevents them from being identical.
4	7.G.B.6	Student incorrectly calculates the percentage of the shaded area due to a combination of at least one conceptual and one calculation error or due to more than one conceptual or calculation error.	Student incorrectly calculates the percentage of the shaded area due to one conceptual error (e.g., taking the incorrect values by which to calculate percentage), but all other supporting work is correct.	Student incorrectly calculates the percentage of the shaded area due to one calculation error (e.g., not summing both shaded areas), but all other supporting work is correct.	Student correctly finds $55\frac{5}{9}\%$ as the percentage of the shaded area, and complete evidence of calculations is shown.
5	7.G.B.6	Student incorrectly calculates the shaded area due to a combination of at least one conceptual and one calculation error OR due to more than one conceptual or calculation error.	Student incorrectly calculates the shaded area due to one conceptual error (e.g., taking the incorrect fraction of the area of the circle to add to the area of the square), but all other supporting work is correct.	Student incorrectly calculates the shaded area due to one calculation error (e.g., making an error in taking a fraction of 4π), but all other supporting work is correct.	Student correctly finds the shaded area to be either $4\text{ cm}^2 + 3\pi\text{ cm}^2$, or approximately 13.4 cm^2, and complete evidence of calculations is shown.
6	7.G.B.6	Student incorrectly finds the volume due to one or more calculation errors or a combination of calculation and conceptual errors.	Student incorrectly finds the volume due to one conceptual error (e.g., calculating the volume of the hollow as a cube rather than as a rectangular prism), but all other supporting work is correct.	Student incorrectly finds the volume due to one calculation error (e.g., an arithmetic error), but all other supporting work is correct.	Student correctly finds the volume of the bead to be 640 mm^3, and complete evidence of calculations is shown.

7	7.G.B.6	Student incorrectly finds the volume due to one or more calculation errors or a combination of calculation and conceptual errors.	Student incorrectly finds the volume due to one conceptual error (e.g., using the wrong formula for the volume of a trapezoidal prism), but all other supporting work is correct.	Student incorrectly finds the volume due to one calculation error (e.g., an arithmetic error), but all other supporting work is correct.	Student correctly finds the volume of the trapezoidal prism to be 225 cm^3 and the volume of the triangular prism to be 75 cm^3, and the larger piece is shown to be 3 times as great as the smaller piece.
8	7.G.B.3 7.G.B.6	Student incorrectly finds the number of bricks due to one or more calculation errors or a combination of calculation and conceptual errors.	Student incorrectly finds the number of bricks due to a calculation error (e.g., using the volume of water rather than the volume of the unfilled tank), but all other supporting work is correct.	Student incorrectly finds the number of bricks due to one calculation error (e.g., a rounding error), but all other supporting work is correct.	Student correctly finds that 85 bricks can be put in the tank without the tank overflowing and offers complete evidence of calculations.
9	7.G.B.6	Student does not appropriately sketch two relative trapezoids according to their relative positions on the pyramid, and an isosceles triangle is not made for the slice through the vertex.	Student sketches two relative trapezoids appropriately but does not sketch an isosceles triangle as a slice through the vertex.	Student makes three sketches but does not indicate lengths of equal measure.	Student makes three sketches that indicate appropriate slices at the given locations on the pyramid and indicates the lengths of equal measure.
10	7.G.B.6	Student incorrectly finds the surface area due to more than one calculation error or a combination of calculation and conceptual errors.	Student incorrectly finds the surface area due to a conceptual error (e.g., mistaking which measure to use in computation).	Student incorrectly finds the surface area due to a calculation error, such as an arithmetic error.	Student correctly calculates the surface area to be 276 in^2, and complete evidence of calculations is shown.

Name ______________________________ Date ________________

1. In the following two questions, lines AB and CD intersect at point O. When necessary, assume that seemingly straight lines are indeed straight lines. Determine the measures of the indicated unknown angles.

   a. Find the measure of $\angle XOC$.

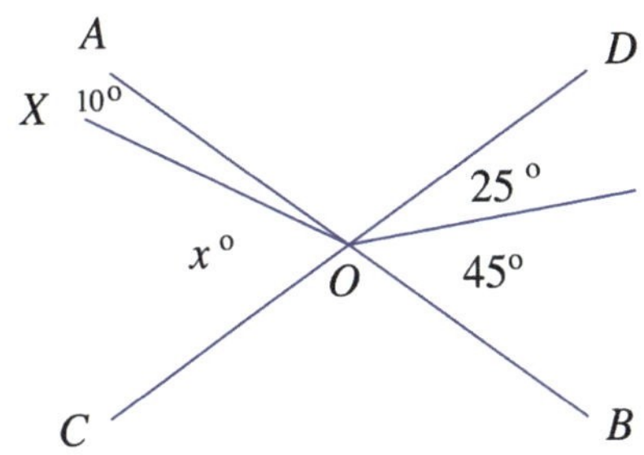

$$x + 10 = 25 + 45$$
$$x + 10 - 10 = 70 - 10$$
$$x = 60$$

$\angle XOC = 60°$

   b. Find the measures of $\angle AOX$, $\angle YOD$, and $\angle DOB$.

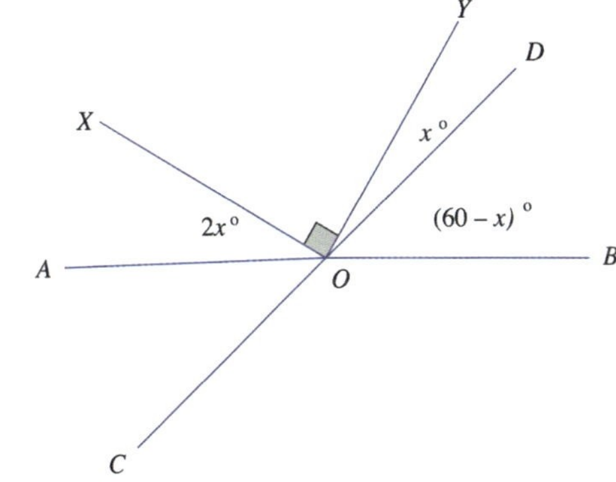

$$2x + 90 + x + (60 - x) = 180$$
$$2x + 150 - 150 = 180 - 150$$
$$2x = 30$$
$$\frac{1}{2}(2x) = \frac{1}{2}(30)$$
$$x = 15$$

$\angle AOX = 2(15)° = 30°$

$\angle YOD = 15°$

$\angle DOB = (60 - 15)° = 45°$

2. Is it possible to draw two different triangles that both have angle measurements of $40°$ and $50°$ and a side length of 5 cm? If it is possible, draw examples of these conditions, and label all vertices and angle and side measurements. If it is not possible, explain why.

One possible solution:

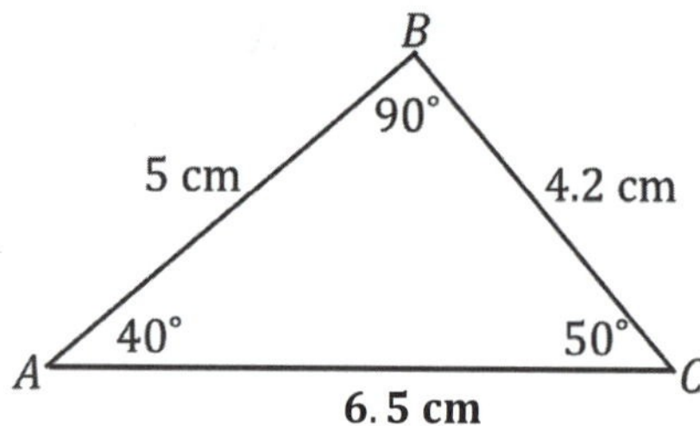

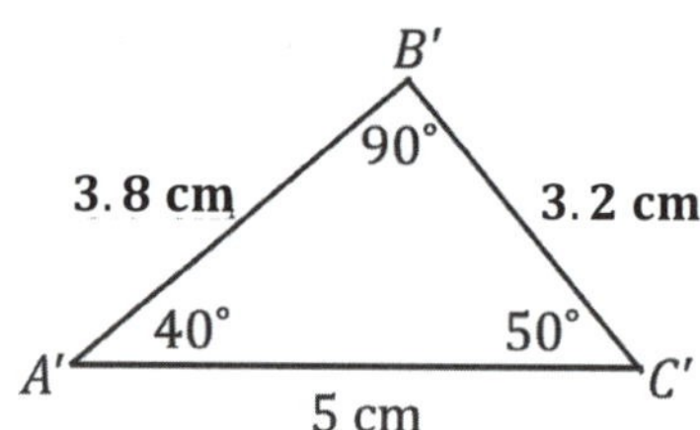

3. In each of the following problems, two triangles are given. For each: (1) State if there are sufficient or insufficient conditions to show the triangles are identical, and (2) explain your reasoning.

a. *The triangles are identical by the three sides condition.* $\triangle ABC \leftrightarrow \triangle SRT$

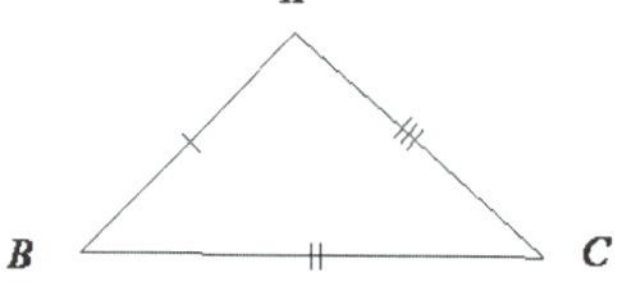

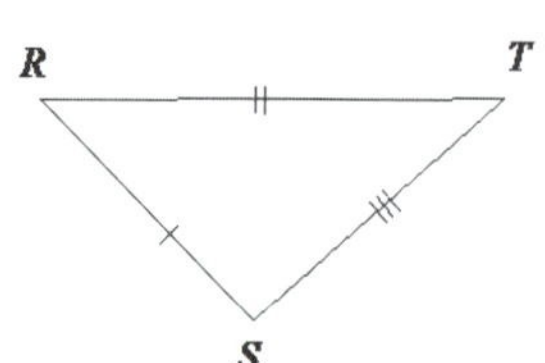

b. *The triangles are identical by the two angles and included side condition. The marked side is between the given angles.* $\triangle MNO \leftrightarrow \triangle RQP$

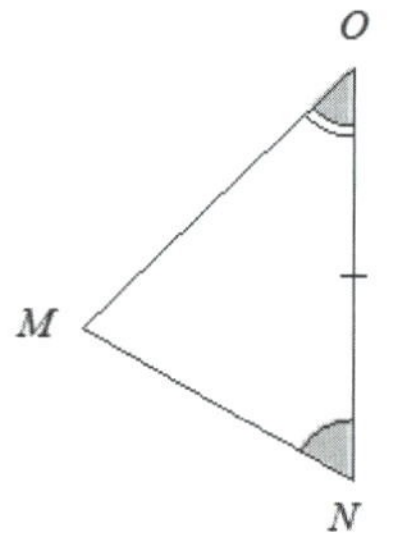

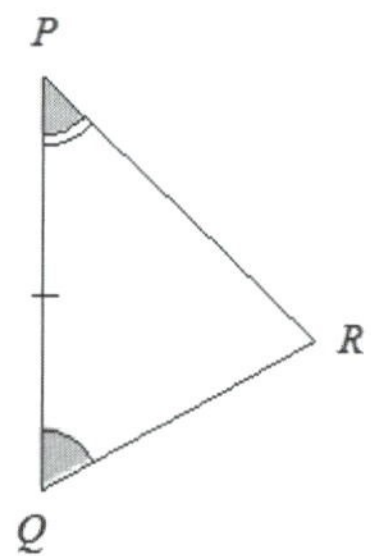

4. In the following diagram, the length of one side of the smaller shaded square is $\frac{1}{3}$ the length of square $ABCD$. What percent of square $ABCD$ is shaded? Provide all evidence of your calculations.

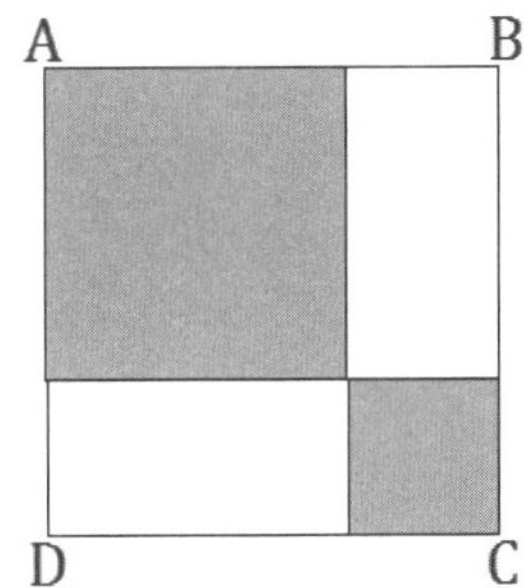

Let x be the length of the side of the smaller shaded square. Then $AD = 3x$; the length of the side of the larger shaded square is $3x - x = 2x$.

$\text{Area}_{ABCD} = (3x)^2 = 9x^2$

$\text{Area}_{\text{Large Shaded}} = (2x)^2 = 4x^2$

$\text{Area}_{\text{Small Shaded}} = (x)^2 = x^2$

$\text{Area}_{\text{Shaded}} = 4x^2 + x^2 = 5x^2$

$\text{Percent Area}_{\text{Shaded}} = \frac{5x^2}{9x^2}(100\%) = 55\frac{5}{9}\%$

5. Side $\overline{EF}$ of square $DEFG$ has a length of 2 cm and is also the radius of circle F. What is the area of the entire shaded region? Provide all evidence of your calculations.

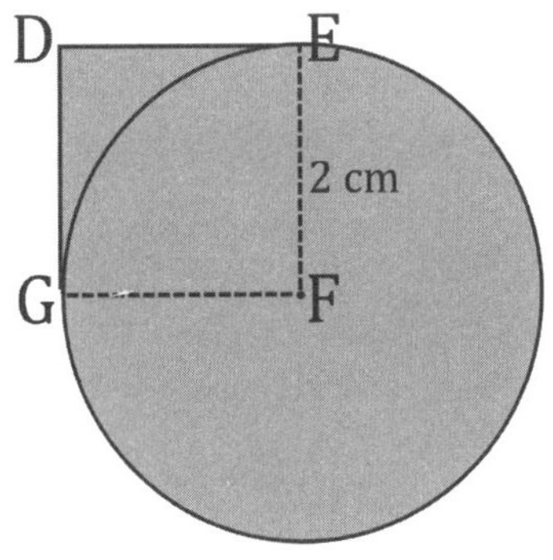

$\text{Area}_{\text{Circle }F} = (\pi)(2\text{ cm})^2 = 4\pi\text{ cm}^2$

$\text{Area}_{\frac{3}{4}\text{Circle }F} = \frac{3}{4}(4\pi\text{ cm}^2) = 3\pi\text{ cm}^2$

$\text{Area}_{DEFG} = (2\text{ cm})(2\text{ cm}) = 4\text{ cm}^2$

$\text{Area}_{\text{Shaded Region}} = 4\text{ cm}^2 + 3\pi\text{ cm}^2$

$\text{Area}_{\text{Shaded Region}} \approx 13.4\text{ cm}^2$

6. For his latest design, a jeweler hollows out crystal cube beads (like the one in the diagram) through which the chain of a necklace is threaded. If the edge of the crystal cube is 10 mm, and the edge of the square cut is 6 mm, what is the volume of one bead? Provide all evidence of your calculations.

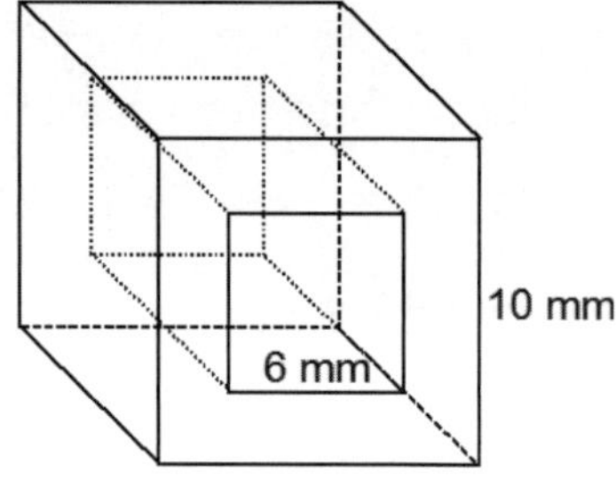

$Volume_{Large\ Cube} = (10 \text{ mm})^3 = 1{,}000 \text{ mm}^3$

$Volume_{Hollow} = (10 \text{ mm})(6 \text{ mm})(6 \text{ mm}) = 360 \text{ mm}^3$

$Volume_{Bead} = 1{,}000 \text{ mm}^3 - 360 \text{ mm}^3 = 640 \text{ mm}^3$

7. John and Joyce are sharing a piece of cake with the dimensions shown in the diagram. John is about to cut the cake at the mark indicated by the dotted lines. Joyce says this cut will make one of the pieces three times as big as the other. Is she right? Justify your response.

$Volume_{Trapezoidal\ Prism} = \frac{1}{2}(5 \text{ cm} + 2.5 \text{ cm})(6 \text{ cm})(10 \text{ cm}) = 225 \text{ cm}^3$

$Volume_{Triangular\ Prism} = \frac{1}{2}(2.5 \text{ cm})(6 \text{ cm})(10 \text{ cm}) = 75 \text{ cm}^3$

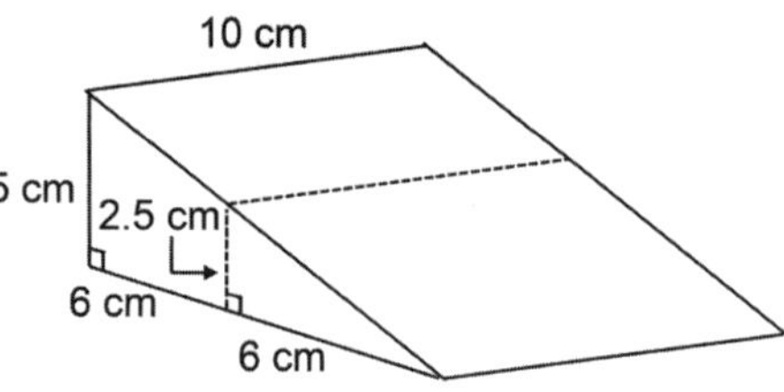

Joyce is right; the current cut would give 225 cm^3 of cake for the trapezoidal prism piece and 75 cm^3 of cake for the triangular prism piece, making the larger piece 3 times the size of the smaller piece ($\frac{225}{75} = 3$).

EUREKA MATH™

G7-M6-TE-B6-1.3.0-10.2015

8. A tank measures 4 ft. in length, 3 ft. in width, and 2 ft. in height. It is filled with water to a height of 1.5 ft. A typical brick measures a length of 9 in., a width of 4.5 in., and a height of 3 in. How many whole bricks can be added before the tank overflows? Provide all evidence of your calculations.

Volume in tank not occupied by water:

$V = (4 \text{ ft.})(3 \text{ ft.})(0.5 \text{ ft.}) = 6 \text{ ft}^3$

$\text{Volume}_{\text{Brick}} = (9 \text{ in.})(4.5 \text{ in.})(3 \text{ in.}) = 121.5 \text{ in}^3$

Conversion (in³ to ft³): $(121.5 \text{ in}^3)\left(\frac{1 \text{ ft}^3}{12^3 \text{ in}^3}\right) = 0.0703125 \text{ ft}^3$

Number of bricks that fit in the volume not occupied by water: $\left(\frac{6 \text{ ft}^3}{0.0703125 \text{ ft}^3}\right) = 85\frac{1}{3}$

Number of whole bricks that fit without causing overflow: 85

9. Three vertical slices perpendicular to the base of the right rectangular pyramid are to be made at the marked locations: (1) through $\overline{AB}$, (2) through $\overline{CD}$, and (3) through vertex E. Based on the relative locations of the slices on the pyramid, make a reasonable sketch of each slice. Include the appropriate notation to indicate measures of equal length.

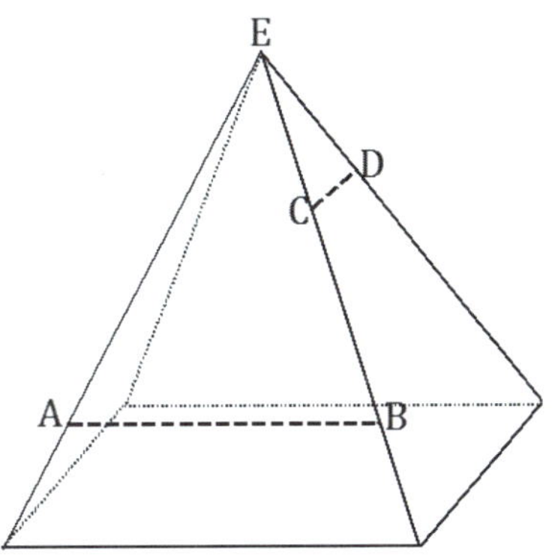

Sample response:

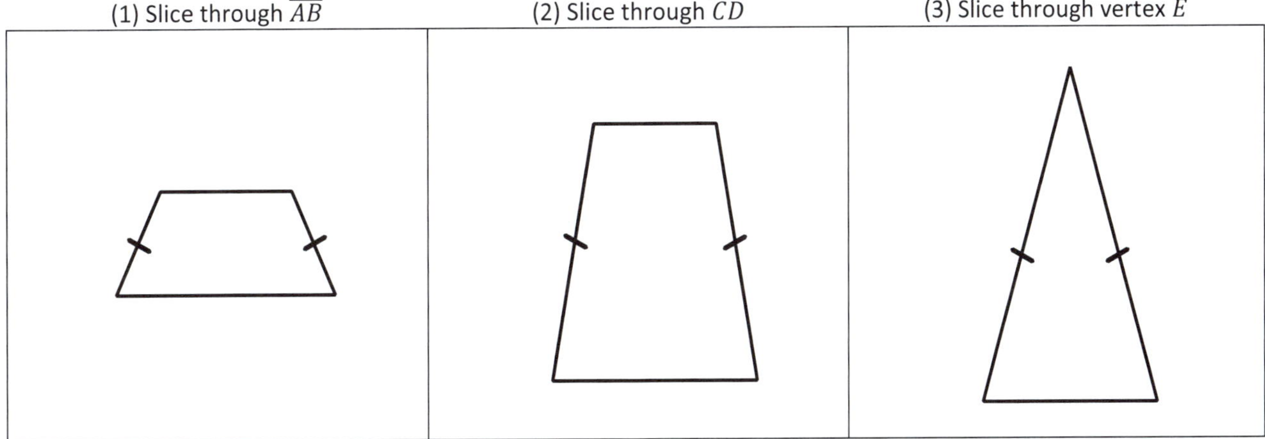

10. Five three-inch cubes and two triangular prisms have been glued together to form the composite three-dimensional figure. Find the surface area of the figure, including the base. Provide all evidence of your calculations.

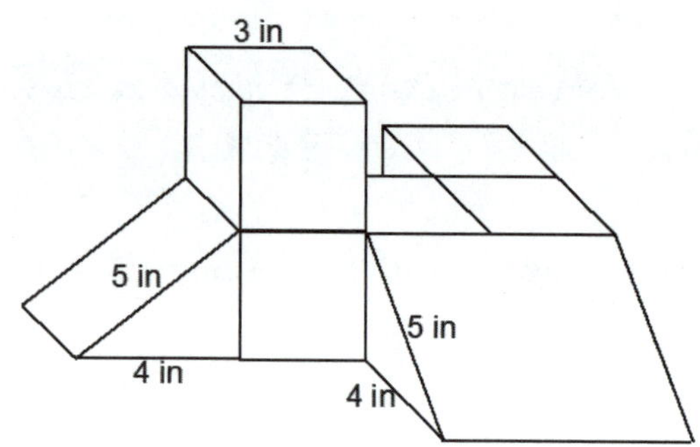

19 square surfaces: $19(3\text{ in.})^2 = 171\text{ in}^2$

4 triangular surfaces: $(4)(\frac{1}{2})(3\text{ in.})(4\text{ in.}) = 24\text{ in}^2$

3×5 rectangular surface: $(3\text{ in.})(5\text{ in.}) = 15\text{ in}^2$

3×4 rectangular surface: $(3\text{ in.})(4\text{ in.}) = 12\text{ in}^2$

6×5 rectangular surface: $(6\text{ in.})(5\text{ in.}) = 30\text{ in}^2$

6×4 rectangular surface: $(6\text{ in.})(4\text{ in.}) = 24\text{ in}^2$

Total surface area: $171\text{ in}^2 + 24\text{ in}^2 + 15\text{ in}^2 + 12\text{ in}^2 + 30\text{ in}^2 + 24\text{ in}^2 = 276\text{ in}^2$

EUREKA MATH™

G7-M6-TE-B6-1.3.0-10.2015